Remote Sensing of Landscapes with Spectral Images

Spectral images, especially those from satellites such as Landsat, are used worldwide for many purposes, ranging from monitoring environmental changes and evaluating natural resources to military operations. In a significant departure from standard remote-sensing texts, this book describes how to process and interpret spectral images using physical models to bridge the gap between the engineering and theoretical sides of remote sensing and the world that we encounter when we put on our boots and venture outdoors.

Remote Sensing of Landscapes with Spectral Images is designed as a textbook and reference for graduate students and professionals in a variety of disciplines including ecology, forestry, geology, geography, urban planning, archeology, and civil engineering, who want to use spectral images to help solve problems in the field. The emphasis is on the practical use of images rather than on theory and mathematical derivations, although a knowledge of college-level physics is assumed. Examples are drawn from a variety of landscapes and interpretations are tested against the reality seen on the ground. The reader is led through analysis of real images (using figures and explanations), and the examples are chosen to illustrate important aspects of the analytic framework, rather than simply how specific algorithms work.

This book is supplemented by a website hosting digital color versions of figures in the book as well as ancillary color images (www.cambridge.org/ 9780521662215).

Remote Sensing of Landscapes with Spectral Images

A Physical Modeling Approach

John B. Adams
and
Alan R. Gillespie
Department of Earth and Space
Sciences
University of Washington

CAMBRIDGE
UNIVERSITY PRESS

University Printing House, Cambridge CB2 8BS, United Kingdom

One Liberty Plaza, 20th Floor, New York, NY 10006, USA

477 Williamstown Road, Port Melbourne, VIC 3207, Australia

314-321, 3rd Floor, Plot 3, Splendor Forum, Jasola District Centre, New Delhi - 110025, India

79 Anson Road, #06-04/06, Singapore 079906

Cambridge University Press is part of the University of Cambridge.

It furthers the University's mission by disseminating knowledge in the pursuit of education, learning and research at the highest international levels of excellence.

www.cambridge.org
Information on this title: www.cambridge.org/9781108462778

First published 2006
First paperback edition 2018

A catalogue record for this publication is available from the British Library

ISBN 978-0-521-66221-5 Hardback
ISBN 978-1-108-46277-8 Paperback

To Caryl and Karen

Contents

About the authors

John B. Adams

John Adams worked on early geological exploration of the Moon, planets, and asteroids, and was instrumental in developing reflectance spectroscopy as a remote-sensing method for mineral identification. As a member of the scientific team that studied the first lunar samples, he demonstrated that Earth-based telescopic spectra could be used to identify and map rock types on the Moon. He has used spectroscopy-based remote sensing of Earth to study geomorphic processes in arid regions and to interpret changes in land use in temperate and tropical landscapes around the world. Adams established the Remote Sensing Laboratory at the University of Washington in Seattle in 1975. He presently is Emeritus Professor in the Department of Earth and Space Sciences at the University of Washington. He spends as much time as possible in the North Cascade Mountains of Washington.

Alan R. Gillespie

Alan Gillespie is a Professor of Earth and Space Sciences at the University of Washington in Seattle, where he has been since 1987. He has been involved with remote sensing since joining the Mars Mariner project at Caltech's Jet Propulsion Laboratory late in 1969. With the launch of ERTS-1, Gillespie switched to terrestrial remote sensing, focusing first on image processing and then on applications, with emphasis on the thermal infrared. He has been a member of the US Terra/ASTER team since 1991, and is responsible for the temperature/emissivity separation algorithms and standard products. Gillespie is also a glacial geologist with a strong interest in paleoclimate. He graduated from Caltech in 1982 with a Ph.D. in geology and a thesis on the glacial history of the Sierra Nevada. Since 1991 he has been using remote sensing and field studies to elaborate the regional variations in the glacial history of central Asia. With Barry Siegal, Gillespie contributed to and edited *Remote Sensing in Geology* (1980). Gillespie is currently the editor of *Quaternary Research*.

Preface

This book is about how to process and interpret spectral images of land-scapes. It is designed for students and professionals in a variety of disciplines who want to do hands-on computer analysis of spectral images to solve problems in the field. The term "landscape" conveys the idea of a part of the land surface that we can see from a vantage point on the ground. It is at landscape scales, rather than at regional or global scales, that we are able to make the most convincing links between remote-sensing measurements and the materials and objects that we see on the ground. In this book, the approach to the study of spectral images at this scale is unconventional. Most other texts on spectral remote sensing emphasize radiative transfer, engineering aspects of remote-sensing systems, and algorithms for manipulating digital images. Indeed, these subjects already have been covered so well that we see no reason to replicate this work. We have taken a new approach that focuses on the intersection of photo interpretation and spectroscopic modeling.

There are practical reasons for shifting the emphasis to physically based models. People in a variety of disciplines use spectral images to obtain information that is important for their landscape-scale projects. For many of these projects it is essential that the extracted information is correct and that it makes sense in the given context. Investigators also want spectral images to give them new insights into materials and processes so that there is the possibility for discovery. However, text books that treat the fundamentals of remote-sensing science and engineering rarely take the next step to explain how actually to go about interpreting spectral images. In our experience, students who have taken the basic remote-sensing courses, when faced with a practical problem that requires interpreting the data, often express the same frustration: "What do you actually DO?" they ask. They know how to invoke algorithms, but they do not yet have a clear understanding of how to tease the desired information out of the data. More importantly, many are not yet familiar with the spectral properties of materials on the ground.

Our objective is to bridge the gap between the more theoretical and engineering side of remote-sensing science and the world that we encounter when we put on our boots and venture into the field. A basic premise is that remotely sensed spectral images are a proxy for observing directly on the ground. Therefore, to interpret spectral images we need to understand the spectral behavior of natural materials, and we need to be able to think like field investigators. This is a tall order, and it may stretch the borders of most remote-sensing curricula. The payoff is a new ability to extract information from spectral images. By applying physical models to remote sensing at landscape scales, we also enhance our ability to "scale up" interpretations to regional and global scales.

To achieve our objectives, we have limited the scope of the book to spectral images in visible and near-infrared wavelengths, and, to a lesser extent, to thermal-infrared images. This emphasis reflects the reality that the most readily available spectral images to support field work are the ones that are being most widely used. Many of our examples are based on Landsat Thematic Mapper (TM) images of areas that we personally have studied on the ground. Landsat images happen to be ideal to illustrate the methods that we discuss, and they comprise a nearly continuous (and under-utilized) 30-year global data base for on-going studies of environmental and land-use changes. The field of view of Landsat TM images is large enough to encompass most field-study areas, and the pixel footprint is small enough to recognize where you are on the ground. The six visible and near-infrared TM Bands provide useful spectral information, but the quantity of data is relatively easily managed. Nevertheless, the analytical and interpretive approach that we describe applies to all spectral imaging systems, including imaging spectrometers having hundreds of spectral channels.

In order to reach a broad, multidisciplinary audience, our explanations are light on equations and heavy on visual examples. The book is suitable as a text for classes in remote sensing and as a professional reference; how-ever, we do assume that our readers already have some background in the fundamentals of remote sensing. For those who need to brush up, there are several excellent basic texts in print. We elaborate on certain topics that only are treated briefly, or are omitted from most texts, and that we feel are essential for image interpretation. These topics include spectral contrast, basic spectral components of scenes, spectral unmixing, the application of fraction images, detectability of targets, and physically based thematic mapping.

Illustrations

The website (www.cambridge.org/9780521662214) hosts digital color versions of the figures in the book and allows us to include more color pictures than the ones in the color plates that are bound within the text. For those who do not have convenient access to a computer when they are reading, a black and white version of each color figure on the Web, with its caption, is included in the text. In addition, the website has reference images that elaborate on the properties of four Landsat TM images that are used frequently in the illustrations.

We constructed a 20×20-pixel synthetic image to assist in explaining some of the methods for exploring spectral images. A particular advantage of the synthetic image is that it allows us to compare the results of various algorithms and operations. First, we know exactly how the image was constructed; therefore, we already know the correct answers, and can evaluate how well various algorithms and image-processing techniques work. Second, the simple pattern of the image avoids the contextual distractions of real images of landscapes. The structure of the synthetic image is defined in Chapter 3.

Terms and definitions

Remote sensing as a discipline has accumulated a substantial amount of jargon. With those of you in mind who care more about applications than about remote-sensing technology, we have attempted to minimize the jargon and to be careful about how terms are used. Many of the technical terms in the text are defined in marginal notes and in the Glossary.

Acknowledgments

Ideas for organizing materials and for explaining arcane aspects of
remote sensing were tempered in the forge of various classes taught
by both authors at the University of Washington over the past 25 years.
We especially benefited from the wide range of backgrounds of our
students in these classes, both beginning and advanced. In addition
to geologists and geophysicists we commonly instructed, and learned
from, students in forestry, geography, urban planning, oceanography,
fisheries, botany, archeology, civil engineering, astronomy, and
statistics.

Our approach to spectral imaging has been strongly influenced by
our colleagues and graduate students in the W. M. Keck Remote
Sensing Laboratory at the University of Washington who have
worked with us over the years on a variety of projects. Research
topics ranged from the composition of the surfaces of the Moon and
Mars, to processes in terrestrial deserts, to changes in Amazon and
northwest forests. The Department of Geological Sciences (now the
Department of Earth and Space Sciences), the home for the Remote
Sensing Laboratory, provided support and encouragement over the
years to what must have appeared at times to be a maverick unit that
crossed all disciplinary lines.

We have benefited enormously from the exchange of ideas with
our many colleagues across the nation and around the world. We owe
a special debt to Milton Smith, who for many years was a key part
of the intellectual brew of the laboratory, and with whom we shared the
excitement and frustrations of developing spectral-mixture analysis.
In preparing the book, we benefited from discussions with Stephanie
Bohlman, Roger Clark, Don Sabol, and Robin Weeks. Thanks to Elsa
Abott for finding long-lost images, files and references. Helpful
reviews of the manuscript were made by Caryl Campbell, Laura Gilson,
Karl Hibbitts, Jim Lutz, and Amit Mushkin. Technical and facilities
support was provided by Bill Gustafson, Tom McCord, Marty Nevdahl,
Sally Ranzau, and Karin Stewart-Perry.

Special thanks are due to Laura Gilson who drafted the illustrations,
and who kept us organized throughout the long process of writing and

revisions. Her skills in technical illustration and image processing emboldened us to produce a much more ambitious book than we originally visualized.

We are grateful to the W. M. Keck Foundation for grants to the Remote Sensing Laboratory over a period of several years that enabled us to explore innovative ways of extracting information from spectral images, and in particular, spectral-mixture analysis and related techniques. In an era when most research grants are awarded with the expectation of quick results, it has been difficult to stay focused on the hard problems that require several years, if not decades, of work. The long view and light hand of the W. M. Keck Foundation has been of immense value in developing and testing our ideas about spectral remote sensing.

We also acknowledge the National Aeronautics and Space Administration for supporting various research projects over many years. This support, along with funding from other governmental and private organizations not only facilitated research, but it enriched the experiences of our students, and provided the crucible out of which this book emerged.

Our thanks to Matt Lloyd, Susan Francis and Wendy Phillips at Cambridge University Press for guiding us through the administrative and technical thicket that lies between a manuscript and a published book. And special thanks to Zoë Lewin, our excellent copy editor, who gently made us realize that we were not quite through yet, and that there always is room for improvement.

Chapter 1
Extracting information from spectral images

North Cascades, Washington. Photograph by D. O'Callaghan

In this chapter

Spectral information, when placed in a photo-interpretive context, is a powerful tool for understanding landscapes. We obtain the most reliable spectral information by invoking physical models that link image measurements with materials and processes on the ground.

1.1 Introduction

Seeing is believing

We are a visual species that relies heavily on images to understand the world around us. The ability of humans to extract information from landscapes through visual images has evolved over millions of years. Of necessity, our hunter–gatherer ancestors learned how to "read" landscapes to find shelter and resources and to avoid danger. Today, with the increasing demands that are placed on the world's ecosystems by a growing population, we find ourselves reading landscapes not only to find more resources and to cope with danger, but for signs of strain

> **Landscape**
> A portion of the land that can be seen from a single vantage point.

in our planet's life-support systems. Ironically, though, our species evolved by paying the most attention to short-term events on local scales. Until now, we have had little need to develop the skills to notice decadal or longer changes or those changes that occur on regional or global scales (e.g., Ehrlich and Ehrlich, 2004). No wonder it is so hard to attract people's attention to issues of environmental change, such as depletion of ground water, deforestation, expansion of deserts and retreat of glaciers. These, and many other changes, are largely ignored by the world's increasingly urban population. Out of sight, out of mind.

Aerial photographs and satellite images now allow us to visualize landscapes in entirely new ways. We can see vast areas all at once. We can construct images that measure diverse properties of landscapes, and we can compare images taken at different times to discover changes. The power of these remote measurements lies in the fact that we can relate them to the familiar realm of what we see on the ground. Astonishingly, within a few decades humans have developed the means to visualize the world in ways that our ancestors never dreamed of. And just in time. Now, more than ever, we need to be able to *see* what is going on in a rapidly changing world where human influences are felt everywhere on the planet. Remote sensing is not just an interesting technical phenomenon, it has become an essential part of an urgent quest to understand our changing environment.

Pretty pictures are not enough

The issues of world resources and environmental change demand society's best efforts, including our most perceptive and accurate assessments from remote sensing. However, in pursuing this goal, we need to be aware that there is an opposite side to "seeing is believing," namely that it is easy to be misled by images. In remote sensing, it is all too common to accept images at face value without understanding how they were made and what they really show, especially when they are dramatic and esthetically pleasing. We all think that we know how to interpret images; after all, it comes naturally, right? Well, not really, at least in the case of remote sensing with modern instruments that operate beyond the range of human vision. Depending on our objectives, processing and analyzing spectral images can be a daunting task – one that often is left to technicians to do the hands-on computer work. In our experience, many, perhaps most, of the people who make use of spectral images do not have the time or inclination to analyze the data themselves, and many do not fully understand what happens inside the computer. Even skilled technicians are not immune. Let's admit it: at heart we are all button-pushers. There is something seductive about clicking on an algorithm and seeing an image appear on the computer screen. With lots of choices of algorithms, it is tempting to try a few to see what happens. Of course, this leaves us vulnerable to generating erroneous or irreproducible interpretations when our real aim is to produce consistently reliable results that help us to solve problems in the field.

One of the best ways to improve the quality and quantity of information that we extract from spectral images is to apply physically based models that connect the spectroscopic and optical properties of known materials with remote measurements. Imaging and spectroscopy were joined in the 1980s for remote-sensing applications when it became technically feasible to acquire simultaneous images in hundreds of narrow wavelengths with spectral resolutions comparable to those of laboratory spectrometers. Spectroscopists were quick to use the new Airborne Imaging Spectrometer (AIS) and Airborne Visible and Infrared Imaging Spectrometer (AVIRIS) to identify materials remotely by the characteristic wavelengths of recognizable absorption bands (Goetz *et al.*, 1985). Actually, though, it has *always* been possible to apply spectroscopy to spectral images – even to images that sample only a few, broad wavelength regions, such as those acquired by Landsat. Perhaps the best example is the widespread use of vegetation indices such as the normalized-difference vegetation index (NDVI), because these indices are firmly linked to the spectroscopic characteristics of green vegetation. Spectroscopy and physical models form a basis for interpreting all spectral images, regardless of the number and the resolution of the wavelengths sampled. We have powerful visual interpretive skills, and we have powerful analytical tools to interpret spectral data. The challenge is to combine the two in order best to extract information about landscapes.

1.2 Field studies and spectral images

Remotely sensed spectral images are a proxy for being in the field. True, there are important differences, such as not being able to sample materials, viewing outside of the visible range of the eye, and observing large areas at once. But there are important similarities between image analysis and field work. Spectral images and the field are data-rich. Skilled analysts, like field observers, can return again and again to an image and find new information each time. In contrast, as beginners, we are apt to turn away from an image after a few moments with the impression that there is not much there. With experience, however, we learn to ask questions of the data. "What do you want to know?" is equally relevant in the field and with images.

Thematic maps of land cover and land use are fundamental products of field studies (Chapter 7). By focusing on particular topics such as plant species, soils, or geology, field observers are able to sort through

---

Field work and image analysis

Both involve sampling the landscape. In field work, a researcher can examine evidence at all scales. In images, the smallest sample of a landscape is determined by the size of the pixel footprint.

the complexities of the natural world to record, organize and interpret relevant information. To maintain focus, the observations and measurements are guided by questions and working hypotheses. Questions determine what to measure. Are there more trees on one part of a hill slope? Is the red soil younger than the brown one? Are the rocks in the distant outcrop the same as the ones we see here? Hypotheses are used to formulate testable ideas about processes that have shaped a landscape. Hypotheses provoke new questions. New questions lead to new observations, measurements and yet more hypotheses.

In the field, we do not expect that the answers to our questions will be revealed everywhere, or at a single spatial scale. We know from experience that critical information, perhaps the presence of a certain type of plant or the evidence for the relative ages of two rock types, is likely to be found in only a few places, and that an essential part of field work is assembling bits of specific, and perhaps scattered, information into a larger and more complete picture. This process is not amenable to a mechanical or automated approach, because the person doing the field work must continually evaluate what is important and what is not. Skilled field workers, therefore, continually make new discoveries and keep seeing a landscape in new ways. Because images are a proxy for being in the field, skilled image interpreters proceed in much the same way.

Our day-to-day visual observations depend heavily on an ability to derive meaning from the shapes, sizes and textures of the objects and patterns in the world around us. We make relatively little use of spectral information. Color-blind people function quite normally in this visual world of ours, which is a reminder that, although the sensation of color enriches our lives and is essential in some instances, our visual experience is dominated by spatial, not spectral, information. It is natural that we bring this same emphasis to our interpretation of spectral images. Furthermore, most of our visual experience has been gained by viewing the world more-or-less horizontally from a few meters above the ground under ever-changing conditions of illumination. We do our best job of image interpretation with familiar subjects such as other people. Viewing the ground surface from aircraft or satellite altitudes introduces an unfamiliar perspective, significant changes in scale and resolution and, commonly, new subjects such as roof tops and the tops of forest canopies (Figure 1.1).

A change in perspective alters shading and shadowing clues, even to the extent of sometimes causing us to perceive topography as inverted. Nevertheless, with experience, we have learned to interpret patterns in images taken from overhead of extended areas of the land surface. An advantage still belongs to those who have experience in the field and in interpreting landscapes at different scales. For example, if we recognize a moraine from a long-vanished glacier in an image it is because we

Color and pattern
Color enriches our interpretations, but interpretations usually are based on spatial patterns and temporal change.

(a)   (b)   (c)

Figure 1.1. Image series taken north of Manaus, Brazil showing the primary forest at different perspectives, scales and resolutions: (a) from the ground; (b) from a 40-m research tower; (c) from an aircraft 200 m above the ground, showing the top of the tower.

Figure 1.2. Oblique aerial photograph of glacial moraine near Bishop, California (see also Figure 1.13). Sinuous, sub-parallel ridges are unsorted debris left behind after alpine glaciers retreated. Moraine is ~1.5 km across. Photograph by D. Clark.

have seen a similar one before, or we have learned from some source what one looks like (Figure 1.2). If we have never heard of a moraine it is unlikely that we could recognize one, or even notice it. Apparently we compare what we see with many stored patterns (image or otherwise) in

(a)   (b)

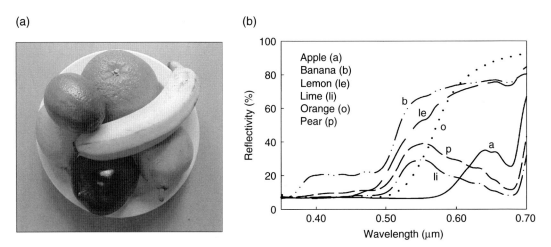

Figure 1.3. (a) Panchromatic photograph of a bowl of fruit. (b) Reflectivity spectra of items in the photograph.

the brain until a good match is found; however, when we interpret visual information we invoke a far more complicated process than matching and naming patterns. Interpretation explores the interrelations between patterns, and it involves a variety of judgments as to the significance of features in a broader context. A glaciologist, for example, can visualize the processes that produce moraines, and thus view all of the topography in an image from that perspective.

Spectral and spatial information

Our skills in spectral recognition are underdeveloped compared with our spatial and pattern-recognition abilities. Consider a bowl containing several different kinds of fruit (Figure 1.3). Most of us could recognize the familiar items in a black-and-white photograph by their shapes, sizes, and textures, but we would be less sure of our choices if we only could see swatches of color samples that matched the objects. And very few of us could identify them just using a plot of their spectral reflectance. Experience has taught us to be wary of relying on perceived colors to identify most objects. We know that different materials may have the same colors. Then too, the same material can have different colors, as is shown by fruits and vegetables that change as they ripen. Once we recognize something by shape, size and texture we can use color to gain further information. We tend to do the same thing with color images, that is, to rely heavily on the spatial information and to use the colors occasionally for help. Modern instruments have extended our visual spectral range by orders of magnitude. Until a few decades ago we never had "seen" the world at 1 μm or at 10 μm, nor at any other wavelengths outside the narrow range of about 0.4 to 0.7 μm. We have had to develop our experience from scratch in each new wavelength range, without the benefit of daily practice, to say nothing of millions of years of evolution.

Now that images are not limited to visible wavelengths, the overall availability of spectral information about landscapes has been enlarged enormously, and with it the potential for discriminating and identifying materials on the ground. Still, the typical approach to spectral images has continued to rely heavily on the more familiar spatial information. In addition, few users of remotely sensed images have had training in spectroscopy. As a result, standard, largely empirical methods for analyzing spectral images have evolved, and are the most widely used analytical tools in remote sensing today. Standard methods (Chapter 3) have been used successfully for a wide variety of applications; nevertheless, the spectral content of images has been, and continues to be under-utilized. To extract the maximum amount of information from spectral images we need to understand the physical connection between the data (encoded radiance) and materials on the ground, a topic that we discuss in Section 1.4.

1.3 Photo interpretation of spectral images

1.3.1 Image context

Context is critically important to be able to interpret spectral images. A rich spatial context already exists for field investigators who want to make use of spectral images. Furthermore, field investigators are likely to approach remote-sensing data with objectives and questions in mind. Photo interpretation is an essential link between field observations and spectral analysis, because it places spectral measurements in a spatial context. For this reason, before unleashing the power of a spectral image-processing program, we should take time to do a careful reconnaissance using our photo-interpretation skills.

> **Image context**
> The context of an image and the goals of the analyst are crucial factors in extracting useful information from images.

Photo interpretation, like learning to ride a bicycle, is not easily described in words. Besides having a strongly experiential component it requires practice. Many people think that they already are good photo interpreters, because they know how to interpret the everyday images formed by the eye–brain system. In the context of remote sensing, however, photo interpretation can be more difficult due to the unfamiliar perspective and scale. Textbooks on photo interpretation generally explain about cameras and photogrammetry, and then give examples of images where the reader either is told what they are seeing or is asked to render an interpretation. Textbook examples are a good way to build experience, but seldom are quite the same as the interpretive problems facing us in our own applications of remote sensing. One way to improve our photo-interpretation skills is to use an hypothesis-testing approach (Section 1.4). It also is helpful to understand the attributes of an image that govern visual perception.

1.3.2 Illumination and viewing geometry

Shadows and shading provide visual information that is needed to understand topography and texture, and being able to interpret topography and texture is important for discriminating and recognizing objects on the ground. To interpret patterns of shadows and shading in images correctly we need to be aware of how a scene is illuminated and viewed. Everyone knows from experience that the land surface varies in appearance during the day depending on Sun azimuth and elevation. Surfaces also vary in appearance depending on the direction from which they are viewed, for example, whether one is looking toward the Sun or away from it. When analyzing an image it is essential to know whether it was taken toward the nadir (straight down) or whether the view of the surface was oblique to the nadir direction. We also would like to know the following for the time the image was acquired:

- The direction toward the Sun (azimuth).
- The zenith angle (the angle between the direction to the Sun and the normal to the mean global surface).
- The Sun-elevation angle (the complement of the zenith angle, providing the orientation of the mean surface of the imaged area is coincident with the mean global surface).
- The angular field of view.

Information on Sun azimuth and elevation generally accompany images that are standard products of remote-sensing surveys and programs. Otherwise, the angles can be computed from an ephemeris if the location, date and time are known. The angular field of view of an image depends on the configuration of the sensor system, and, along with the altitude at the time the data were taken, determines the field of view of the image. For public-domain satellites this information is published. The illumination and viewing geometries are relatively simple for nadir-viewing satellites (Figure 1.4); however, these parameters can vary substantially for some satellites, planetary spacecraft and aircraft cameras that view surfaces obliquely.

1.3.3 Shadows and shading

Shadows, shading and inverted topography
By knowing the Sun azimuth we can predict the orientation of shadows that are cast by topographic or other features. The relationship of

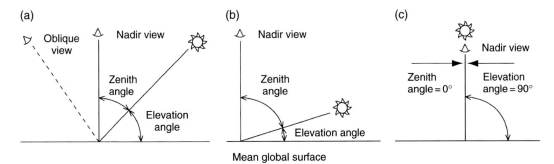

Figure 1.4. Sketch defining zenith angle, Sun-elevation angle, nadir view and oblique view relative to the mean global surface. Notice that these angles are *not* measured relative to local topographic surfaces. (a) Moderate zenith angle. This is the most common geometry for remote-sensing images. (b) Large zenith angle, low Sun-elevation angle. Images show strong shadowing, even by low objects, but shadows obstruct view of much of the surface. (c) Zero zenith angle. When the Sun and the observer are close to the same angular position, objects on the ground block the observer's view of shadows. Sketches are in the "principal plane" that includes the Sun, ground and observer. For simplicity, the angular field of view of the observer or camera is assumed to be zero; and "rays" from the Sun are considered to be parallel.

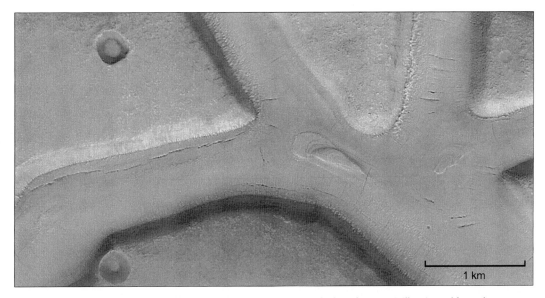

Figure 1.5. Image that may cause the topography to appear inverted when the page is illuminated from the top. Nilosyrtis, Mars. Mars Global Surveyor; by permission of NASA/JPL/MSSS. Rotate the page 180 degrees and look again.

shadows to the objects that cast them is perfectly obvious in many images. The eye–brain system is so practiced at making sense out of shadows and shading that we take the ability for granted. But the same ability plays tricks on us sometimes by inverting topography (Figure 1.5). The main reason for inversion is that we are accustomed to seeing landscapes illuminated from above – a natural consequence of where solar illumination comes from when we are standing on the ground. However, a satellite view of a landscape toward the nadir need not obey this convention. For example, on a winter morning in the northern hemisphere a landscape will be illuminated from the southeast, and, if the image, as is customary, is oriented with north at the top, the illumination will come from the bottom right. For some, the eye–brain system insists that the illumination must actually be coming from the top of the image and, if so, the ridges must then be valleys and vice versa. Sometimes it is possible to change this perception of inversion by looking at the image while keeping in mind the correct Sun azimuth. A more powerful remedy in the case of hard copy is to rotate the image 180°. Obviously, rotation is more difficult with a computer monitor, and analysts have been known to resort to the indignity of inverting themselves. Whatever method we use to solve this problem, it is essential for further analysis of the image to able to see the topography correctly.

Shadows, exposed and hidden

Many of the widely used images of Earth (e.g., Landsat, SPOT, air photos) are acquired in mid morning. It is the time when shadows cast by topography and roughness elements are not too long and not too short for optimum photo interpretation of most scenes. (Morning images avoid the clouds that build during the day in many areas, and, therefore are more popular than afternoon images.) Everyone knows that shadows are prominent in images that are taken near sunrise or sunset when the Sun is low in the sky. It is less widely recognized that significant shadowing occurs anytime that the Sun grazes a local surface such as a hillside. Spatially resolved shadows often reveal subtle features of topography and roughness that otherwise would go unnoticed, because small objects can cast large shadows. Low-Sun satellite images of Earth are not familiar to most image analysts, but they are more common for moons and planets. A drawback of low-Sun images is that at extremely low Sun elevations shadows may obscure a significant portion of the surface, resulting in a loss of information. However, when a sensor is directly in line with the surface and the Sun, shadows disappear, because they are hidden behind the objects that cast them. This is the so-called "zero-phase angle" geometry at which images appear "washed out," because it is difficult or impossible to see topography and roughness (Figures 1.6 and 1.7). Although these conditions make photo interpretation difficult, they are optimum for measuring the spectral properties of surface materials, precisely because the confusing effects of shadows and shading are minimized.

Lightness gradients can occur from one side of an image to another as a result of variations in shadowing that are caused by changes in

---

Brightness and lightness

We use the term *"brightness"* for illumination sources (e.g., Sun, light bulbs and computer screens); and the term *"lightness"* to describe reflecting surfaces (e.g., Earth, Moon and hardcopy images).

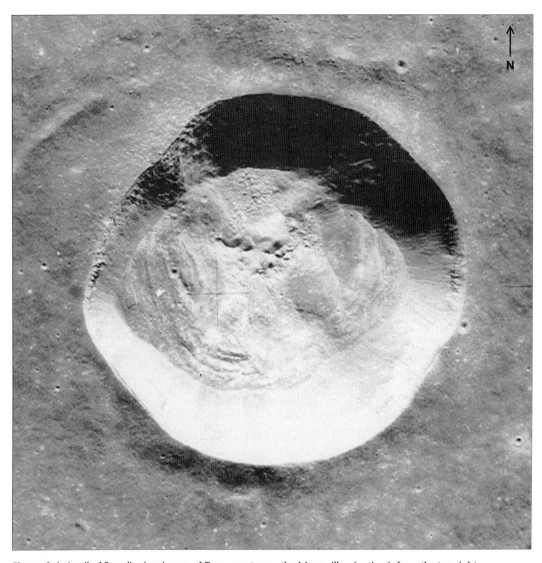

Figure 1.6. Apollo 12 nadir-view image of Dawes crater on the Moon. Illumination is from the top right. Topographic detail is lost along the fully illuminated inner wall where the phase angle is small and in the area of complete shadow. The optimal lighting geometry for interpreting surface topography and texture is in the left and right segments of the crater wall where the local incidence angle is small and details such as rock layers are revealed by their shadows. Dawes crater is approximately 18 km in diameter. NASA photograph.

a camera's angular field of view (Figures 1.7 and 1.8). Such gradients are most noticeable in images taken from low-altitude aircraft, but across-scene gradients also are measurable in some satellite images of rough surfaces. In these images, shadows are more hidden in the down-Sun direction compared with the up-Sun direction. The loss of shadows down-Sun may be difficult to separate from the loss of contrast caused

Figure 1.7. Sketch of an aircraft sensor view in the principal plane. To the reader's right side, the sensor views toward the Sun; on the other side, the sensor views away from the Sun. The phase angle is the angle between a line from a point on the ground to the Sun and a line from the same ground point to the sensor. The sketch shows the special case where the angular field of view just includes the zero-phase angle.

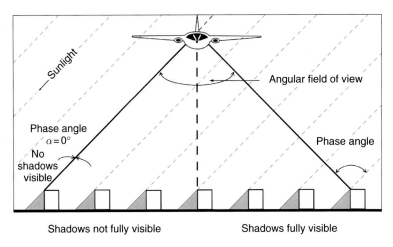

Figure 1.8. Kilauea caldera, Hawaii. Gradient across a NASA NS001 aircraft image. Details of the surface are obscure in the down-Sun direction of the image (left side), primarily because of a loss of shadows. North is up. Image by NASA/JPL.

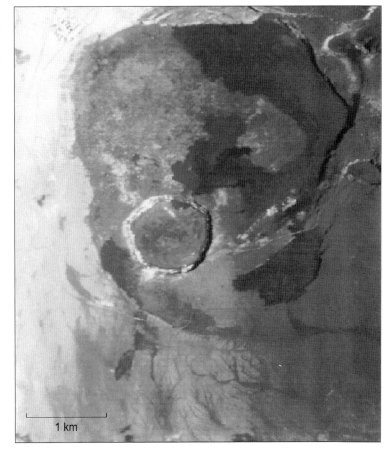

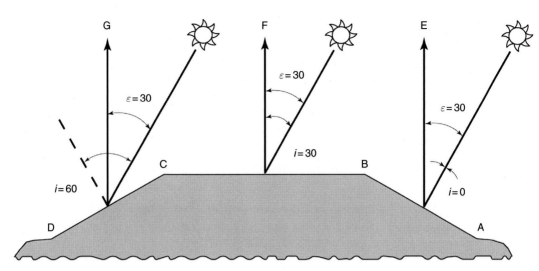

Figure 1.9. Sketch in the principal plane through a hypothetical hill having the photometric properties of a perfect, diffuse (Lambertian) reflector. The hill is constructed such that the incidence angles for the three slopes are: $i_{AB} = 0°$; $i_{BC} = 30°$; $i_{CD} = 60°$. Normalizing to the radiance received at AB, the measured radiances at overhead sensors E, F, G are cosine functions of the incidence angles, and: $E = \cos 0° = 1$; $F = \cos 30° = 0.87$; $G = \cos 60° = 0.5$. The exitance angle (ε) is the same (30°) for all slopes. On an image taken at high altitude, AB would have maximum lightness, BC would be slightly shaded, and CD would be half as light as AB.

by atmospheric backscattering (Chapter 3). A special case of the hidden-shadow effect occurs in oblique images when a camera views away from the direction to the Sun ("down-Sun"). Shadows also can be hidden from view simply by being too small to be spatially resolved. Indeed, most of the shadowing in a scene may occur at a sub-pixel scale. Sub-pixel shadows darken a pixel, as does shading (see section below) and the presence of dark materials on the ground.

Shading

The term "shading" describes the variation in light reflected from a given scene element in a given direction as a function of the *local* incidence angle. The local incidence angle is the angle between the direction to the Sun and the normal to the local surface, where the surface is defined at a convenient spatial scale (Figure 1.9). Materials on the ground reflect more light when illuminated from a direction normal to the surface, and less as the angle to the normal (the incidence angle) increases. At zero-incidence angle a surface receives the maximum illumination flux (radiant power) per unit area. Thus, slopes that face the sun directly are more powerfully illuminated than ones oriented at larger incidence angles. Using this simple relationship our eye–brain systems are able to interpret tonal variations in a two-dimensional

Figure 1.10. Photometric
functions measured in the
principal plane for three types
of surfaces: (a) diffuse reflector
(matte paper); (b) back-
scattering surface (dark
moss); (c) specular reflector
(polished aluminum).
Incidence angle ($i = 35°$) is
relative to the normal to the
mean surfaces of the samples.
The exitance angle is
from $-90°$ to $+90°$.
Measurements for each
sample are normalized to the
maximum reflectivity.

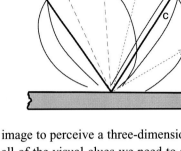

image to perceive a three-dimensional surface. Shading alone gives us all of the visual clues we need to see topography and roughness in an image over a wide range of scales.

Shading occurs for all wavelengths of reflected light. In images having evident topography, the contribution of shading in each spectral channel can be shown by highlighting parts of the data histogram, and observing where the corresponding pixels on the image are located (Section 3.4.1). Within each wavelength channel, Sun-facing slopes will be lightest (ignoring compositional differences), and, as the local incidence angle increases, slopes will be progressively darker, eventually becoming shadowed where the incidence angle is $90°$. There still can be wavelength-to-wavelength differences in lightness for any given pixel, even in shadow, because a shadowed slope may be illuminated by skylight or indirect light from surrounding objects (Section 3.2.1). Shading is manifested in "scatter plots" of one spectral channel against another (Section 3.4.1) by a roughly linear clustering of data points that extends from near the zero point. Highlighting the points on a scatter plot and observing the corresponding pixels on an image will show how lightness is correlated with the orientation of topographic slopes.

Lambert's law

A Lambertian surface scatters light in all directions uniformly. Therefore, $L_i = L_o \cos(i)$ describes the radiance L_i measured from a surface when the sun has an incidence angle of i; L_o is the measured radiance when the Sun is overhead. For Lambertian surfaces the measured radiance is independent of the exitance angle.

Photometric function

The full description of how a surface reflects light in all directions for each possible illumination configuration is known as the photometric function or the bi-directional reflectance distribution function (BRDF) (e.g., Hapke, 1993). It is not a simple matter to predict quantitatively how a topographic surface will become darker or lighter as the illumination geometry is varied. Although the orientation of the local topography dictates the power of the incident light per unit area, the power of the light *leaving* a surface depends on its angular direction (Figure 1.10). For example, perfectly diffusing (Lambertian) surfaces scatter light equally efficiently in all directions, whereas everyone knows that a specular (mirror) surface returns light preferentially in one direction that is determined by the incidence angle. Certain dark materials that have rough surfaces can scatter a substantial portion of light back in the direction to the Sun. Examples include dark conifer trees and opaque, rough soils.

Complete photometric functions are known for a few materials that can be arranged and measured under laboratory conditions. For remote sensing at landscape scales, though, only partial, bi-directional, photometric measurements can be made, because of the limited angular ranges of solar illumination and sensor viewing. It is not practical to measure complete photometric functions for extended natural surfaces. For example, imagine trying to measure from a satellite the radiance from the surface of the Earth at all possible angles for each position of the Sun for all seasons, and trying to correct the measurements for variations in the atmosphere. To complicate this picture further a photometric function can vary with wavelength. Thus, even if a perfect description of the structure and composition of the scene is available, it still may be difficult to predict exactly how shading will be manifested for each pixel footprint. Not surprisingly, natural landscapes, most of which reflect more or less diffusely, usually are assumed to be Lambertian for remote-sensing applications, and wavelength effects usually are ignored (Section 3.2.1).

One reason it would be helpful to know the photometric functions of natural surfaces would be to remove the shading effects of topography. This is important for photo interpretation, because it may be ambiguous whether an area on an image has intrinsically dark materials, or whether it is dark because of the illumination geometry (Figure 1.8). Methods for correcting topographic and other illumination effects are discussed in Chapter 3. Another use of the photometric function is to infer structure or composition. Telescopic measurements of the Moon in the days before spacecraft exploration showed evidence of diffuse scattering combined with an unusual, highly backscattering photometric function. Laboratory experiments and optical theory suggested that the lunar surface was covered with a dark, loosely packed powder (e.g., Hapke, 1967). Apollo astronauts and returned samples of Moon dust confirmed this interpretation. Back on Earth, theoretical studies indicate that it may be possible to invert bi-directional, remote measurements of forests to infer canopy structure and stand attributes. For some landscape surfaces, modeled bi-directional reflectance functions agree well with remote-sensing measurements; however, definitive tests of this approach await more experience interpreting multi-angle measurements from spacecraft sensors such as NASA's Multi-angle Imaging SpectroRadiometer (MISR).

> **Pixel footprint**
> A pixel footprint ("instantaneous field of view" or IFOV) is the area in the scene from which light is averaged by a detector. The size of a pixel footprint affects the spectroscopic information that can be retrieved.

Figure 1.11 and Web color Figure 1.11. Death Valley, California. (See also Plate 1.) (a) Digital-elevation model with superposed Landsat Thematic Mapper (TM) color-composite image of Band 5 (red), Band 4 (green) and Band 3 (blue); (b) DEM with superposed Thermal-Infrared Multispectral Scanner (TIMS) image of channel 5 (red), channel 3 (green) and channel 1 (blue). Scene is ~40 km across. Images by NASA/JPL.

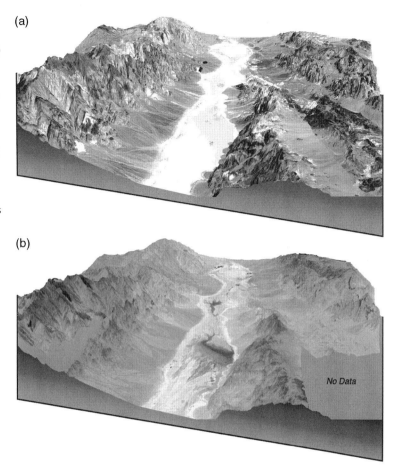

(a)

(b)

Images of topography

Images made from measurements of the elevation of the land surface can be extremely useful as an aid in photo interpretation. Digital elevation models (DEMs) derived from topographic maps are commercially available for many areas worldwide. In some cases, such as the European Space Agency's Mars Express, elevation is calculated directly from a stereoscopic model. Other elevation data are derived from RADAR or LIDAR data. Using DEMs, most image-processing programs can create shaded-relief images, images of elevation intervals, and three-dimensional grids of topography. Interpretation of spectral images can be enhanced by registering them with DEMs; the visual effect is that of draping the spectral data over the topography (Figure 1.11 and color Figure 1.11; see Web).

In recent years, RADAR and laser altimeter measurements acquired by aircraft and spacecraft have become available of some planetary and

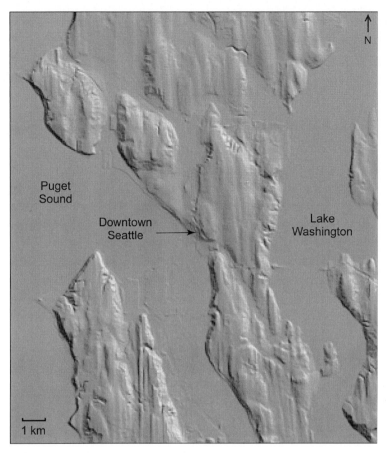

Puget
Sound

Downtown
Seattle →

Lake
Washington

1 km

Figure 1.12. Shaded-relief image of Seattle, Washington derived from aircraft LIDAR topographic data that has been rendered as a shaded-relief image using a cosine function and illumination from the upper right. North–south trending grooves and ridges that were produced by continental glaciation are evident in the LIDAR image, but they are difficult to see in air photos or satellite images because of urban structures and vegetation. Adapted from a digital image provided by R. Haugerud, USGS.

terrestrial regions. Shaded-relief images made from these data often capture fine topographic details that facilitate photo interpretation (Figure 1.12). Wherever possible, it is advantageous to incorporate topographic information into our analysis of spectral images of land-scapes. A detailed treatment of the analysis of topographic data is beyond the scope of this book; however, there is further discussion of how to isolate topographic information from spectral information in Chapters 3 and 7.

1.3.4 Spectral patterns

Up to this point we have discussed illumination geometry and how shadows and shading influence tonal variations in reflected-light images. These are critical perceptual clues for interpreting topography and texture. On the other hand, spectral patterns are produced by fundamental differences in the capacity of surface materials to absorb,

Table 1.1. *Darkening mechanisms for reflected-light images.*

Cause of darkening	Explanation	Interpretation	
		Topography	Composition
Shadows		Yes	No
Resolved	Illumination is blocked by opaque object, and image of blocking object is projected on the ground		
Un-resolved	Signal from sub-pixel shadow is mixed with signal from illuminated areas		
Shading		Yes	No
Incoming irradiance is spread over larger area	Angle of incidence (i) >0 (cosine effect)		
Outgoing radiance is diminished or not detected	Radiance is directed out of the view of the detector by specular reflection or scattering		
Absorption	Photon energy is absorbed by surface materials by vibrational, electronic, or charge-transfer resonances	No	Yes

reflect and emit radiant energy. In the human visual range we would describe these variations as different colors. To make use of spectral patterns in images it is not necessary to have a background in spectroscopy. After all, we do not need a spectroscopic explanation of the physical phenomena to discriminate and interpret visual colors in the world around us (Figure 1.3).

When we use only one spectral channel (or a wide range of wavelengths) it may be difficult to determine which processes are responsible for tonal variations in a given image, because all three processes, shadowing, shading and absorption by materials can influence place-to-place (spatial) variations (Table 1.1). By using two or more channels, however, we can extend our observations into another dimension, the spectral dimension, which allows us to look for channel-to-channel variations in image lightness of a given area. From experience we do not expect that shadowing and shading will cause much variation in the spectral dimension; therefore, areas in an image that are mainly influenced by shadows and shading will appear similar from channel to channel, for example in a color composite. Patches of ground that are not strongly affected by shadowing and shading will reveal the intrinsic capacity of the surface material(s) to absorb (and emit) light. For many

Channels and bands
We refer to discrete wavelength regions measured by spectrometers and spectral cameras as "channels," rather than "bands." To spectroscopists, bands have specific meaning as wavelength regions in which physical processes of absorption or emission occur. Since our subject is the meeting ground of spectroscopy and imaging, we need to use terms that are clear in both contexts.

materials, absorption and emission vary with wavelength, and this is what lets us discriminate and identify materials spectroscopically.

There are circumstances when the spectral properties of surface materials mimic those of shadows and shading. Some materials, for example rocks that contain metals, metal oxides, carbonaceous material, or dark glass, can be strongly absorbing in visible and near-infrared (VIS-NIR; 0.48 to 1.2 μm) wavelengths. We commonly can resolve ambiguities between dark surface materials and shadow/shading by context, because shadows have a consistent orientation that is dictated by the direction of illumination, unlike patterns of surface materials. We could solve the problem directly if we had another image taken at a different time in which the shadow pattern had changed, or taken from a different angle (see ahead to Figure 8.8), or if we had DEM or other topographic information from which we could predict the locations of topographic-scale shadows. It also is possible to identify shadows using thermal images. In the daytime, shadows have relatively low radiant temperature, whereas fully illuminated, dark surfaces generally have higher temperatures.

1.4 Spectral analysis of images

1.4.1 Empirical approach

Recognizing spatial and spectral patterns

At the most basic level we begin to extract information from images by looking for patterns. In the context of spectral images, we define a pattern as "a regularity in the data." We can find regularities in image data without understanding what they represent on the ground. At this point we are looking for spatial patterns and spectral patterns, and for combinations of the two; in Chapter 8 we bring in temporal patterns as well. Patterns can be described, measured, and correlated with one another and with other information. Patterns in the data stand in contrast to random and non-random noise, and, therefore, give potential information about the ground. Patterns of unwanted noise, such as image defects, often can be recognized visually. We have noted that we already are skilled at finding visual spatial patterns, but that we have far less experience with the spectral properties of the world, especially outside of the visible wavelengths. However, humans are much better at pattern recognition than computers, especially when the subject is viewed or illuminated in various ways, or when the spectral quality of the illumination changes.

> **Empirical**
>
> Derived from or guided by experience; depending upon experience or observation alone, without using scientific method or theory.

Relating patterns and experience

The next step in the process of extracting information occurs when we relate patterns in images to experience. This is the same empirical approach that we all use daily to recognize objects in the world around us. To use a remote-sensing example, at an appropriate scale it is easy to recognize an agricultural area in a black-and-white image by the familiar patterns of the crop fields. We could discriminate the crop fields further by making color images. It would not even be necessary to know what wavelength channels were used, as long as some of the fields had different colors on the image. A skilled interpreter or a local farmer might be able to use this information to discriminate or identify crops based on experience. A computer also might be able to "learn" to recognize crops by their spectral properties in a multispectral image. Although it may be tempting to conclude that this process leads to identification, actually, it only establishes the similarity of spectral patterns.

Information from correlations

Once we have correlated patterns in the image data with attributes of the ground surface, we can begin the process of extracting information. Subjective judgments are involved here. Unimportant clutter to one person may be useful information to another. What is defined as important depends on the objective of our study, and there is no standard measure of the quality of the evidence that is needed to arrive at information. We also need to keep in mind some human behavioral quirks. For example, once we have found a good match we are tempted to end a search, rather than to consider other possible solutions. Double-checking is boring, so we generally do not like to explore an image carefully to evaluate how good our solution really is. We like to find quick answers and move on. Skilled interpreters learn to resist this bad habit.

The empirical approach to image analysis described above is based on correlations, and not on physical models of how the radiance recorded by the image is related to the ground surface. Correlation is an essential first step in the process of image interpretation; however, it is entirely possible to correlate patterns, such as the shapes of crop fields, spectral signatures and agricultural practices without knowing how or why the correlations occur. Lacking this vital information, we can never be sure that our analyses are valid anywhere other than where our ground control – our experience – was established.

Physical model
A simplified description of one or more physical processes that allows one to make predictions.

1.4.2 Physical models

We also can relate patterns by means of physics-based, or physical models. The descriptions themselves typically are (or can be) expressed mathematically, and are derived from controlled, repeatable experiments and from theory. Models are simplified versions of reality, because no one has all of the information necessary to describe the full detail of the workings of the natural world, especially at landscape scales. Physical models are considered robust when they are highly effective at predicting the function and results of physical processes. Especially important in remote sensing, physical models are robust when they can be applied anywhere, far from the geographic source of the original observations.

Robust physical model
A model that is insensitive to noise or undesired or uninteresting clutter.

When doing field work we naturally organize information using physical models in order to understand processes and to be able to make predictions. Therefore, we can use physical models to interpret spectral images whenever we are able to understand the image data in terms of processes on the ground, particularly those processes that affect the measurements themselves. We especially need the physical-modeling approach to interpret spectral data for the non-visible wavelengths, for it is here that our visual, empirical tools are least effective.

Physical modeling is a "surface-up" approach that begins with consideration of what is on the ground and how materials will be expressed in spectral images. In contrast, the empirical approach is "image-down," in the sense that analysis is done first on the image, and then an attempt is made to interpret the results in terms of what is on the ground. In a physical model the spectral properties of the ground surface are predicted, based on knowledge of surface materials, and then these "forward-modeled" properties are compared with the measured spectral data in an image. A model is evaluated by measuring the residual between the predicted and the actual measurement. If the "fit" is good, a model is consistent with the image data and can be considered as a basis for interpretation. If the fit is poor, we need to construct a new model and try again.

Testing models
A model makes a prediction. The residual between the prediction and the actual measurement determines the accuracy of the model.

In spectral remote sensing, physical models entail some form of spectral matching. In the simplest case, we compare the spectrum of a well-understood training area with the spectra of other areas on the same image. Comparisons of laboratory spectra with image data require calibration (Chapter 3). Models become more complex when they include mixtures of spectra of multiple materials (Chapters 2 and 4). For spectral matching to be convincing, we need to be able to formulate equations that link the ground and the image data. To do this, we must understand the processes that give rise to the features in the spectra of the materials on the ground, and we must understand the factors, such as atmospheric effects and illumination geometry, that alter spectra in a remote-sensing context. Although we want equations that take everything into account, in fact, in the real world we usually are satisfied to make a good estimate.

Physical models

Physical models commonly are described by an equation that relates attributes of the ground surface to attributes of a spectral image. By inverting the equation, spectral image data can be interpreted in terms of physical processes on the ground. Under unusual circumstances, some inversions can produce a unique interpretation. More often, there are several possible interpretations that must be evaluated further in the context of spatial and other information. Usually this occurs because there are too many important characteristics of a scene, and too many important processes, all to be deduced from a limited set of spectral observations. A straightforward way to evaluate an estimate is to measure the difference (channel by channel) in the value of a model spectrum and the image data. For all channels together we can express "fit" as a root-mean-square (RMS) error. If the error is large we may need to revise our equation. Even if the RMS term is small, we may not have a good fit if a single channel, or a small number of channels, have large differences that are masked by the overall good fit. Examination of the RMS error is, however, a first step. The equation that describes spectral mixing, Equation (4.2), is an example of a physical model applied to images.

To illustrate the general idea of how a physical model works let us return to the previous example of a spectral image of an agricultural area with fields of crops. Consider a situation where we are searching for a particular type of crop, for example, wheat. We have little information about the actual location of wheat fields at the time the image was acquired and, although we can use various standard techniques to discriminate many different fields, we are not sure which fields contain the sought-for wheat. We cannot go to the field to make an empirical correlation with our image, because the seasons have changed since the image was acquired, and the crops have been harvested.

We approach the problem from the ground up by finding field or laboratory spectra of the crop of interest. If we have not made these measurements ourselves, we may be able to find published spectra. Knowing the spectrum (or range of spectra) we estimate how it would be distorted by a measurement made through the atmosphere and by differences in illumination, and we predict the spectrum that would be measured in our image. Any pixels that have the predicted spectrum are candidates for the target crop. If our crop has a distinctive spectrum (which will depend on the material itself, the background, and the spectral resolution of our image) we will be able to narrow our search, or perhaps even make a confident identification using the image.

Of course, we will want to evaluate our physical model in a spatial context using the empirical method of photo interpretation. For example, we might not be pleased if all the pixels that fit our spectral model were sprinkled about the image like random noise. In this case we probably would revise our model, because experience tells us that the crop we are looking for always is planted in cultivated fields, not scattered in small patches. It would be more consistent with our understanding of agriculture in the area if our target pixels occurred in large clusters that have regular outlines. We then could interpret the image in light of the spectral evidence.

There is one more dimension that we can use to strengthen a physical model: time. We can use a series of spectral images obtained over time to test physical models of processes that change the land surface. Patterns of temporal change, like spatial patterns and spectral patterns, can be used to interpret processes and to help discriminate and identify materials. Consider, for example, that our target crop (wheat) has a spectral signature nearly identical to another crop grown in the area – barley, perhaps. Our spectral analysis would locate all of the crop fields that could be either wheat or barley. However, if the two crops were planted, matured and harvested at different times, and if we had a time series of images, we could resolve the ambiguity. The time series could consist of just two images, providing that they were obtained when differences between the two crops were evident.

Time series
Several spectral images acquired over a span of time strengthen image interpretations, especially for land use.

Because there are many ways to extract information from spectral images, we usually have the problem of deciding what analytical methods to use under what circumstances. There is no single, best method, for the same reasons that there is no one way to approach field work. It depends on what we want to find out, and on the nature of the data. Some scientists characterize field work as "messy," because there are so many uncontrolled variables. Images that fuse spatial, spectral and temporal information also are messy, and we will need a flexible strategy that lets us use a variety of analytical tools.

1.4.3 Thematic maps

Some of the information that we extract from spectral images can be displayed in the form of newly created images that convey certain specific ideas. These new images commonly are known as "thematic maps." Thematic maps are interpretive, and by definition they are limited to certain themes. In this book we further restrict the use of the term "thematic map" to those images that demonstrate a physical basis for interpretation. One kind of thematic map, a *physical-attribute map*, portrays continuous properties of the scene, for example, radiant temperature (Figure 1.13) or relative absorption by chlorophyll. The other kind of thematic map, a *unit map*, is a physical-attribute map that has been taken one step further in terms of interpretation by defining the spatial limits of the described properties, as well as the nature of adjacent units. Thus, boundaries exist between units, and each unit has a different

Thematic maps
Thematic maps that are derived from spectral images convey certain ideas about the nature of a landscape.

Figure 1.13. Thematic maps of Owens Valley, California that are interpretations of physical properties of the scene. (a) Radiant temperature; lighter tones are warmer. Remotely measured radiances are interpreted as radiant temperatures by assuming emissivities. From Landsat TM Band 6, October, 1983. (b) Thematic map with units. Vegetation map of Owens Valley. Vegetation units are interpretive, and are based on compiled data from ground measurements, aerial photography and satellite images. Modified from Smith *et al.* (1990) with permission from Elsevier.

(a)

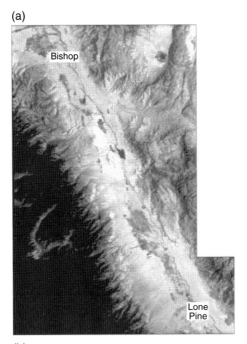

(b)

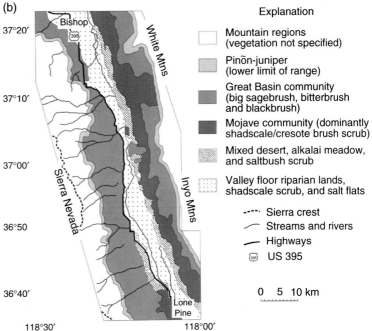

interpretive label (Head *et al.*, 1978). For example, Figure 1.13b is a unit map of vegetation types that is based on interpretation of aerial photographs, spectral images and field measurements. Other kinds of unit maps exist for the same area, including soil maps and geologic maps. Thematic mapping is discussed further in Chapter 7.

Proxies

The basic measurements in spectral remote sensing are of reflected or emitted radiance from the ground surface. We can improve the accuracy of these measurements by correcting the data to remove or diminish the unwanted instrumental and atmospheric effects (Chapter 3). Aside from augmenting photo interpretation, radiance is difficult for most of us to convert into useful information without invoking physical models. Consider thermal-infrared (TIR) measurements. If we make the assumption that a surface radiates like a blackbody, we can convert radiance measurements into brightness temperature (Chapters 2 and 3). If we know or assume how parts of a surface depart from blackbody behavior (emittance) we can create a map of radiant temperature. Temperature is a physical attribute that we can understand and interpret. In this example, radiance is a substitute, or proxy, for temperature. In fact, extracting spectral information from images consists largely of finding ways to use radiance measurements as proxies for other, more interesting physical attributes of the ground.

> **Radiance proxies**
> Image analysis involves using radiance as a proxy for other properties of the ground surface that we are interested in.

Proxies need to be robust. Unless radiance, and calculations derived from radiance, are convincing substitutes for other attributes, we may be unable to extend our interpretations beyond simple empirical observations or correlations, and we run into problems validating our results (Section 1.5). Physical models provide the most solid basis for extracting information from images (Section 1.4.2), but because physical models are a simplified version of reality, it is not surprising that proxy measurements have to be applied carefully. For example, when we use radiance as a proxy for radiant temperature, it is easy to forget that the physical model has a built-in assumption about emissivity that we know is far too simplistic for real landscapes.

Vegetation indices (Chapter 3) provide another familiar example of how we use proxies. A vegetation index is based on the physical model that radiation at red wavelengths (about 0.66 μm) is strongly absorbed by chlorophyll, whereas radiation in the near infrared (NIR) (from about 0.8 to 1.1 μm) is strongly reflected (and scattered) by typical leaf structures. In the simplest application, approximate boundary conditions are defined by (1) a surface having no green vegetation, and by (2) another surface entirely covered with green vegetation. The same ratio also may vary, although less dramatically, across surfaces such as bare soil or rock that have no chlorophyll, in which case the proxy breaks down, and it is meaningless to use the term "vegetation index" for an algorithm when there is no vegetation. When the index is used as a proxy for percent cover of vegetation, as is common, it no longer applies to variations across a landscape having 100% vegetation cover. In both of these cases, other physical models must be invoked. We have to remember that just because something is called a "vegetation index" does not mean that it always is measuring vegetation. In fact, it is easy to forget that we are not actually measuring vegetation in the first place, only a radiance substitute for the real thing. This observation applies to many of the algorithms that we use in image processing to extract information. The lesson is that any algorithm that invokes a physical model must be evaluated critically, and the main assumptions and limitations need to be understood.

Radiance and vegetation

It is easy to forget that we are not actually measuring vegetation directly, only a radiance substitute for the real thing.

Linked models and derived attributes

It is common in remote sensing to link physical models together to develop interpretations. The basic idea is that remote measurements of radiance are linked to one physical model, and the first model is linked to another physical model, and so on. An example is the use of satellite images (AVHRR) to predict outbreaks of Rift Valley fever in eastern Africa (Linthicum *et al.*, 1999). A modified outline of the story goes like this. (1) Radiance is measured at two wavelengths that are sensitive to green foliage, as in the examples above. (2) A normalized difference of the radiance measurements is used as a proxy for the amount of green-vegetation cover. (3) Changes in the amount of green-vegetation cover, adjusted for seasonal variations, are used to show regions of recent greening. (4) Greening that is attributed to heavy rainfall during times of El Niño southern oscillation (ENSO) predicts a subsequent increase in the population of disease-bearing mosquitoes. (5) Where mosquitoes proliferate, there is a health risk to livestock and to the human population from Rift Valley fever. The product of the analysis is an image showing the final attribute: areas of predicted increased health risk. The logic is compelling. When it rains, dry areas become green, and wet areas breed disease-bearing mosquitoes. It is reasonable to predict, therefore, that the local incidence of disease would increase in areas that have recently turned green, and, indeed, the authors present evidence that this actually happens.

Not all linked models work as expected, however, and caution is warranted. The final conclusion will be wrong if one model is wrong. Furthermore, derived attributes become less convincing when we try to press this or any other chain of models for accurate, quantitative results. Recalling that physical models are simplified versions of reality, we know that each model contains assumptions and uncertainties. The errors and uncertainties in one model usually propagate into the next one linked to it. The longer the chain, the greater the opportunity for accumulated error. The process is especially vulnerable to weak links that are not based on sound physical models but instead on correlations where cause and effect are unclear.

1.5 Testing and validating results

1.5.1 Accuracy, error and validity

Why results go wrong

Whenever we extract information from spectral images, we would like to have assurance that our results are *valid*, meaning that they are sound, well-founded, logical and convincing. Results of remote-sensing analyses are considered valid if they can be confirmed by an independent means, usually by observations and measurements made on the ground. The land surface is our standard frame of reference, and the observations and measurements that are made in the field are different from the ones made with spectral images (Chapter 7). The simplest test of validity is whether or not a label that we apply to part of an image agrees with one that we would apply to the corresponding area if we were in the field. When labels agree, we judge that our image results are valid; when they do not agree, we have made an error. On a per-pixel basis, labels are either right or wrong, but when we consider aggregates of pixels, we are faced with the question of how many pixels, and which pixels, we are willing to tolerate being in error. Unfortunately, to validate results, we cannot just rely on some generally accepted score of right vs. wrong pixels. Our own judgment is required, because our tolerance for wrong pixels depends on *which* pixels are wrong and what we are trying to do. Similarly, our judgment is needed when we try to understand why labels are not correct, for we become drawn into an analysis of where the unacceptable errors originated. Judgment also is required when we try to validate per-pixel, quantitative measurements; for example, percent vegetation cover or the proportion of a mineral species. In these cases, valid results depend on whether the image estimates are within allowed error tolerances, which themselves depend on the objectives of each study.

> **Accuracy and validity**
> Terms related to accuracy and validity have been used in many different ways in different contexts. Accuracy and validity are different concepts, and accuracy does not imply validity. *Valid*: well-founded; producing the desired result; effective. *Accurate*: free from error. The extent to which a given measurement agrees with a standard value.

Incorrect results and results that do not meet our error tolerance can arise from inaccurate measurements, errors in analytical procedures, or faulty interpretations. Intuitively, some minimum level of accuracy is required to be able to produce valid interpretations from spectral images. If our image measurements and calculations are seriously in error, our results are not likely to be valid. How accurate do measurements and analyses have to be to give valid results? There is, of course, no simple answer to this question. The same measurement may be quite valid for one purpose but not valid for another. Consider a hypothetical case where we process a spectral image to study the cover of green vegetation. Imagine that we have two objectives in mind. First, we want to have a generalized picture of the distribution of green vegetation at a particular season to compare with other seasons. Second, we want to know the percent cover of green vegetation at each point in the image. If the *pattern* of vegetation cover "makes sense" and is consistent with the time of year, we may make the judgment that these results are valid; that is, the pattern is sufficiently accurate for our purpose of comparing seasonal cover of vegetation. However, on a *per-pixel* basis, we may be unable to confirm our estimates of percent vegetation cover simply by photo interpretation, and we would need independent measurements from the ground.

For simplicity in developing this example further, let us assume that we can access the field under the same conditions that the image was taken. Because it is impractical to visit every pixel footprint on the ground, we also will assume that we have captured the reality of the field conditions by acquiring a statistically robust sample. And we will assume that the accuracy of the measurements that we make on the ground is adequate for the objectives of our study. Next, we need to decide on the per-pixel accuracy that we require for our study. Consider two extremes: At one extreme we require very high accuracy, say, 2% of the amount of vegetation cover actually present. This means that if the actual cover is 42%, our image estimate has to be 41, 42, or 43%, otherwise the results are not valid. Such high accuracy is elusive even with careful field work in a 1-m^2 plot, and it is not realistic for larger field areas or at remote-sensing scales. If we demand such high accuracy, we probably will find that our image results are never valid (perhaps excepting rare pixels where image and ground values just happen to coincide). At the other extreme, let's say that we can tolerate 50% error and still achieve our study objectives. This means that although the actual value is 42% cover, we would be satisfied if we got numbers from 21 to 63%. In this case, we might be able to validate the estimates of nearly all of the pixels. Realistic cases probably will lie somewhere between the extremes in our hypothetical examples. The point we are making, though, is that to discuss validation we need to consider the degree of error that can be tolerated and still meet the objectives of a study. In fact, error tolerance is a useful thing to know at the start of a study. There may be no point in doing an analysis if we know from the beginning that we cannot validate our results. (Analyzing an

Validation of models

Validation pertains to models, not data. A model is valid if its predictions are found to be within a desired level of accuracy.

To discuss validation, we need to consider the degree of error that can be tolerated and still meet the objectives of a study.

image for a study that demands better than 1% accuracy in vegetation cover is an example.) Then there is the price in time and effort devoted to the analysis that must be paid to achieve a given level of accuracy. It may be technically feasible, but is it worth the cost? Curiously, tolerance for error rarely is specified, or at least reported, by image analysts.

High analytical accuracy does not necessarily mean that our results will be valid as judged by field observations. It is quite possible to make very accurate measurements using image data that do not make sense in the context of our objectives. The measurements and analyses may be accurate, but the interpretation that follows may not be valid unless we have established a physical connection between the remote measurements and the materials on the ground. Consider another hypothetical example in which we process a Landsat TM image using a maximum-likelihood classifier (Chapter 3) to map areas of deer habitat. A training area known to be inhabited by deer is selected to guide the classification. Internal statistical checks on the robustness of the classification show that all areas like the training set (Chapter 3) can be discriminated from the rest of the image with high confidence. Unfortunately, field checking by wildlife biologists reveals that the classified image has no predictive value. We are puzzled. After all, the training area is known to be a good one. (Stay with us. This is not that far-fetched. We have seen it happen.) The answer, of course, is that we did not establish a physical connection between the image training area and deer. Thirty-meter TM pixels cannot resolve the individual animals, so there is no point in even thinking about the spectrum of deer hair. What is it about deer habitat that might be definitive? A wildlife biologist would recognize a number of attributes in the field such as availability of food (of many types), distance to water, and areas of cover and shelter. But these attributes are not *spectrally* specific. Then there is the problem of scale. Specific types of food for the deer that would be apparent on the ground probably would not be detectable at the Landsat scale. Thus, the training area that was selected was only a key to finding other spectrally similar areas. In this example, our analysis is flawed because of the lack of a physical link between spectral images and the animal habitat. It does not matter whether the image classification accurately finds all other areas that have the same spectral properties as the training area, because high classification accuracy per se does not guarantee results that make sense.

Another hypothetical example may help to illustrate the importance of framing the objectives and the analysis realistically so that the results can be validated. An analyst has just compared a 200-channel image and a geologic map of the same desert region. Only a few of the high-resolution spectra from the image can be interpreted in terms of the rock compositions shown on the geologic map. Most of the

> **Accuracy and meaning**
> Just because measurements are accurate does not mean that they are useful or sensible.

spectral data, though, do not make sense in the context of the mapped geologic units. To field geologists, the reasons for the discrepancies are evident. A geologic map is an interpretation, whereas a spectral image records the surface exactly as it is – including alluvium, colluvium, soils, weathering products, and vegetation. Field geologists commonly ignore surficial materials and concentrate on fresh, un-weathered rocks. A geologic hammer is used to break away surface weathering, and scattered outcrops may be adequate to interpret the subsurface distribution of unseen materials. Nevertheless, even when little or no fresh rock is revealed at the surface it sometimes is possible to identify rock types in spectral images by secondary properties. For example, some of the rock units may have distinct mineralogical properties resulting from chemical weathering. Other units may show distinct textures resulting from physical weathering. Others might have characteristic vegetation. Notice, however, that in our example the geologic field map specifically left out "overburden" (alluvium, colluvium, soils) and vegetation, whereas the image included this unwanted information. Even where fresh rock is present, there is no guarantee that a unit mapped by a geologist could be identified spectrally. Most geologic units, in fact, are defined in the field by non-spectral attributes such as grain size and texture, internal structure, chemical composition or fossils (Chapter 7).

The examples above illustrate the importance of defining realistic objectives based on an understanding of the physical processes that connect an image to the ground. Unfortunately, there often is a gap between the objectives of a user of spectral images and the realities of what actually can be determined from spectral measurements. As discussed in Section 1.4.3, it is sometimes possible to achieve our objectives by linking radiance to one or more well-founded physical models, but we certainly are not justified in thinking that we can always use radiance as a proxy for whatever we want to measure. If mapping deer via habitat with Landsat TM is likely to be a lesson in frustration, it would be good to be aware of that in advance (especially if you are funding the project).

Tradeoffs between accuracy and spatial scale
We know from a practical point of view that spatial resolution and spatial coverage are inversely related; if we want very high spatial resolution of objects on the ground we usually will have to accept less spatial coverage, and vice versa. In a similar manner, spatial resolution influences our ability to validate the accuracy of spectral images by field measurements (Section 1.5.3). If we consider radiance measurements alone, it is more difficult to compare ground and image measurements at coarse spatial scales where individual pixel footprints are

relatively large (e.g., kilometers) than in cases where pixel footprints are relatively small (e.g., meters). At fine scales, it is easier to locate the corresponding area on the ground and to make representative ground measurements, whereas it is more difficult and costly to study large regions on the ground. We accommodate to these realities of scale by adjusting our expectations for accuracy. For example, we might not expect (calibrated) radiance measurements for half-kilometer-sized MODIS pixel footprints to agree as closely with ground measurements as would be the case for 17-m AVIRIS pixels. Even if we are willing to pay the price required to do extensive field work, accuracy still depends on the extent to which a measurement agrees with a "standard value." Standard values of measurement are most common at fine scales (e.g., meters and below) where we can make laboratory measurements under controlled conditions.

> **Expectations**
> We adjust our expectations for accuracy according to spatial scale. Expectations are different for small pixel footprints than for larger ones.

Tradeoffs between accuracy and detectability
When we talk about "extracting information" from spectral images there is a connotation that the information that we want is imbedded in a background of other, unwanted data. Sometimes, desired information "pops out" of the background; other times it has to be teased out; and sometimes we cannot find what we want among the clutter. In this book, we use the term "detectability" in discussing the degree of ease or difficulty that we encounter in extracting a given piece of information from a given background, and we frame the overall problem as one of separating a target from a background (Chapter 6).

Accuracy and detectability are related in complex ways. From experience, we know that whenever something stands out spatially or spectrally from its background it is easier to identify it or to make accurate measurements of it. Conversely, whenever something is hard to distinguish (spatially or spectrally) it means that the information that we are looking for is so similar to the unwanted information that we have trouble telling them apart. In the simplest case, we assess the detectability of one object set against the background of another, say object A against object B. Spectrally, there are straightforward ways to compare the contrast of A vs. B at each wavelength, and to determine whether measurement accuracy and precision are sufficient to tell the two apart. The situation becomes more complex when we try to detect pixels of type A in a matrix of A, B, C, and D pixels. The more other types of pixels that are present, the greater the likelihood that there will be some that have attributes similar to, or even the same as, type A. Accuracy is at stake, because we introduce error into our image analysis each time we confuse some other spectral type with type A. The opportunity for error due to mimicry is further increased when spectral component A is not spatially resolved, but instead is imbedded in a spectrally mixed pixel that also contains components A, B, C, and D. This, of course, is what happens in remote sensing when a pixel measurement integrates the signals from multiple materials within a pixel footprint on the ground. The general conclusion at this stage in our discussion is that spectral detectability of any one component in a landscape becomes more difficult, and accuracy in measuring component attributes decreases, as the complexity of the background increases. This topic is discussed further in Chapter 6.

Replicating results

Consider a situation where we have made a thematic map from a spectral image. We do a field check, and our interpretation is valid in every area that we visit. Then we are asked how we conducted the image analysis. Unfortunately, because we processed the image by trial and error and did not keep a record, we cannot remember what we did. Worse, we cannot replicate our results. Are our results still valid? Well, one definition of "valid" includes the words "sound," "well-founded," "authoritative," and "convincing." It is a matter of judgment whether our results are still valid, but we certainly do not present a convincing case when we are unable to demonstrate how we arrived at the results and cannot replicate them. Keeping careful records would appear to be desirable and even necessary to permit validation (e.g., Jensen, 1996, p. 11); nevertheless, details of analytical procedures in image processing are not always reported in published articles, and they hardly ever accompany unpublished images.

1.5.2 Errors in the acquired data

Those of us who use spectral images to address field problems are concerned with testing the *interpretations* that are derived from the remotely sensed data. This job is distinct from that of remote-sensing specialists who use laboratory and field tests to verify the radiometric and geometric accuracy of the *data* acquired by satellite and aircraft systems to see how well the systems function. Field testing the accuracy of measurements of radiance over pixel footprints on the scale of meters to kilometers is a formidable challenge – one that the practically oriented field observer can be glad to sidestep. Typically this involves calibrating the detector system under laboratory conditions, making measurements of the atmosphere at the time of overflight, characterizing test areas in the field, and estimating the sources of error one at a time.

Remote-sensing textbooks generally partition data errors in spectral images into those that are internal to the detector system itself and those that are external, such as atmospheric and topographic effects. Calibration is the process by which both internal and external errors are corrected – or at least minimized (Section 3.2). Fortunately, images from many of the standard sources already have been corrected for the most obvious internal geometric and radiometric errors by the time they reach the analyst. External errors usually are the most difficult to fix, because they were produced at the moment of data acquisition, and it is unlikely that anyone was on the ground at that time making all the needed measurements of the true conditions. To the field investigator, though, data errors are only a problem to the extent that they limit or cause mistakes in interpretation. As discussed in Section 1.5.1, error tolerance depends on the field application and the nature of the information that is being sought. Much work has gone into developing formal ways to detect and remove errors from spectral images, but there are no guidelines regarding how much error is allowable for specific applications.

> **Error tolerance**
>
> Error tolerance depends on the field application and the nature of the information that is being sought.

1.5.3 Testing and validating in the field

If observations and measurements made in the field are our standard frame of reference for testing and validating interpretations derived from spectral images, what do they entail, and how difficult are they to make? Field validation often is visualized by image analysts simply in terms of "checking" classification images or thematic maps, where each pixel has certain data attributes or a label. At first glance it would seem that all we have to do is locate a pixel footprint on the ground and decide whether that pixel is right or wrong. In reality our task is not that simple. In Section 1.5.1 we pointed out that we have to adjust our expectations for accuracy according to spatial scale. In Chapter 7 we discuss the differences between what an observer on the ground can see and what is recorded by remote sensing. For field validation, we need to address the following problems: (1) locating pixel footprints on the ground, (2) making field measurements of attributes, and (3) evaluating per-pixel labels.

> **"Ground truth"**
>
> The term "ground truth" is misleading, although it is commonly used. The problem is that the "truth" is not extracted from the field, only observations and measurements that will vary from user to user. Some researchers have abandoned remote sensing prematurely, because they did not understand why radiance measurements and ground-based observations disagreed.

Locating pixel footprints

Field investigators routinely use aerial photographs to locate themselves relative to resolved ground features, but at coarser resolutions the choices of recognizable spatial patterns become narrower. At the scale of nominal Landsat TM or ASTER pixels (30×30 m), familiar objects such as trees or small buildings no longer can be resolved, although high-contrast, sharp edges such as roads, streams and boundaries of cultivated fields are still recognizable. At coarser scales, such as that of MODIS (500×500 m) and AVHRR (1×1 km), it is not realistic for most applications to define pixel footprints in the field using photo interpretation alone. In the cases of ASTER or Landsat 7, the availability of co-registered panchromatic channels with 15-m pixels has made it easier to interpret the spatial patterns in the multispectral channels. In principle, one can co-register any high-resolution image, including aerial photographs, with coarser-scale multispectral data, but there are practical limitations on the time and effort required. Considering the limitations of photo interpretation, the most attractive alternative for locating pixel footprints is to use global positioning systems (GPS) in conjunction with rectified and geo-coded multispectral images. This subject is outside the scope of this book; however, it is covered in several remote-sensing textbooks.

Validating attributes

> **Validating attributes**
>
> Validating image attributes can be more difficult than validating classification images or thematic maps.

Attributes of landscapes can be more difficult to validate than per-pixel labels, because we need to measure continuous, variable parameters on the ground. For example, when we produce an image that we interpret in terms of percent cover of vegetation, we can test our results in the field by measuring transects (the amount of vegetative cover intercepted along reference lines) or quadrats (the area occupied by vegetation within defined rectangles). Assuming that we actually can make accurate measurements on the ground, we now must decide whether the values for the corresponding pixels are correct, and this requires us to set a threshold value for tolerable error.

The issue of whether we can make accurate measurements on the ground that are representative of pixel footprints is not trivial. For example, in the case of vegetation cover, in many areas it is difficult to make field measurements at all, because of factors such as rough terrain, thorny shrubs or indistinct canopy boundaries, to say nothing of making enough measurements to be statistically robust. In addition, there is the fact that standard intersection and area methods do not take into account small gaps in which branches and substrate are exposed when seen from above. In this case, spectral images of desert shrublands will underestimate vegetation cover relative to the field measurements that consider each shrub as an unbroken entity. In principle, both methods can be accurate but still not agree. Other difficulties occur in forested areas, where it is impossible to measure canopy attributes such as the proportions of green leaves, branches and flowers from the ground, and one must turn to very high-resolution (low-altitude) aerial measurements, or, in special test areas, research towers that rise above the canopy (Figure 1.1).

> **Accuracy and agreement**
>
> Field and image measurements can both be accurate but still not agree.

Validating per-pixel labels

We can relax the requirement to locate individual pixel footprints by validating representative areas. For a large block of pixels having the same label, it often is sufficient to verify the label by inspecting a representative part of the ground area. In practice, this is done routinely for well-understood, homogeneous parts of landscapes such as bodies of water, some forests and grasslands. If we already are sure that labels are correct (or wrong) for representative pixels, there is no need to try to find and visit any particular pixel footprint. Near the boundaries between types of land cover, though, it is more important to validate pixel labels individually. This means that spatial patterns affect the strategy for field validation, because there is more field work to be done when there are many small patches than when there are just a few large ones. Unfortunately, the spatial properties of types of land cover usually are not reflected in statistical classifications and estimates of their accuracy. For example, it is common to see a single value, say, 89.5% accuracy, attached to a classification image. This number implies a high level of accuracy, but it would be misleading if, in fact, most of the correct pixels were in one homogeneous class that occupies most of the image. Statistical sampling strategies and the use of error matrices are discussed in detail in the technical literature.

On the ground, checking per-pixel labels can range from straightforward to an exercise in frustration. In some cases it is simply a matter of experience. For example, an agronomist presumably would be able quickly to identify all of the different crops on the ground, in contrast with a less experienced observer. In other cases, even skilled observers on the ground may not be able to validate pixel labels if they are based on land use rather than on physical or spectral attributes (e.g., cow pasture vs. horse pasture; retail vs. wholesale facilities). In Chapter 7 we further explore the ambiguities inherent in interpreting labels on thematic maps that are derived from spectral images.

1.6 Summary steps for extracting information

Evaluate the data

In Chapter 3 we discuss the importance of always evaluating a data set before beginning to work on an image. We recommend using a checklist to make sure that the structure of the data is understood and to reveal any problems.

- **Do a photo reconnaissance**. An important part of the process of evaluating a spectral image involves photo interpretation; therefore, data evaluation blends with the need to become familiar with the context of the image. Field investigators presumably already will be familiar with the landscape on the ground. This also is the time to gather together topographic maps, thematic maps, air photos, stereographic images, ground photos, sketches, reports, or any other relevant information. It is highly unusual to be faced with a spectral image of an unknown area on Earth that comes with no additional information. Planetary scientists, though, are more apt to encounter images of objects in the Solar System for which the context is meager.

- **Specify objectives and required accuracy.** Our objectives may be as general as finding out all we can about an area on the ground – a reconnaissance – or they may involve very specific goals. In either case, the objectives and the requirements for accuracy will determine how to carry out the analysis of an image and how to validate the results.

- **Expand from small to larger areas**. Skilled image analysts often spend much of their time on one or more small subsets of a larger image in order to acquire key information about the ground surface. If we can interpret sub-images of small areas of a landscape, we can extend our interpretations to a larger scene, in much the same way that field investigators assemble detailed observations from a few places into a broader interpretive context.

- **Use a flexible approach**. At this point we will have done a basic photo reconnaissance, taken into account what already is known about the study area, and considered our objectives in light of physical processes that link an image to the ground. Ideally, the choice of analytical tools will be guided by an understanding of spectroscopy. But, realistically, we learn as we go, and no matter how much background and experience we have, there may be surprises along the way. Indeed, this is how we discover new and unexpected things in spectral images. The analytical process is not linear. Expect to try out different methods and refine results by several rounds of iteration. In the process, as we learn more about a scene, our objectives may evolve, or change significantly. Keep in mind the hypothesis-testing approach and the similarities between image analysis and field work.

- **Move from simple to complex**. We concluded earlier that the way to extract the most information from spectral images was to integrate spatial, spectral and temporal data using physical models; however, this also is the most difficult path, one that often requires calibration and correction for variations in illumination. We do not want to engage in a difficult analysis if we can achieve the desired results by one or two simple procedures. An important ground rule, then, for working with a new image is to start with simple operations and, as necessary, to move to more complex ones. There is more to this ground rule than just avoiding unnecessary work. By starting with simple operations and building each analytical step on the results of the previous ones it will be easier to sustain an understanding of the physical connection between the image and the ground surface. The quirk of human behavior that urges us to seek quick answers means that we often are tempted to unleash the latest algorithms on our images in the hope that amazing results will appear like magic. This is a gambler's approach. It may pay off, but the odds are stacked against us. When we begin our analysis by applying a complicated procedure that we do

Flexibility
The advantage of the trained human analyst over an "expert system" algorithm is the flexibility of the former. The advantage of the algorithm is its consistency.

not fully understand, we also defeat the problem-solving process, because we may not see what to do next if we get stuck.

- **Document all work**. After working on a spectral image for a few hours or days it is easy to get lost and to forget exactly what we have done. Documentation is the key to being able to replicate results. If we cannot replicate our results we probably do not understand what we did, and we certainly will not be able to convince anyone else that we do. And, without documentation we may not be able to validate our results. Unfortunately, the tendency to seek quick answers conflicts with our need to record carefully what we do, and for some analysts it may even be more difficult to keep notes than to solve the analytical problems.

- **Validate the results**. Image analysts who do not have the experience, skills, or resources to do field work must rely on others to validate their results – or else omit validation entirely. Field investigators, on the other hand, are in a better position to specify the degree of error that is tolerable in the image analyses and interpretations, and to verify the image results by measurements on the ground. We owe it to those who use our results to state explicitly how validation was done and how we judge the soundness of the information that we have extracted.

- **Troubleshooting**. What if our initial attempts to extract information from spectral data do not seem to work? Before abandoning the image or yelling at the cat it may be worthwhile to consider some common stumbling points.

Unrealistic expectations

We need to make sure that what we want to see in an image is physically realistic. Just because we really want to find something, or have been told to, does not mean that it can be done. As discussed earlier, if we understand the physical connection between image and ground we are less likely to be sidetracked by unrealistic expectations.

Using the wrong tools for the job

When we understand what the image-analysis tools can and cannot do we can stay focused on physical processes. Selection of inappropriate spectral channels or methods for a task may signal that we are using a hit-or-miss approach. However, if we experiment with a tool that apparently gives good results, but we do not really understand how it works, we need to be extra careful, because we may not be able to replicate our results.

Calibration problems

For some purposes, instrumental and atmospheric effects must be removed to achieve sufficient spectral fidelity. Check to make sure that you have used an analytical tool that is minimally sensitive to calibration problems, or that you have made the necessary corrections to the data (Section 3.2).

Variable illumination problems

Spectral properties across an image are difficult to compare when the illumination of a scene is not uniform. Topographic variations in lightness are an especially common cause of problems for spectral analysis. Use tools that minimize the effects of variable illumination (Section 3.2).

Next

What is the spectroscopic basis for interpreting spectral measurements at landscape scales? How do we bridge the gap between the scale at which we measure spectra in the laboratory and the scales of aircraft and satellite images? If we are interested in solving problems in the field, do we really need to understand spectroscopy? These are some of the topics that we address in the next chapter.

Chapter 2
Spectroscopy of landscapes

The Needles, Utah. Photograph by D. O'Callaghan

In this chapter

Spectroscopic methods can be applied to most spectral images. Resolved resonance bands, when present, are the first choice for remote identification of materials. But, in an image context, resolved resonance bands typically are not needed to determine the basic spectral components of a landscape.

2.1 Basics of spectroscopy for field investigators

Spectral imaging systems are specifically designed to measure wavelength variations in the way landscape materials reflect or emit radiant energy. Although spectral images can be used simply for photo interpretation, their primary purpose is to make it possible to extract compositional information that otherwise would not be available from black-and-white images. Spectroscopy is the basis for understanding why materials vary in the way they reflect or emit energy from one wavelength to another. Reflectance and emittance spectroscopy are

treated thoroughly in textbooks and professional publications. The basic
physics is well understood, and spectroscopy is routinely applied in
many fields, ranging from astronomy to the analysis of industrial
chemicals. There also is an extensive literature on the spectroscopy
of natural and man-made materials, including some of the things that
make up the surfaces of landscapes. In this chapter we address only
those aspects of spectroscopy that are most relevant to an investigator
who is trying to use spectral images to address field problems.

2.1.1 Laboratory frame of reference

Those who are eager to proceed with image analysis in order to assist
with investigations on the ground may be impatient with the prospect
of pausing to learn about spectroscopy and the arcane details of
remote-sensing measurements. There are indeed many subjects that
a field investigator can postpone for another time or leave to those who
specialize in remote sensing; however, there are a few basic ideas that
are especially relevant for interpreting spectral images. One of the
most basic notions is that spectroscopic measurements, like all measurements, require a standard frame of reference. The ability of a
material to reflect or emit radiant energy at any given wavelength is
defined relative to a reference material that is measured under the
same conditions. Measurement conditions, including factors such as
sample purity, roughness, particle size, illumination source, geometry,
and temperature are best controlled in a laboratory. And for practical
reasons of sample purity and laboratory space, samples measured for
reflectivity or emissivity usually are on the size scale of millimeters to
centimeters.

> **Spectral data**
>
> Spectroscopic measure-
> ments require a standard
> frame of reference.

Reflectivity and emissivity are measures of how samples depart from
being perfect reflectors or perfect emitters, respectively. Thus, reflectivity
is measured relative to a standard material that reflects efficiently at
wavelengths of interest, and emissivity is measured relative to a standard
that emits efficiently at wavelengths of interest. In the laboratory, reflectivity typically is measured relative to stable, "white" materials such as
halon or spectralon, whereas emissivity is measured relative to a visibly
dark, cavity-filled "blackbody" of known temperature.

> **Terminology**
>
> The terms *reflectivity* and
> *emissivity* refer to measure-
> ments made in the labora-
> tory on small, smooth or
> polished samples.
> The terms *reflectance* and
> *emittance* apply to textured
> or rough surfaces although
> usage is not at all standard-
> ized in remote-sensing lit-
> erature. Most of the natural
> samples studied for optical
> remote sensing are not
> polished; hence, *reflectance*
> is the preferred term.

In a typical laboratory measurement, the radiant energy (radiance)
that is received from the sample at each wavelength interval is registered by a detector and recorded as a voltage or current; then the field
of view is changed and the standard is immediately measured using the
same instrument system. The sample measurement at each wavelength
is then divided by the corresponding measurement of the standard.
This means that reflectivity and emissivity values are dimensionless

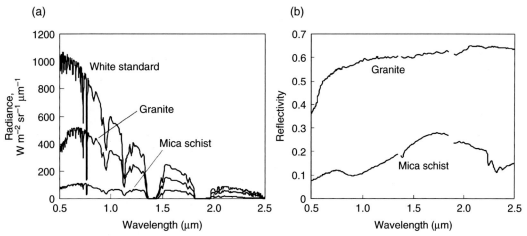

Figure 2.1. Comparison of radiance and reflectivity spectra measured by a field spectrometer. (a) Radiance spectra. Top curve is radiance of the standard, a white surface. The lower two spectra are of smooth, centimeter-sized rock samples, granite and mica schist. All three radiance spectra are dominated by the spectrum of the illumination source (in this case, the Sun) and by atmospheric attenuation. (b) Reflectivity spectra obtained by dividing the radiance spectra of each of the rock samples by the spectrum of the white standard. Reproduced from the ASTER Spectral Library through the courtesy of the Jet Propulsion Laboratory, California Institute of Technology, Pasadena, California. Copyright 1999, California Institute of Technology. All rights reserved.

and range from 0 to 1. By making ratio measurements it is easy to compare how a sample differs from being a perfect reflector or emitter and to see how the properties of any given sample vary from wavelength to wavelength. Furthermore, we can compare the spectra of samples that have been measured at other laboratories, as long as the same standard was used. Reflectivity spectra, for example, can be compared with one another even though the sources of illumination and the detectors used in the measurements were not the same. This is because, as shown in Equation 2.1, in the process of calculating reflectivity (ρ) the multiplicative factors are lost:

> **Spectra**
>
> Reflectivity spectra are independent of illumination conditions; radiance spectra are dominated by lighting.

$$\rho = \frac{L_s}{L_{ref}}\rho_{ref}; \quad \rho = \frac{\left(\rho\frac{I}{\pi}\cos(i)\right)}{\left(\rho_{ref}\frac{I}{\pi}\cos(i)\right)} * \rho_{ref} \tag{2.1}$$

where I is irradiance, L_s and L_{ref} are the measured radiances from the sample and reference, respectively, ρ_{ref} is the reference reflectivity, and (i) is the incidence angle.

The difference between radiance and reflectivity is illustrated in Figure 2.1. The radiances of the samples and the standard (a white surface) are dominated by the spectral properties of the light source and by atmospheric absorption. The important differences between the two rock samples are not apparent until the spectra are converted to reflectivity by dividing the radiance values at each wavelength by those of the standard.

Thermal IR measurements in the laboratory are made relative to a perfect emitter (blackbody). At each wavelength, the radiance of a blackbody is the maximum possible for a given (radiant) temperature. The shape of the spectral curve of a blackbody as a function of temperature can be calculated from Planck's law. This means that as long as we know the temperature of the sample and the standard we can divide the emitted radiance (L) of the sample by that of the standard to produce emissivity (ε). Alternatively, emissivity can be estimated indirectly by

$$\varepsilon = L_{\text{sample}}/L_{\text{blackbody}} \qquad (2.2)$$

measuring reflectivity and applying Kirchhoff's law that says: $\varepsilon = 1 - \rho$, where ρ is the total reflectivity measured from all directions (hemispherical reflectivity).

Hemispherical reflectivity

Surfaces that reflect isotropically are called Lambertian (Section 1.3.3). Many natural surfaces are sufficiently Lambertian that Kirchhoff's law may be applied.

2.1.2 Interpretation of spectra

Visible and near-infrared reflectance spectra

The part of the electromagnetic spectrum that is important for remote sensing of reflected light extends from about 0.4 μm to about 3 μm. This wavelength range commonly is divided into two segments: visible (VIS) from 0.4 to 0.7 μm; and near-infrared (NIR) from 0.7 to 3.0 μm. There are extensive collections of laboratory reflectance spectra at VIS-NIR wavelengths, and experimental and theoretical studies of reflectance spectroscopy are discussed in detail in the literature (e.g., Hapke, 1993; Clark, 1999).

The most interesting parts of a reflectance spectrum are the wavelengths where light has been absorbed. For absorption to occur, light must enter a sample, and not just bounce off the first surface that is encountered. Therefore, in the laboratory, spectra are measured in ways that minimize surface (specular) reflections, and that enhance the fraction of the incident light that has traveled through part of a sample, scattered internally, and exited to a detector. In this context, the photon energy is absorbed by activating certain electronic or vibrational processes. Absorption only occurs for those wavelengths (frequencies) at which the energy "resonates" with the energy needed to trigger these processes in a given sample. Processes of particular interest for spectroscopic identification of materials are electronic transitions within certain ions (e.g., displacement of d-orbital electrons in transition-metal ions such as Fe^{2+}), transfer of electrons between ions (e.g., between metals and ligands such as Fe^{3+} and O^{2-}), and assorted molecular vibrations and rotations (Burns, 1970). In materials where only a narrow range of wavelengths activates one of these processes, the

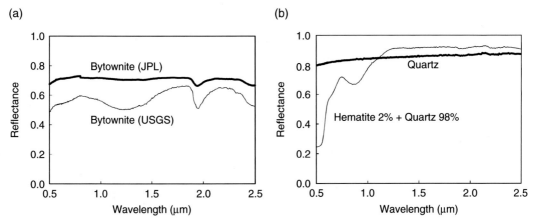

Figure 2.2. Reflectance spectra of plagioclase feldspar (bytownite, $NaAlSi_3O_8$–$CaAl_2Si_2O_8$) and quartz (SiO_2). (a) Pure bytownite (JPL), and bytownite containing trace amounts of Fe^{+2} that produces the broad, electronic (crystal-field) absorption band that is centered near 1.3 μm (USGS). (b) Pure quartz (heavy line), and 98% quartz + 2% hematite powder (Fe_2O_3) that coats the quartz grains (light line). Several types of electronic processes involving Fe^{3+} produce the absorption bands in the hematite at the short-wavelength end of the spectrum. Spectra from USGS; also reproduced from the ASTER Spectral Library through the courtesy of the Jet Propulsion Laboratory, California Institute of Technology, Pasadena, California. Copyright 1999, California Institute of Technology. All rights reserved.

absorbed wavelength interval often serves as a unique identifier. Some materials, such as pure samples of the common rock-forming minerals, quartz and feldspar, absorb little photon energy in the wavelength region from 0.4 to 2.5 μm, and their reflectance spectra are relatively flat and featureless. However, spectra of the same minerals show absorption at certain wavelengths when small amounts of impurities are present (Figure 2.2).

Spectra of natural materials

Spectral libraries may be found at http://speclib.jpl.nasa.gov/ and http://speclab.cr.usgs.gov/

Some common minerals, such as magnetite and ilmenite, absorb strongly throughout the VIS-NIR, but, because their spectra have little structure, they are difficult to use for identification. Other materials, though, have absorption "bands" that can be characterized in terms of the wavelength of maximum absorption, the band depth and width. Absorption bands may reveal information about the composition and structure of a sample. In minerals, bands that are produced by electronic transitions often can be assigned to specific elements having specific valence states and having a particular bonding configuration with neighboring ions. Other absorption bands are caused by vibration modes (e.g., stretching, bending, rocking, rotating)

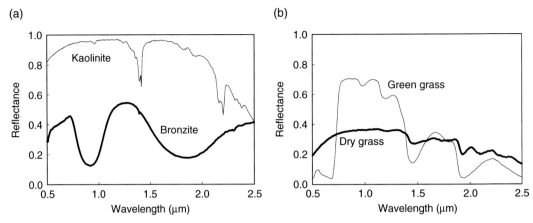

Figure 2.3. Examples of types of absorption bands in reflectance spectra. (a) Narrow vibrational bands in the mineral kaolinite [$Al_2Si_2O_5(OH)_4$] (light line). The band near 1.4 μm is produced by O–H stretching; the band near 2.25 μm involves Al–O–H bending. Broad electronic-transition bands in the mineral bronzite [$(Mg,Fe)SiO_3$] (heavy line). The two bands near 0.9 and 1.9 μm are caused by Fe^{2+} at different crystallographic sites. (b) Green grass (light line). The strong absorption bands below 1 μm are produced by electronic processes in organic pigments, primarily chlorophyll; the bands at longer wavelengths are vibrational bands involving H–O–H as liquid water. Dry grass (heavy line). Chlorophyll bands are absent, and water bands are weak, revealing faint vibrational bands produced by cellulose and lignin. Spectra from USGS.

of specific molecules in specific arrangements in a material. Examples of different types of absorption bands are shown in Figure 2.3.

TIR radiation

Thermal radiation emitted from the Earth's surface is commonly strongest near 10 μm, because the surface temperature is about 300 K. The peak wavelength can be found by differentiating Planck's law.

Thermal-infrared (TIR) spectra

The wavelength region from about 3 μm to about 20 μm is divided into the midwave infrared (3 to 5 μm) and the longwave infrared (5 to 20 μm). The term "thermal infrared" is used because at these wavelengths radiant energy is emitted by materials as heat. The total amount of energy emitted by a sample is determined primarily by its temperature. Spectroscopists, though, are mainly interested in variations in radiance from one wavelength (or frequency) to another and the deviation of a spectrum from that emitted by a blackbody. Heat energy causes atoms and molecules to vibrate and rotate at frequencies that are determined by the composition and structure of a sample. In the absence of any resonance, all of the heat energy radiates from the sample and the radiance is that of a blackbody. Emissivity spectra reveal the wavelengths at which energy has been absorbed in the process of activating molecular vibrations and/or rotations. As in the case of reflectance spectra, the wavelength position, depth and width of resonance bands in TIR spectra can be interpreted in terms of the composition and molecular structure of a sample. Examples of TIR spectra are shown in Figure 2.4.

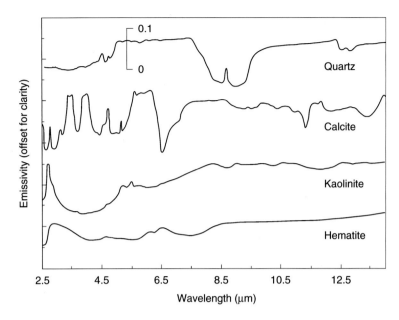

Figure 2.4. Laboratory thermal-infrared (TIR) spectra from 2.5 to 14 μm of common minerals, quartz [SiO_2], calcite [$CaCO_3$], kaolinite [$Al_2Si_2O_5(OH)_4$] and hematite [Fe_2O_3]. Hemispherical reflectivity spectra of powders (125 to 500 μm) were converted to emissivity using Kirchhoff's law. Spectra by Salisbury *et al.* (1991). Figure modified from Hook *et al.* (1999), by permission of John Wiley & Sons Inc.

Wavelength-to-wavelength (WW) contrast

At a given wavelength, the fraction of radiant energy that is absorbed by a sample depends on several factors, including the nature of the resonance mechanism, the concentration of the absorbing species and the path length of radiation through the sample. Under ideal measurement conditions, and ignoring scattering and losses at sample surfaces, absorption at each wavelength is given by the Bouguer–Beer–Lambert law, where I_o

$$I = I_o e^{-k_\lambda d} \qquad (2.3)$$

is the intensity of light entering a sample, I is the measured intensity after passing through a sample, k is the absorption coefficient at a given wavelength (λ), and d is the path length through the sample (Figure 2.5). In this ideal case (approximated, for example, by polished wafers of pure mineral grains) we can measure the transmitted power at each wavelength and plot the wavelength pattern of absorption (k). Absorption-coefficient (or absorbance) spectra are measured in the laboratory for certain pure materials by passing light through polished, flat samples (e.g., Burns, 1970), subject to the practical constraints of the required path length (d).

In reflectance spectra of natural samples that are rough, fractured or particulate, the differential absorption that is described by the Bouguer–Beer–Lambert law only applies in principle. Light typically undergoes multiple internal reflections in natural samples, before it exits to a detector; thus, it is not feasible to measure absorption

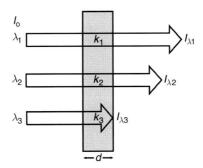

Figure 2.5. Conceptual sketch of light transmitted through a polished sample according to the Bouguer–Beer–Lambert law, Equation (2.3). The WW contrast is governed by the interplay of the absorption coefficients (k) for each wavelength and the path length (d). Wavelengths λ_1, λ_2 and λ_3 have corresponding absorption coefficients where $k_3 > k_2 > k_1$. Given d as shown in the example, contrast is at a maximum between $I_{\lambda 1}$ and $I_{\lambda 3}$ and between $I_{\lambda 2}$ and $I_{\lambda 3}$. A thicker sample (larger d) would decrease $I_{\lambda 1}$ and $I_{\lambda 2}$ but would not affect $I_{\lambda 3}$ that already is zero. A thinner sample would reduce absorption and increase $I_{\lambda 1}$, $I_{\lambda 2}$, and $I_{\lambda 3}$. The practical result is that for a given material, spectral contrast is influenced by the physical properties that govern mean path length such as particle size of mineral grains or thickness of leaves.

WW contrast

Wavelength-to-wavelength contrast is measured on a single spectrum. It is defined as $(r_{\lambda 1} - r_{\lambda 2})/r_{\lambda 1}$. In Section 2.2.1 we discuss pixel-to-pixel (PP) contrast that is a comparison of one spectrum with another.

DN

"Data number," DN, is an anachronistic term left over from the early days of computing and digital image processing. Although generally DN indicates radiance, it can stand for any parameter in an image.

---

coefficients, because we cannot measure the photon paths that actually are followed. Nevertheless, a reflectance (r) spectrum contains information about the relative absorbing power of a sample at each wavelength, and this is the information that we need to identify certain materials remotely. To describe the relative absorbing power from one wavelength to another, we introduce the term "wavelength-to-wavelength (WW) contrast." In spite of the usefulness of such a term in discussing laboratory and remote-sensing spectra, there appears to be no standard way to measure this property. (Indeed, spectral contrast is virtually unmentioned in the literature.) With reference to Figure 2.6, we have found it convenient to adopt the method of Clark and Roush (1984) who define the depth (D) of an absorption band as $(r_c - r_b)/r_c$ where r_c is reflectance measured on the continuum at λ_b, and λ_b is the absorption maximum. Generalizing, we define WW contrast as $(r_{\lambda 1} - r_{\lambda 2})/r_{\lambda 1}$. In remote sensing, contrast in the per-pixel data numbers (DNs) from channel to channel have been treated in a similar way. Ratios or normalized differences such as the normalized-difference vegetation index (NDVI) are relative measures of WW contrast that minimize the effects of topography and other gain factors.

Wavelength-to-wavelength contrast is reduced whenever a significant fraction of the detected radiation has scattered from the surface without entering a sample, or when it follows only very short path lengths within low-k materials before returning to the detector. Low WW contrast is common in the reflectance spectra of particulate materials such as some mineral powders. For this reason, laboratory spectra of natural materials

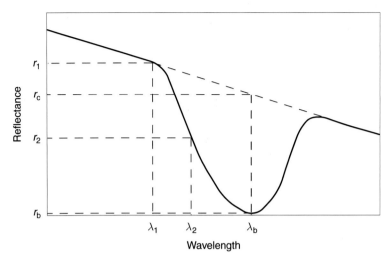

Figure 2.6. Hypothetical reflectance spectrum with an absorption band centered at λ_b. Wavelength-to-wavelength contrast may be defined as $(r_{\lambda 1} - r_{\lambda 2})/r_{\lambda 1}$. The depth ($D$) of the absorption band is defined as $(r_c - r_b)/r_c$.

such as minerals, rocks, soils, and vegetation are made under conditions that optimize WW contrast for each sample. Surface reflections are minimized, as are extraneous sources of radiance from the surroundings. For reflectance measurements, contrast often can be enhanced by selecting a particle size that produces a mean path length that allows maximum absorption. Thermal-infrared spectra of particulate samples typically are obtained by making hemispherical reflectance measurements and converting them to emissivity using Kirchhoff's law. By optimizing WW contrast in the laboratory, we obtain the best possible spectra to use for interpreting spectral images. The corollary is that remote spectra have less WW contrast than laboratory reference spectra.

Analytical tools

Spectroscopists have developed a variety of ways to characterize reflectance and emittance spectra, with particular emphasis on resolved resonance bands. Many of these analytical methods have been applied to remote sensing, and specifically to imaging-spectrometer data. Image-processing software packages typically include tools to fit a continuum to a spectrum, normalize spectra to remove the continuum, measure the wavelength position of resonance bands, and to define other parameters such as band depth, width and symmetry. Specialized programs have been developed to fit modified Gaussian functions to resonance bands, often making it possible to resolve the components of multiple, overlapping bands (Sunshine *et al.*, 1990). These tools are especially helpful to geologists who want to compare imaging-spectrometer data with laboratory reference spectra to identify minerals.

Spectral mixing

Mixtures of pure materials have mixed spectra. It may be evident from a mixed spectrum what the individual components are, but often one or a few components are spectrally dominant, and others are obscure. When there are many components, the resulting spectral "soup" may defy interpretation. Analysis of spectral mixtures is important in remote sensing, because virtually any spectrum of a landscape is a mixture. When we analyze a mixed spectrum we would like to answer two questions. What are the spectra of the individual materials? And, what are the proportions of the individual materials? Determining proportions is more difficult, and, according to Clark's law, generally follows identification.

In Figure 2.7 we give examples of mixed spectra to illustrate some of the issues involved in isolating and identifying components. In the quartz–alunite mixtures (Figure 2.7a), it is clear that the mineral alunite is present, because the several absorption bands are diagnostic, but it is not evident that quartz is a component, because its spectrum is featureless at these wavelengths. The mixed spectrum of olivine (forsterite) and pyroxene (enstatite) (Figure 2.7b), however, has combined features of the two minerals, and we need to know the spectra of the pure minerals to be able to interpret the mixed spectrum. In Figure 2.7d, the mixed spectrum of a desert soil ("CUP-4") is not easily interpreted, because the weak absorption bands are not mineral-specific. However, by dividing CUP-4 by the spectrum of wind-deposited "background" soil, the normalized spectrum reveals a weak band at 2.16 μm that indicates the presence of alunite – a mineral of interest to geologists seeking ore deposits.

Fig. 2.7c is a mixed spectrum of the shadow cast by a green leaf on bare soil. The spectrum of the soil that is illuminated by light transmitted through the leaf is similar to that of the leaf itself, although it is not as light and has slightly less WW contrast. This is a particularly good example of the type of mixing that occurs when light is partially absorbed by one material before entering another material – often referred to as "intimate" mixing. The Bouguer–Beer–Lambert law, Equation (2.2), states that such mixing will be non-linear, and, indeed, non-linear mixing occurs in all of the examples discussed above. For example, if we calculate the mixed spectra of quartz and alunite as non-linear mixtures and as linear mixtures, and compare the results with the spectra of the physically mixed samples, we find that linear mixing does not match the spectra of the physical mixtures well, especially in the absorption bands (Figure 2.8). This discrepancy between the two types of mixing models becomes important when we want to estimate the proportions of the mixed materials.

Clark's law

This law, which has broad applications that include remote sensing, was articulated by our colleague Roger Clark of the USGS. "First find out what is there, then worry about how much."

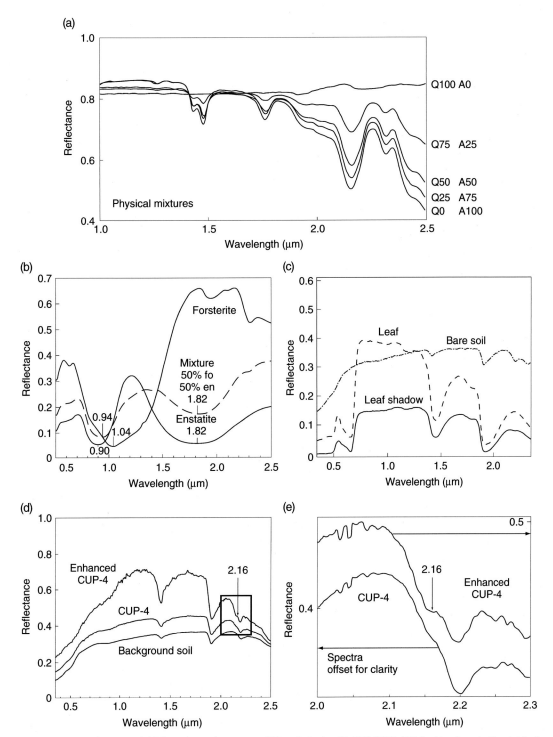

Figure 2.7. Mixed spectra. (a) Mineral powders: quartz (Q) and alunite (A), $KAl_3(OH)_6(SO_4)_2$. Numbers to the right of the figure indicate mixing ratios. (b) Mineral powders: forsterite ["fo": $(Mg,Fe)_2SiO_4$] and enstatite ["en": $(Mg,Fe)SiO_3$]. The "mixed" spectrum differs from the arithmetic average of the "fo" and "en" spectra due to non-linear absorption and the differences in packing of the particulate samples. Adapted from Adams (1974). (c) Green leaf, soil, and leaf shadow on soil. Courtesy of D. Roberts. (d) Desert soils from the Cuprite mining district, Nevada (sample "CUP-4"). The box shows the spectral region enlarged in (e). After Shipman and Adams (1987).

Figure 2.8. Calculated
mixtures of spectra of quartz
(Q) and alunite (A): (a) nonlinear mixing; (b) linear
mixing. A non-linear mixing
model is in better agreement
with spectra of the physical
mixtures of quartz and alunite
powders (Figure 2.7).
Numbers to the right of the
graphs indicate mixing
proportions in percent.

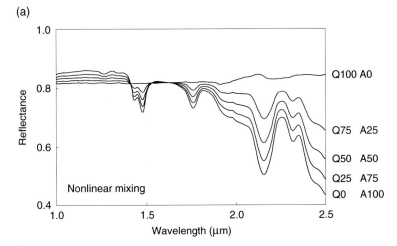

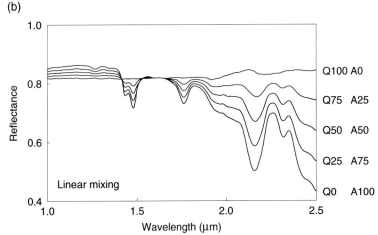

Why, then, would we consider using a linear mixing model at all to estimate proportions of components, aside from the fact that it is computationally simpler? The answer lies in the fact that when we move from the laboratory scale to the scale of remote sensing, linear mixing begins to dominate for most scenes. This is illustrated simplistically in Figure 2.9, where the spectra of adjacent meter-scale patches mix linearly for practical purposes, because the amount of intimate mixing at patch boundaries is negligible. Spectral mixing becomes an increasingly important factor as pixel size becomes larger (Chapters 4, 5, and 6).

2.1.3 Spectroscopy in the field

Measurements that are made in laboratories comprise the main database for spectroscopy, because sample quality and conditions of measurement

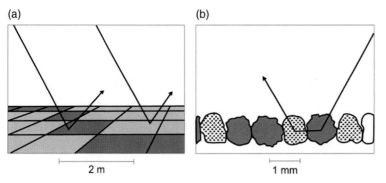

Figure 2.9. Sketch to illustrate spectral mixing at different scales. (a) The spectra of two different materials mix linearly at the scale of meters according to area (perspective view). (b) At the millimeter scale, the spectra of different mineral grains mix in a non-linear way according to the Bouguer–Beer–Lambert law, equation (2.3) (cross-section view).

can be carefully controlled and documented. It is possible, though, to acquire spectra of high quality in the field using portable instruments. Among the advantages, samples can be measured in their natural setting, larger samples can be included, and a greater variety of samples can be accessed quickly. As in the laboratory, field spectra typically are made by alternately measuring sample and reference while maintaining the same field of view. However, there are added complications. For example, although the Sun is the main illumination source for reflectance measurements in the field, there is a contribution from (spectrally different) skylight. And the atmosphere, especially if there are thin clouds present, can vary between measurements of sample and reference unless the spectra can be acquired quickly. In the TIR, emission by the atmosphere and surrounding objects must be minimal to be able to detect the properties of the sample. Furthermore, samples measured in the field generally are not as well characterized as the ones that are selected for laboratory spectra. It can be inconvenient to bring analytical equipment into the field, but an even larger problem is how adequately to describe large, complex samples such as a branch, a tree, a plowed field, or an outcrop of rock. Field spectra, therefore, are particularly valuable for understanding how well relatively pure laboratory spectra represent larger and more complicated samples at coarser scales.

The bridge to remote sensing

In remote sensing, spectral images measure uncorrected or "encoded" radiance. We could make laboratory-like reflectance and emittance measurements remotely if we had suitable standards on the ground in each field of view. Alas, this is not practical for most remote sensing, and this precludes us from comparing remote spectra directly with laboratory and field spectra that comprise the main interpretive database. To work around the problem, there are various methods by which to "calibrate" encoded radiance to reflectance or emittance. This subject is discussed in Section 3.2.

2.2 Spectroscopy at landscape scales

2.2.1 Spectral contrast

The term "contrast" describes a quality of images, and it is widely understood that an image that has bold patches of white and black has high contrast, whereas one that consists of faintly differing tones of gray has low contrast. The contrast ratio (CR) is given in Equation (2.4). Contrast

$$CR = DN_{max}/DN_{min} \qquad (2.4)$$

ratio and other metrics, such as contrast range and DN standard deviation, commonly are applied to digital images that either are panchromatic or are taken in a single spectral channel. In Section 2.1.2 we discussed a different kind of contrast, namely the contrast between the amount of radiant energy that is reflected or emitted at one wavelength vs. another, which we called WW contrast. The idea of WW contrast can be extended directly to spectral images, because, for each pixel, we can easily compare the reflected or emitted radiance in two channels. (Ratio images are an example.) However, the contrast between just two wavelengths does not describe contrast over all of the sampled wavelengths, that is, between two spectra. Accordingly, we introduce a new term, "pixel-to-pixel (PP) contrast," to make clear that this is the overall contrast between pixel *spectra*, and that it is not the same as WW contrast within a single spectrum.

> **PP contrast**
>
> Overall contrast between pixel spectra.

Wavelength-to-wavelength contrast is governed by the interplay of resonance and path length (Section 2.1.2), and laboratory measurements are designed to maximize contrast between wavelengths by controlling experimental parameters such as sample purity, particle size and sample geometry. These controls, of course, are absent with remotely sensed spectra that are measured on natural surfaces over the scales of pixel footprints (meters to kilometers); therefore, WW contrast generally is lower at landscape scales than for laboratory spectra of the same materials (Section 6.3).

> **WW contrast**
>
> Generally WW contrast is lower at landscape scales than for laboratory spectra of the same materials.

Wavelength-to-wavelength contrast also becomes smaller as pixel footprints become larger, because the spectra of pure materials are diluted by the spectra of other kinds of materials and by the spectra of geometrically diverse surfaces. If our objective is to use the spectra of individual pixels to identify materials, the loss of WW contrast with increasing pixel-footprint size affects the amount of compositional information that we can extract from spectral images and especially from imaging-spectrometer data. For example, AVIRIS has approximately a 17-m pixel footprint (depending on the aircraft altitude). At this scale, depending on the scene, relatively small amounts of a material can dominate a pixel spectrum, and the full power of the very high spectral resolution of AVIRIS can be brought to bear to resolve resonance bands. By contrast, the same amount of material that could be identified in one AVIRIS pixel would likely be lost in a 500-m pixel of the Moderate Resolution Imaging Spectrometer (MODIS). The latter spectrometer, therefore, was not designed to have as many channels or to be used for the same degree of detailed spectroscopic identification as AVIRIS.

Pixel-to-pixel contrast

Pixel-to-pixel spectral contrast consists of two components, spectral-length contrast and spectral-angle contrast. In Figure 2.10 the spectra B and C are defined by points (vectors) in the two- and three-dimensional sub-space projections. Spectral length (SL) is the vector length of one spectrum in multi-channel data space. It is a measure of the overall spectrum lightness. Spectral-length *contrast* defines the contrast in SL between two spectra (B and C), and we define it as the normalized difference (Figure 2.10):

$$SL_{contrast} = (SL_C - SL_B)/(SL_C + SL_B) \qquad (2.5)$$

> **Components of PP contrast**
> Pixel-to-pixel contrast consists of two components, spectral-length contrast and spectral-angle contrast.

where $SL_C > SL_B$. Spectra having vectors of equal length have zero SL contrast.

Spectral angle is the angle (in the data space defined by the spectral channels) between two vectors, and it is independent of SL. Thus, SA is inherently a measure of contrast. When $SA = 0$, two spectra can differ only in overall lightness (SL). The spectral angle is described by Kruse (1999) as a means of classifying images (Chapter 3) that is relatively insensitive to a variety of gains, including those imposed by topographic shading (Chapter 1); however, this means that SA also is insensitive to overall reflectance or sample albedo.

> **Spectral data space**
> A Cartesian space in which each axis corresponds to the DN for a single image channel. Data spaces with more than three dimensions may be hard to visualize, but can be manipulated mathematically.

The Euclidean distance between two vectors in the data space combines the SL and SA components of pixel-to-pixel spectral contrast. Euclidean distance is commonly used as a basis for classifying spectral images, and, although it has the advantage of being a complete measure of spectral contrast, it has the disadvantage that the SL component is ambiguous in the presence of topographic shading and shadowing.

Competing effects influence PP contrast. If we hold the size of the pixel footprint (instantaneous field of view, or IFOV) constant, and increase the field of view (FOV) of a landscape, there is the potential for PP contrast to increase, because over a larger area we eventually will

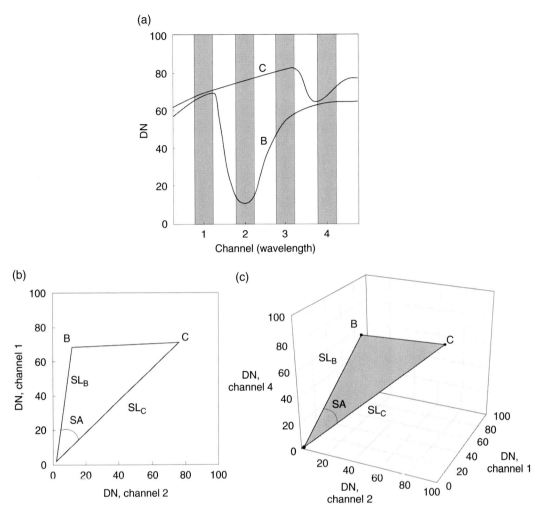

Figure 2.10. Pixel-to-pixel (PP) spectral contrast. (a) Hypothetical spectra B and C, sampled by channels 1, 2, 3, and 4. (b) Spectra (vectors) B and C in the two-dimensional Cartesian space defined by channels 1 and 2; SL_B and SL_C are the spectral lengths (SL) of the respective vectors in this sub-space projection, and are calculated as the square root of the sum of the squares of the channel DNs. Spectral-length contrast between B and C is the normalized difference of the spectral lengths ($(SL_C - SL_B)/(SL_C + SL_B)$); SL contrast between B and C in channels 1 and 2 = 0.20. Spectral angle (SA) contrast between B and C in channels 1 and 2 = 0.64 radians. (c) Spectra B and C in the three-dimensional space defined by channels 1, 2, and 4, where SL contrast = 0.13, and SA = 0.53 radians. The Euclidean distance between B and C incorporates SL and SA contrast.

sample more and more spectral variety. We also know that we can increase PP contrast by making the pixel footprint smaller, because we will be able to spatially resolve more objects that are different from one another. The corollary, of course, is that progressively larger pixel footprints average the finer-scale spatial variability, and PP contrast is diminished. Both WW and PP spectral contrast influence detection thresholds and the compositional information that can be extracted from spectral images. We explore this subject in more detail in Chapter 6.

2.2.2 Reference spectra for landscapes

As discussed at the beginning of this chapter, virtually all of our fundamental understanding of spectra comes from theoretical and laboratory studies. Unfortunately, the spectra that we understand best, the laboratory reference spectra, are made at the wrong spatial scale, on the wrong samples and using incompatible measurement methods for remote-sensing purposes. As a result, we do not have standard (reference) spectra that apply to remote-sensing measurements at landscape scales. Let us review the problems.

> **Reference spectra**
> Reference spectra do not exist for many of the things that we are interested in for remote sensing; for example, whole trees or an alluvial fan.

- **Geometric factors**. Illumination and detection geometry for measuring laboratory reference spectra are standardized from sample to sample, and are designed to optimize spectral contrast. With remotely sensed images, there are variations in solar zenith angle, local incident angle (roughness and topography) and local exitance angle (nadir vs. off-nadir viewing) (Chapter 1). To standardize spectra at remote-sensing scales, one either must correct the measurements to a standard geometry or model the expected effects of geometric variations.

- **Calibration**. Laboratory spectral measurements and remote-sensing spectral measurements are made in entirely different ways (relative vs. absolute). To compare them we must calibrate (Chapter 3); unfortunately, it is not possible to know all of the calibration factors precisely.

- **Atmospheric effects**. The spectral effects of atmospheric scattering and absorption are minimized in laboratory measurements by using short optical path lengths and standard references. Atmospheres, however, can distort remote measurements significantly, and measurements can be difficult to correct, because atmospheric properties may vary in space and time.

- **Spatial resolution**. Samples that are used for laboratory spectra typically are restricted in size to mm or cm, but it is impractical to make controlled laboratory-type measurements of larger entities such as trees, cornfields, buildings, or volcanoes. Remote-sensing measurements generally make poor reference spectra, because pixel footprints incorporate variable spectral mixtures of constituent materials in addition to unresolved shading and shadowing. However, if image spectra can be normalized to standard illumination conditions, it is possible to accumulate standard spectra of some parts of the land surface that are well understood.

- **Spectral resolution**. There is a practical tradeoff between spectral resolution (channel width, spacing and number) and spatial coverage. Images that cover large areas typically have low spectral resolution. Reference spectra at pixel scales have to be expressed at the spectral resolution of each imaging system.

- **Impurities**. Most collections of laboratory spectra are strongly biased toward pure materials, because they are easy to understand and categorize. At the land surface, though, materials typically are contaminated by impurities, stains, and coatings.
- **Spectral variability**. Materials having the same name can be variable in composition and structure, and therefore, in their spectral properties. As examples, we do not expect one reference spectrum to represent all green vegetation or all granites.

In spite of these problems, some remote spectra can be interpreted in terms of familiar laboratory data, but these tend to be either spectra having especially large WW contrast (green vegetation is the best example), or they are the spectra of a few materials that dominate all other materials by area. Field investigators commonly are surprised that so few different materials can be found in spectral images, because they are not aware that many of the things that are of interest to them are not really very abundant or that they are subsumed in spectral mixtures. Alas, in remote sensing, to find what we are looking for we often are forced to analyze carefully the most basic spectral components of landscapes before we can detect the minor components.

> **A few spectrally dominant materials**
>
> In remote sensing we often are forced to identify the common spectral components of landscapes before we can pick out the less abundant components from the spectral clutter.

2.2.3 Basic spectral components

Spectral variability on the ground in most landscapes usually is far greater than can be measured conveniently with a field spectrometer or by collecting samples for laboratory measurements. Each pixel incorporates the spectral variability within a pixel footprint into a single, mixed spectrum; therefore, we generally have no hope of detecting all of the contributing materials. Sometimes, though, it is possible to infer the presence of a few spectrally distinct, and spectrally dominant basic components. For example, widely used vegetation indices (Chapter 3) take advantage of the fact that green vegetation and substrate (litter, soil, rock) are so spectrally distinct in the VIS-NIR that the relative proportions of green vegetation and substrate can be estimated from just two well-placed channels. A vegetation index provides information on only two basic landscape components, but it turns out that the proportions of these two components often are of great interest. Vegetation-index measurements are, of course, approximate, because it is not possible to take into account all of the other sources of spectral variability such as different kinds of vegetation and substrate and the presence of other materials.

> **Basic spectral components**
>
> Basic spectral components consist of a few spectrally distinct materials on a landscape that describe the main sources of spectral variability.

We can generalize the concept of basic spectral components to include most landscapes. A requirement is that we are able to visualize a landscape in a spectral context rather than in just the familiar spatial one. An imaging system, for example, "sees" shading and shadow (hereafter referred to as "shade"; see ahead to Section 4.2) as a landscape component, whereas we ignore variations in illumination in the field. We think of a forest as consisting of individual trees or of different species, whereas most of the spectral variations can be attributed to green vegetation (GV), non-photosynthetic vegetation (NPV), shade, and substrate. Basic spectral components, then, consist of a few (typically two–five) spectrally distinct features of a landscape that describe the main sources of spectral variability. The basic components may or may not be the most interesting features to a remote-sensing analyst or a field investigator; nevertheless, they are the ones that are most readily revealed in a spectral image. Even if the basic spectral components are not the ones that investigators would like most to find, they define the spectral background, or matrix, in which other spectral information is imbedded (Chapter 6).

Table 2.1 lists basic spectral components for several types of landscapes. The ability to discriminate components depends on the properties of each spectral imaging system. For example, dark rocks and shade are mimics in the VIS-NIR, but can be distinguished if a thermal channel is available. Leaf litter and soil can be difficult to distinguish with Landsat TM channels, but may be separable with higher spectral resolution systems such as AVIRIS. In such cases, components remain combined until the ambiguity can be resolved by additional spectral information or context. Materials that are not common constituents of a given type of landscape are not included as basic spectral components.

Representative spectra of two common landscape components, green vegetation and rock/soil are shown in Figure 2.11. The spectra are proxies for basic spectral components, recognizing that the categories are very broad, and that, in reality, no one spectrum can be a perfect representative. The chosen laboratory spectra exhibit high WW and PP contrast; however, some basic components such as soils and NPV have low spectral contrast, and can be difficult to discriminate. Figure 2.11 also shows the wavelength channels of several NASA imaging systems. In general, spectral contrast, both WW within the individual spectra, and contrast between the two spectra, is maximum when sampled by a high-spectral resolution system such as AVIRIS. With fewer and/or broader channels, there is less chance that resonance bands within the individual spectra will be fully resolved (WW contrast) or even detected. And, unless some of the image channels happen to sample wavelengths at which two spectra differ from each other, the

Table 2.1. *Basic spectral components for several types of landscapes.*

Landscape type	Basic spectral components
Closed-canopy forest	GV (green leaves)
	Shade (shadows and shading)
	NPV (woody tissue, senescent leaves)
Open-canopy forest	GV (green leaves in canopy and in understory)
	Shade
	NPV (in canopy and as litter on ground)
	Soil/rock
Savanna	GV (green leaves on trees, shrubs, grass, forbs)
	NPV (woody tissue, senescent grass, litter)
	Shade
Shrub steppe	GV (grass, trees, shrubs, forbs, lichen)
	NPV (senescent grass, litter, woody tissue)
	Soil/rock
	Shade
Warm desert	Soil/rock
	Shade
	GV (shrubs, forbs, lichen)
	NPV (woody tissue, litter)
Agriculture	GV (crops/cover)
	NPV (senescent crops/cover)
	Soil
	Shade
Moon	Mature highland soil
	Mature mare soil
	Highland rock/immature soil
	Mare rock/immature soil
	Shade
Mars	Global dust
	Rocks (un-weathered)
	Shade
	Ice/frost

GV = Green vegetation. NPV = Non-photosynthetic vegetation. Notice that this is not a list of all of the components of each landscape type, but instead is focused on spectrally dominant ones. The list does not include all landscapes. It is intended to be illustrative. The spectral components are in approximate order of relative abundance. Shade varies with illumination conditions.

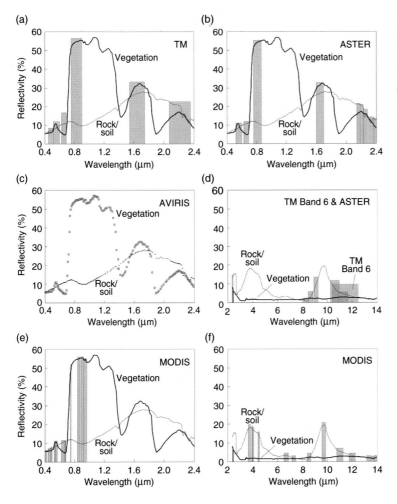

Figure 2.11. Laboratory spectra of green vegetation and rock/soil representing two basic spectral components of certain landscapes. Wavelengths of the spectral channels of different imaging systems are shown to illustrate how each system would respond to the laboratory spectra. (a) Landsat TM VIS-NIR; (b) ASTER VIS-NIR; (c) AVIRIS VIS-NIR; (d) Landsat TM and ASTER TIR; (e) MODIS VIS-NIR; (f) MODIS TIR.

PP contrast will be low. Thus, for a given scene, we are led to the conclusion that fewer and wider channels, along with larger pixel footprints, usually have the effect of lowering spectral contrast.

There are two potential difficulties in drawing on laboratory spectra to model basic spectral components of landscapes. The first problem is calibrating encoded-radiance measurements and converting ground radiance into reflectance or emittance, and the second problem is selecting spectra that are representative. We defer discussion of calibration to Chapter 3, but point out here that approximate calibration is satisfactory for many applications. At first glance, applying representative laboratory spectra to images appears impossible, because these spectra usually are made on small, pure samples, whereas remote-sensing measurements comprise mixtures on coarser scales. In practice, laboratory spectra actually have performed well as proxies for basic spectral components, especially when they are used as endmembers in simple mixture models (Chapters 4 and 5).

Spectra of landscape components also can be derived directly from images. This has the advantage of avoiding calibration and re-sampling laboratory spectra to the channels of a particular imaging system. The main problem is finding pixels in a scene that are pure samples of each basic component, rather than mixtures. Whether or not an image contains "pure" pixels depends on the nature of each scene and the size of the pixel footprint.

2.3 Spectroscopy applied to images

High-resolution telescopic spectra of planetary surfaces and experience with imaging-spectrometer measurements of terrestrial landscapes have shown that certain materials can be identified remotely by resolving diagnostic absorption or emission bands. Remote identifications can be made of many minerals and some of the components of vegetation by their resonance bands, even though they occur in intimate mixtures or make up only a modest fraction of a pixel footprint. However, not all materials exhibit bands, and it is common for bands to be obscured for various reasons.

Why resonance bands may be scarce in remote spectra

- Many materials collected in the field do not exhibit clearly defined resonance bands in the first place.
- Of the materials that have bands, many cannot be identified unambiguously by their spectra, even by laboratory measurements.
- Only a few of the materials that can be identified spectroscopically are interesting for remote-sensing applications.
- There may not be enough of a material present for its bands to be detected, because band strength is proportional to the abundance of the absorbing or emitting component.
- Absorption and emission bands can be masked by the presence of small amounts of some other materials, such as opaque carbon or metallic oxides.
- Absorption and emission bands can be masked by particle size and other geometric effects.
- To resolve a band, a sensor system must have channels at wavelengths near the band center and outside the band. Channels, therefore, must be narrower than the band itself.
- Narrow channels collect less radiant energy, which places constraints on detector sensitivity and the time that the sensor collects energy from the pixel footprint (dwell time).
- Only a few imaging-spectrometer systems have narrow channels that collectively cover a wide wavelength range. Even then, parts of the spectrum are not accessible because of detector and sensor limitations or atmospheric interference.
- When pixel footprints of spectral-imaging systems are applied to landscapes, the spectra generally consist of mixtures. Mixed spectra dilute the signals from resonance bands in any one material, making their detection more difficult.

Background on resonance bands

The significance of resonance bands for remotely identifying minerals and other solid materials was recognized in the late 1960s. The earliest remote measurements that detected absorption bands in the VIS-NIR were telescopic spectra of Mars, the large asteroid, 4-Vesta, and the Moon. In the 1980s, aircraft imaging-spectrometer systems, AIS and AVIRIS, made it possible to acquire images with high enough spectral resolution to detect absorption bands in VIS-NIR spectra of Earth. Imaging spectroscopy now extends to the TIR. Modern imaging spectrometers have spectral resolutions comparable to those of laboratory instruments. Today there is a growing data base of high-spectral-resolution images of Earth and a variety of Solar-System bodies. Naturally, interest has focused on finding resolved bands that can be interpreted in terms of the composition of remote materials. Analytical techniques have been developed to search imaging-spectrometer data for resonance bands, and to match band positions and strengths with those of materials in laboratory spectra of known materials (Chapter 3).

Two special conditions are required to resolve bands in spectral images. The first, and most evident, requirement, is that wavelength channels must be sufficiently narrow and appropriately placed. The second, and less widely recognized, requirement is that the pixel footprints must be small enough to sample at a scale where there is enough spectral contrast to resolve resonance bands. Only the data from a few imaging-spectrometer systems meet both requirements. Most of the commercially available spectral images do not.

Investigators who have access to spectral images that can resolve resonance bands will want to make full use of this powerful method for identifying landscape components. However, if suitable images are not available, or resonance bands are not present in the surface materials to begin with, it is still possible to apply spectroscopic methods.

2.3.1 Images without resolved resonance bands

For many applications of remote sensing, it is quite adequate to discriminate among materials, as long as their spatial patterns also can be mapped in an image. Once materials can be discriminated and mapped, they often can be identified within a broader interpretive context. Indeed, most of the standard methods for analyzing spectral images have been developed simply to discriminate subtle differences among spectra and categories of spectra (Chapter 3). It turns out that within certain contexts resonance bands can be inferred even when there is insufficient wavelength resolution to measure band minimums and band shapes. Indeed, the channels of most spectral-imaging systems are placed to sample, but not resolve, resonance bands that occur in interesting materials. For example, green vegetation has an especially strong chlorophyll absorption band at 0.66 μm that is flanked by a wavelength region where chlorophyll absorption is weak or absent and reflectance by leaf structure is strong. As a result, most terrestrial spectral-imaging systems have channels that are placed at the 0.66 μm chlorophyll-a band and at the adjacent 0.80 μm "plateau." Because no other common, natural materials have the same spectral properties, green vegetation is readily inferred, if not identified, within the context of terrestrial landscapes, simply by comparing measurements in these two channels (Figure 2.11).

Based on the examples in Figure 2.11, it might appear that we need only two channels to identify green vegetation. However, we really have not made a true spectroscopic identification, because a few, broad channels do not define the wavelengths of band minimums and the band shapes. Actually, what we have done is to apply a simple spectroscopic model that is consistent with green vegetation. The vegetation model fits the data when the ratio of the radiances at 0.80 and 0.66 μm is large. To simplify our task, we narrow the context so that it is not likely that any other materials would give the same response. In an entirely different context, though, such as vegetation-free areas on Earth, or the Moon and Mars, the same channels simply respond to the local materials, and low radiance values at 0.66 μm and higher values at 0.80 μm would have nothing to do with green vegetation. An important point here is that context allows us to make spectroscopic inferences from imaging systems that have low spectral resolution.

The spectral properties of interesting materials have guided the channel configurations of many spectral-imaging systems. Landsat MSS, Landsat TM, AVHRR, SPOT, MODIS, and ASTER, for example, all have channels placed in one or two of the main chlorophyll bands and at adjacent wavelengths. The early Landsat TM was given a new channel at 2.2 μm (TM band 7) to coincide with absorption bands in clays and other hydrated minerals that are associated with hydrothermally altered ore deposits. About a dozen of the 36 channels on MODIS are placed to facilitate measurements of clouds, aerosols, and smoke. To interpret measurements in channels that have been selected for specific purposes, we need to know both the context and the relevant spectroscopic model.

Placement of channels

Wags call this the "band-fer" approach – in our terminology, "a channel for this and a channel for that."

2.3.2 Spectral matching and classifying

Pixel spectra, with or without resolved absorption bands, can be compared one at a time with reference spectra in a data base. The objective is to find spectra of known materials in a reference set that most closely match the image spectra, channel by channel. In this approach, no initial assumptions need to be made about the materials on the ground or their spectral properties, although the number and type of reference spectra typically can be narrowed according to the interpretive context. Because of calibration uncertainties, matches with reference spectra often have to be limited to a few pixels of known materials in an image. However, if an image can be calibrated and converted to reflectance or emittance, all pixel spectra can be compared with laboratory reference spectra, thereby enlarging the potential matches. High-resolution image spectra that exhibit resonance bands offer the best opportunities for interpretation (Section 3.5.4). In the absence of resolved resonance bands, spectral matching of this type is not likely to lead to a unique interpretation. Nevertheless, it may be possible to narrow the choices, especially if there is a rich spatial context or information from other sources.

There are a number of analytical methods for categorizing spectra, including those without resolved resonance bands. Most of these methods pay no attention to resonance bands or spectroscopy per se, and are designed instead to compare vectors in a multi-dimensional data space. Classification, one of the most widely used image-processing techniques, compares image spectra with one another and defines groups having similar properties. The number of classes and their nature can be defined automatically or using guidance from an analyst. Spectra categorized in this way are not interpreted spectroscopically, and must be understood in terms of image context. Standard methods for classifying are discussed in Section 3.4.2.

2.3.3 Spectral modeling

We use the term "spectral modeling" to describe a physical modeling approach (Section 1.4.2) whereby many image spectra are compared with a few carefully chosen reference spectra that represent scene components that we know, or can guess, are on the ground. In spectral modeling, we start with a physical interpretation (a model) that is based on field observations or other prior knowledge, and we test how well image spectra agree with our model (Section 1.4.2). The process is iterative. If a model does not fit the data, we learn something about what is not in the image, and we try other models until we have narrowed the possible interpretations. In contrast, spectral matching and categorizing methods such as classification algorithms do not require any initial interpretation. Interpretation comes later – after the image spectra and reference spectra have been arranged taxonomically by one or more spectral-matching algorithms.

> **Spectral modeling**
> Physical interpretation comes first, and is tested in light of the data.

> **Spectral matching**
> Categorizing comes first, and interpretation follows.

The modeling approach can use a few or all spectral channels, and can be applied equally well to imaging spectrometer data with resolved absorption bands or to image data without resolved bands. If image spectra fit a model, it may be possible to infer the presence, and even the proportions, of certain materials, especially if the choices can be narrowed by prior knowledge or ancillary information. Certain channel ratios, such as the NDVI, can be used as simple spectral models. Match filtering and spectral-mixture analysis are examples of multi-channel techniques that apply spectral models (Chapters 3 and 4).

Next

How does one actually go about analyzing spectral images, keeping in mind the spectroscopic properties of a landscape? What algorithms work under what circumstances, and why? What are the pitfalls to avoid?

Chapter 3
Standard methods for analyzing spectral images

Methow Valley, Washington State. Photograph by D. O'Callaghan

In this chapter

We discuss and evaluate standard methods for analyzing spectral images, with the objective of physically modeling landscapes. Many of the algorithms in image-processing programs are useful for enhancing images for photo interpretation and for data reconnaissance, but only a few invoke physical models.

3.1 Initial evaluation

3.1.1 Pedigree

Ideally, every spectral image should be accompanied by basic information on ground location, time acquired, spatial coverage, spatial resolution, spectral channels, illumination and viewing geometry, and detector characteristics. Recommending that one should know the nature of an image may sound like gratuitous advice, but it is common for investigators who are not familiar with remote sensing to work with images that are not fully characterized. This may not matter for a few applications of photo interpretation that depend only on recognizing objects, but for serious interpretive work, including extracting spectral information, it is essential to know the properties of the data.

Modern spectral images that are purchased from the public domain or commercial sources include metadata that document acquisition and processing information, but these valuable data may have been lost from older, archived images. Sometimes, things have been done to images in addition to the documented protocol. Second-hand digital images are especially notorious for having undocumented histories, and the higher the level of processing the more obscure the trail. A friend who passed along an image may have thought it perfectly reasonable to modify the original data, but, if the changes were not recorded, unexpected problems may show up during subsequent analysis. Hard-copy images typically are manipulated to achieve the most visually satisfying result for the intended study. Without a well-documented pedigree it may not be possible to replicate such images (Section 1.4.5).

3.1.2 Geometric factors

Illumination

Spectral information is affected by variations in illumination geometry.

In Section 1.3 we emphasized the importance of illumination geometry in photo interpretation. Spectral information, also, can be affected significantly by variations in illumination geometry within a scene. There are two main things to look for when doing an initial evaluation: (1) spectral variations caused by topography, and (2) spectral variations caused by changes in the view angle from the nadir. Usually the main problem is with topography, because surface materials having the same inherent spectral properties will range from light to dark, depending on the solar incidence angle at each local surface. If important parts of an image are affected by shading or shadow, we may need to pay further attention to calibration (Section 3.2.1) and methods for suppressing topographic effects (Section 3.3.2).

Two main factors can cause spectral changes when view angles depart from the nadir: (1) variations in atmospheric scattering and absorption, and (2) variations in shadowing (Section 1.3.3). Subtle across-image gradients may be present even in satellite images that have a narrow angular field of view (e.g., Landsat); however, they are more noticeable in images having relatively wide fields of view such as those acquired by some planetary spacecraft, AVHRR, and most aircraft, including those made from TIR scanners (Figure 1.7). Recall that for common cross-track scanner systems, the view angle is measured in the cross-track direction. Typically, a scene will appear lighter in the down-Sun direction (i.e., the direction away from the Sun). If this is the case, and the gradient is larger for shorter wavelengths, it is probable that the gradient is dominantly caused by atmospheric path radiance (scattering). A similar down-Sun gradient also can be caused by variable shadowing of a rough surface, independent of any atmosphere, but in this case the gradient does not change significantly with wavelength. If a DN gradient is less than 1 or 2 percent it may not be necessary to attempt to remove the effect, depending on the nature of further analysis. A spurious gradient in a thematic classification or other spectral product signals the need to correct the channel data by an appropriate algorithm (Section 3.2.1).

3.1.3 Atmospheric effects

Naturally, we select images that do not have visible clouds obscuring areas of interest. Field investigators, however, sometimes are surprised to find that there also are other kinds of more subtle atmospheric interference that may significantly affect photo interpretation and spectral analysis. Potential problem areas usually can be revealed simply by examining a visible-channel image. For example, Landsat TM Band 1 often is ignored when making color composites or other products, because it appears "hazy." However, sometimes this is the best TM channel to show the edges of clouds, thin layers of cirrus, aircraft contrails, fine aerosols, fog, and smoke. By identifying problem areas in a visible channel, we can be alert to atmospheric variations that might not otherwise be recognized in longer-wavelength channels, but that nevertheless might affect a spectral analysis.

3.1.4 Defects

It is best not to assume that everything is "normal" about image data. A basic starting point in an initial evaluation is to display black-and-white images for each spectral channel for visual comparison. The first objective is to look for any channels that are grossly defective. This includes channels for which no image was recorded, channels that have dropped lines or pattern noise, and channels where clouds or other atmospheric effects mar the image. Major defects commonly can be seen even without contrast-stretching the image. More subtle problems can be revealed by stretching each image individually (Section 3.3.1). Grossly defective channels should be eliminated from further analysis. At this point, be cautious about fixing defects in the images. Although there are a number of techniques for "restoring" data lost due to dropped lines and striping, they alter the spectral information.

In most image-processing programs the contrast-stretch tool displays a histogram of the DN values of the pixels alongside an image.

> **Initial inspection**
>
> The advantage of doing an initial inspection is that you can anticipate which channels may adversely affect the results of further processing. In some cases, flawed channels can be deleted.

Examine the histograms and the range of DN values covered. Look for two potential problems. (1) If the DN values for a channel are all confined within a narrow interval (say 5 to 20 DNs out of a possible 255), the data have a low dynamic range, and spectral contrast will be low. (2) If the DN values for a channel extend to the top of the scale, the channel is "saturated," meaning that any higher DNs were not measured. These problems usually stem from gain settings on the detector that are not optimum for the measured scene. For example, Landsat TM Band 3 (red) may have the gain turned down as it passes over highly reflective snow-covered North America, but the same gain renders the Amazon forest too dark, producing a narrow dynamic range. Although these types of problems cannot be fixed after the fact, it is essential to be aware of them when analyzing images for spectral information.

With channel images and histograms displayed, it may be possible to make a rough estimate of the threshold of system noise. Select a dark area on the image and display its histogram. Contrast-stretch the DNs of the dark-image subset until stripes or other patterns appear that obviously are not part of the real scene. The DN values that are contributing the patterns can be measured by adjusting the contrast stretch and noting the DN readout. For example, in Landsat TM VIS-NIR channels it is common for noise to contribute DNs at about one percent of the dynamic range. At this level, noise generally is not a problem, but it can become a problem if it is important to extract spectral information from the darkest areas on the image. Depending on the objectives of the study, and the way in which the image is processed, noise may show up as unwanted patterns in the final products.

3.1.5 Channel-to-channel registration

Registration
In registered images the pixel footprints overlap perfectly.

It is common for image analysts to assume that each pixel footprint in each spectral channel is from exactly the same part of a scene. After all, that is the whole idea of a multispectral data set. However, even with commercial data (e.g., Landsat), misregistration by a fraction of a pixel is typical, and by one, two, or even more pixels, is not unusual in older data. Channel-to-channel registration can be especially difficult if the camera system acquired data sequentially in separate spectral channels as it moved along its path, thereby creating different "look" angles. This is a familiar problem for those who work with images of the Moon, Mars and other Solar-System bodies, which were acquired by so called "pushbroom" scanners that consist of CCD line detectors. Orbital and system constraints do not always allow spectral images to be acquired simultaneously in all channels.

Depending on the objectives of the analysis, misregistration by a few pixels can be a serious problem for extracting information from spectral images. Homogeneous areas in a scene present the least difficulty, but at sharp boundaries, the measurement at any one wavelength for any one area on the ground is being compared with a measurement at another wavelength that includes part or all of an adjacent pixel footprint (Section 3.2.1). The resulting pixel spectra are spurious, and may even be misinterpreted as evidence for non-existent materials on the ground. Registration errors will propagate through any spectral analysis, and may be especially detrimental in thematic classifications.

Misregistration may stem from the way images were acquired by a satellite or aircraft system, from the way the data were sampled (e.g., non-integer differences), or they may be introduced inadvertently during processing or handling of the data. It may not be possible for the analyst to discover the cause, but detection of misregistered channels can be relatively straightforward. Image-processing programs typically have software that makes it possible to link two images of different channels, such that a small area of one image can be moved across the other "background" image. Where the linked images are superposed, sharp boundaries, such as those along roads, ridges and streams, or between types of land cover, the edges of features will be appear to be repeated or interrupted if the channels are not properly registered. Misregistration also can be recognized visually in color-composite images (Section 3.3.2) by the presence of color artifacts along the edges of spatially resolved objects. Correcting the problem may be as simple as adjusting the (x,y) coordinates on each channel until the images coincide with a reference channel. In some cases, this can be done visually by trial and error using spatial features with sharp edges, or by using a program that entails finding reference points (tie points). However, if different channel images were taken from even slightly different "look" angles, or the landscape has significant topographic relief, the geometric errors may vary strongly across the scene, and the task of registration can be formidable.

Other types of geometric adjustments to images have been devised to remove a variety of spatial distortions and to make images correspond to conventional map coordinates. These types of geometric rectification are important for mapping and GIS applications, although they do not directly affect the fidelity of spectral information. The topic of geometric rectification is extensively covered elsewhere in the remote-sensing literature. For example, see Schowengerdt (1997) and references therein.

Misregistration

Depending on the objectives of the analysis, misregistration by a few pixels can pose serious problems for extracting information. Errors will propagate through spectral analysis, and may affect thematic classifications.

Rectification

Distortion of the image geometry to force it to coincide with a map.

Checklist for evaluating a spectral image

1. Know the basic properties of the image and how, when, and where it was acquired.
2. Find out how the image has been processed. Be alert for cosmetic enhancements that may have altered the spectral fidelity.
3. Know the illumination and viewing geometry.
4. Examine the images and histograms of each channel for defects, dynamic range, saturation, gradients, atmospheric effects, and noise.
5. Visually verify channel-to-channel registration using color composites or ratios.
6. Prepare a well-stretched color-composite image to use as a photo-interpretive reference during further analysis.

3.2 Calibration

3.2.1 Calibrating VIS-NIR images

In order to extract high-quality information from spectral images we need to be certain that the DN values for each pixel in each channel correctly represent the spectral nature of the land surface within each pixel footprint. Spectral images never capture the true radiance of a surface, owing mainly to various distortions introduced by sensors and the atmosphere; thus, we need to correct the measurements by a process that is generally known as "radiometric calibration." Sensor calibration and atmospheric correction comprise the main objectives of radiometric calibration. Once the pixel DNs have been radiometrically corrected, we need to make further modifications of the data to convert them to reflectance if we want to compare pixel spectra with laboratory or other reference spectra. This process sometimes is described as "calibration to surface reflectance" and it entails normalization of the land-leaving radiance by the solar spectral irradiance reaching the ground, adjustments for solar-illumination geometry, and further adjustments for topography and the photometric function of the surface.

Calibration of spectral images can be a difficult task if we demand highly accurate results. But before discussing calibration further, recall our suggestion in Chapter 1 about having a clear idea of our objectives in analyzing a spectral image (Section 1.4.3), and how accurate we want our results to be (Section 1.4.4). For many applications, particularly those that depend mainly on photo interpretation, it is not necessary to calibrate at all. When study of a landscape requires discrimination among spectral properties, or comparison of image spectra with laboratory spectra, it usually is sufficient to make an approximate calibration. Actually, all calibrations end up being approximations. In this section we focus on the essentials of the calibration process and those procedures that are readily

---

Image calibration

Generally there are two levels of calibrated image data – "radiance at sensor" and "land-leaving radiance."

accessible with most image-processing software. Conventional calibration procedures are discussed in considerable detail in other texts. Our objective here is to provide only a simplified explanation of what calibration entails. In Chapter 4 we describe how mixture models can give information on calibration, and in Chapter 7 we discuss how miscalibration defines predictable patterns in spectral data space.

Sensor calibration

Before sensors are flown, they are tested in the laboratory to measure how they respond to a standard light source. The output of the light source is radiance (radiant flux), in units related to power: $W\ m^{-2}\ sr^{-1}\ \mu m^{-1}$. Radiance received at a detector is encoded as byte or half-word integers ("DNs" to the image analyst), and the result commonly is referred to as "encoded radiance." From an engineering perspective, measuring the responses of the sensors is the correct sense of the term "calibration." One task is to measure any non-zero DN values that are measured even when the light source is turned off. Such DNs are spurious and must be subtracted (an "offset" or "bias") to get a correct DN proxy for radiance. Another task is to measure the multiplicative factor ("gain") by which the known radiance values have been amplified by a sensor, assuming a linear-response function. The results typically are published as offsets and gains for each channel in units of DN per unit radiance (e.g., Schowengerdt, 1997). Image analysts, however, want to calibrate image DNs in each channel so that the encoded radiance correctly represents the received radiance (radiance per DN). Image analysts, therefore, calibrate the encoded radiance by "inverting" the calculations; subtracting the laboratory offsets and dividing by the laboratory gains. Fortunately, this process often is made easy by the availability of image-processing software that includes sensor-calibration factors for the widely used satellite systems.

> **Units of calibration**
> Data numbers in calibrated images have equivalent physical units such as $W\ m^{-2}\ sr^{-1}\ \mu m^{-1}$.

Sensor-calibration factors change as instruments age, and sometimes detectors deteriorate significantly over time. Only rarely do spacecraft sensors have the capability of continuous in-flight calibration. Pre-flight calibrations, therefore, are likely to be most accurate for images taken soon after launch. Sensor calibrations should be applied before correcting for atmospheric and other effects, following the logic that the sensor is the last thing to affect the measured signal and should be the first removed. Generally, this correction has less impact on image data than does an atmospheric correction (discussed below). Calibration problems with imaging-spectrometer data often can be detected visually by plotting DNs as a function of wavelength. Data numbers are expected to vary smoothly as a function of wavelength at the resolution of most instruments, even where there are strong absorption bands that arise from surface materials or the atmosphere. "Spikes" in DN values (up or down) at only one wavelength imply instrumental or other artifacts. (However, with ultra-high-resolution imaging spectrometers, some spikes resolve very narrow bands in atmospheric gases.)

Atmospheric correction

Radiant energy that is scattered into detectors by an atmosphere, and that does not interact with the ground surface, is known as path radiance. This usually is unwanted information for a field investigator, because it reveals nothing about the nature of the land surface. Furthermore, path radiance reduces contrast in individual channel images (e.g., contrast ratio), and it reduces the spectral contrast (WW contrast) between channels (Section 2.1.2).

Correcting radiance values

A simplified view is that remote measurements have lost DNs and/or accumulated unwanted DNs in each wavelength channel owing to several possible processes. The effects of some processes are multiplicative, others are additive.

$$DN_m = G * DN_c + O \quad \text{and} \quad DN_c = 1/G * (DN_m - O) \qquad (3.1)$$

DN_m is the measured value, DN_c is the correct (calibrated) value, and G and O are, respectively, the cumulative gains and offsets. Each gain and each offset is the result of a separate physical process. Thorough discussions of these processes and the physics of radiative transfer are left to several excellent textbooks such as Schott (1997) and Schowengerdt (1997).

Atmospheric path length

Atmospheric path length varies with view angle and scene elevation. Path radiance and atmospheric transmissivity additionally vary with atmospheric humidity and aerosols.

Atmospheric scattering increases with shorter wavelengths, and in spectral images it is most apparent in visible-wavelength channels. The gross effects of path radiance generally are evident by visually examining the histograms of individual channel images. When applying corrections for atmospheric path radiance, it commonly is assumed that the effect is uniform across an image. This is not strictly true, of course, because path radiance depends on atmospheric path length which depends on view angle. Images such as those acquired by AVHRR or aircraft require more path-radiance correction at maximum view angles than at nadir. Furthermore, real atmospheres are spatially variable, particularly in mountainous terrain, and wherever particulates such as aerosols, smoke, and dust are important. Calibration of a variable atmosphere is a daunting undertaking, because there hardly ever are sufficient measurements available for the time of image acquisition.

A simple way to estimate and remove the main effects of path radiance is to subtract the DNs (in each channel) of a dark object or dark area in an image from all pixels. Figure 3.1 shows histograms for the six VIS-NIR Landsat TM Bands of a sub-scene of the Seattle area.

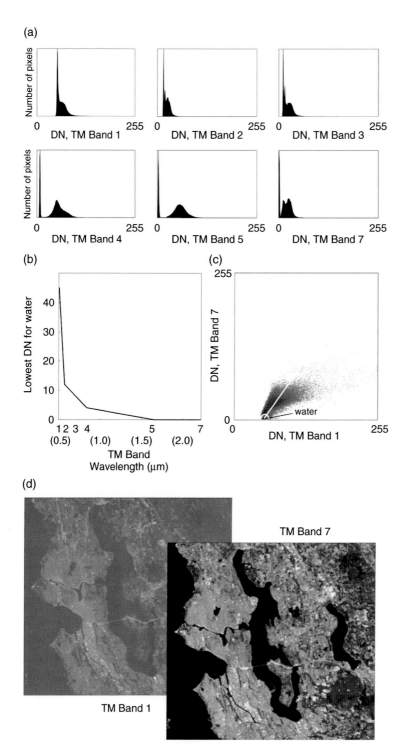

(a)

Number of pixels

0   255   0   255   0   255
DN, TM Band 1   DN, TM Band 2   DN, TM Band 3

Number of pixels

0   255   0   255   0   255
DN, TM Band 4   DN, TM Band 5   DN, TM Band 7

(b)

Lowest DN for water

40

30

20

10

0
  1 2 3 4   5   7
 (0.5)   (1.0)   (1.5)   (2.0)
        TM Band
    Wavelength (μm)

(c)

255

DN, TM Band 7

0
 0   DN, TM Band 1   255

water

(d)

TM Band 7

TM Band 1

Figure 3.1. (a) Histograms for the reflective channels of a sub-scene of an uncalibrated August 1992 Landsat TM image of Seattle, Washington. The tall, narrow peak at the left side of each histogram represents pixels of dark water. The lower and broader peaks at higher DNs represent pixels of land areas. (b) Plot of the lowest DN for water vs. TM Bands. An approximate correction for atmospheric path radiance can be made by subtracting the offset from zero in each TM Band. (c) Scatter plot of TM bands 1 and 7. A least-squares fit to the data follows a mixing line between the lightest and darkest scene elements, urban buildings, and deep water, respectively. The line intercepts the TM Band 1 axis at the value of darkest water (40 DN). (d) Images of TM Bands 1 and 7. Both images are contrast-stretched from 0 to 255 DN between the values for darkest water and the peak of the land part of the histograms. The lower contrast in TM Band 1 is caused mainly by atmospheric scattering (path radiance). See also Figure 3.4, Plate 2, and reference images on the Web.

Measured at the surface, deep water typically has low reflectance in all TM Bands; however, the DNs for Bands 1–3 appear anomalously high. The pattern of descending lowest DNs (for pixels of water) with longer-wavelength channels is consistent with a physical model that predicts the greatest amount of scattering by the atmosphere at shorter wavelengths (Figure 3.1b). Presuming that our physical model is correct, we estimate the DNs contributed by path radiance as the difference between the expected near-zero DN value and the measured one. We recognize that the expected DN value of water in each channel is a guess because we do not have access to actual measurements of the water surface at the time the image was taken. Furthermore, we know that the reflectance of deep water, although typically low, is not actually zero at all wavelengths. Nevertheless, the example illustrates a simple and widely used in-scene method to correct for path radiance, namely, to use a dark area that is expected to have near-zero DNs in all channels. Of course, there has to be a suitably dark area in the scene.

If pixels of dark materials are not resolved in an image, spatially resolved shadows can be used for calibration, although shadows do not have the same spectra as intrinsically dark materials. Shadows are not entirely dark, at least on Earth and on Mars, because they are illuminated by scattered skylight and indirect sources such as nearby sunlit objects. In some images, though, resolved shadows cannot be located, and it may be necessary to define a "virtual" dark pixel by projecting a mixing line in the data space to the channel axes. The offset from zero DN can be used as an estimate of the path-radiance correction, as is shown in the scatter plot in Figure 3.1c. This method often works, because pixels usually contain some fraction of unresolved shadow, and an assemblage of pixels consists of the spectra of the fully illuminated materials mixed in various proportions with "shade" (as defined in Chapter 4).

Some fraction of the solar irradiance reaching the ground and of the radiance coming from the ground is absorbed by an atmosphere. Unlike path radiance that adds unwanted radiance, absorption selectively diminishes DN values at certain wavelengths. The effects of absorption are difficult to isolate using image data alone; however, the empirical line method (discussed below) produces a correction that combines absorption (gain) and scattering (offset) effects. Commonly, corrections for atmospheric absorption are based on radiometric models such as MODTRAN (e.g., Berk et al., 1998). The accuracy of models, of course, depends on the quality of the input data. It usually is not possible to measure atmospheric properties during image acquisition; therefore, the coefficients that describe the state of the atmospheric column typically are estimated or assumed. Imaging-spectrometer data may need particular care to correct atmospheric absorption effects, for the

Triangulation of shade

Radiance from a black object is mostly the atmospherically scattered path radiance, and can be estimated from the intersection of mixing lines passing through distributions of partially sunlit and shadowed materials, even if totally dark pixels are not found in an image, or if candidate pixels do not have zero DN in any channel.

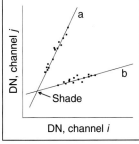

reason that some channels extend partially or entirely into absorption bands caused by atmospheric water, CO_2, and other components. On the other hand, images acquired in or near atmospheric absorption bands potentially can be used for pixel-by-pixel in-scene corrections. The model MODTRAN and other radiometric models are available in some commercial image-processing packages, and there is an extensive literature on atmospheric modeling.

Calibration to surface reflectance

In Chapter 2 we described how spectral reflectance and emittance measurements are made in the laboratory, and contrasted them with the way remote-sensing measurements are made. Recall that laboratory VIS-NIR spectra are normalized *relative* to the spectrum of a white standard, where both sample and reference are measured under identical conditions. In the laboratory there is the enormous advantage that sample and reference can be measured with the same illumination geometry, and the geometric, instrumental, and atmospheric effects cancel, because they are the same for both sample and reference, Equation (2.1). In the case of images, we can easily apply gains to correct radiance values for variations in solar elevation and calculate "apparent" reflectance by dividing by the solar spectral irradiance. Fortunately, solar irradiance is well known. Small seasonal changes in solar distance from Earth and variations in solar-energy output usually are ignored, except, of course, that distance to the Sun is important when doing remote sensing of other objects in the Solar System. But we cannot directly compare apparent reflectance with laboratory spectra ("true" reflectance) unless the illumination geometry and the photometric properties of the surface are the same. It is unlikely that there will be a pixel-footprint-size reference material in the scene, and it is unrealistic to expect one for each spatially resolved topographic slope and aspect. Therefore, we need to model the radiance (in DN) of the same reference material that is used in the laboratory.

> **Apparent reflectance**
> The term "apparent" is used to distinguish reflectance in calibrated images that has not been corrected for macro-topography at the scale of a DEM.

> **True reflectance**
> Reflectance measured under conditions where the sample and the reference have the same illumination and detection geometry.

The correction of radiance or apparent reflectance for topography typically is done by applying a DEM (Section 1.3.3). A DEM specifies slope and aspect of each surface element of a landscape at a particular scale – usually tens to hundreds of meters. The reflecting ability of surface elements depends on two things: the composition of the surface material(s) and the surface geometry at all scales. By correcting radiance or apparent reflectance for each DEM element we can remove the effects of the macro-topography at the DEM scale; however, this does not account for geometric effects that occur at finer scales. Furthermore, it is unlikely that we will know the true photometric function of the surface. Digital elevation models, therefore, commonly are applied assuming a Lambertian (cosine) function (Figure 1.9). This assumption works well enough for many applications and is computationally convenient. More accurate results can be obtained with other photometric models, depending on the nature of the landscape surface. For example, various models have been developed for forested landscapes, including the SAIL model (e.g., Verhoef, 1984), hybrid models (e.g., Li *et al.*, 1995), and the Sun-canopy-sensor model (Gu and Gillespie, 1998).

Combined calibrations

Before being sent to the user, spectral images commonly are pre-processed to correct for sensor calibration, (global) solar elevation, and solar spectral irradiance. At this stage, the data are corrected to "top of the atmosphere" (TOA) reflectance (also known as "exoatmospheric reflectance"). If this has not been done already, many image-processing packages have programs that allow these corrections to be made individually or in combination. Selecting only a sensor calibration produces TOA radiance. Selecting a sensor calibration combined with corrections for Sun angle and solar spectral irradiance gives TOA reflectance. Top of the atmosphere radiance or TOA reflectance values still need to be corrected for atmospheric and topographic distortions. Therefore, another set of offsets and gains needs to be found to correct for atmospheric scattering and absorption, and further gains derived from DEMs need to be applied, pixel by pixel, to correct for topography. In reality, it is not possible to determine all of these gains and offsets with assured accuracy. Calibration, therefore, relies on best estimates, often using a combination of methods.

> **TOA reflectance**
>
> Commercially available images commonly have been corrected for sensor calibration, sun angle and solar spectral irradiance; therefore, the DNs represent TOA reflectance. The user still needs to correct for atmospheric effects and macro-topography to arrive at surface, "true" reflectance.

Empirical calibrations

One way to calibrate remote measurements to reflectance is to have standard reference targets on the ground. However, placing light and dark reference panels the size of a pixel footprint or larger on the ground before over-flights is, to say the least, not a practical approach for most applications. There is an alternative when field investigators know enough about materials on the ground itself to be able to estimate the reflectance spectra of the footprints of a few pixels. When image spectra and laboratory or field spectra (reference spectra) of the same area are known, we can calculate the offset and gain factors that reconcile them. Assuming that the atmosphere is spatially uniform, these empirically derived calibration factors then can be applied to all other pixels.

In a method known as an "empirical-line" calibration (e.g., Conel, 1990), light and dark pixels are selected for which reflectances are known from field information. In the simplest form of the empirical-line calibration that is available in commercial image-processing programs, reference spectra are re-sampled by the integrated wavelength functions of the image channels, and the image data simply are forced to fit the reflectance data, channel by channel, using a linear regression. Pixels that are included in an empirical-line calibration should not sample significant topographic slopes; alternatively a topographic correction (Section 3.3.2) should be applied to an image after calibration. Commonly, though, an empirical line is selected to pass through the vector of a normally illuminated pixel footprint and either the origin or the vector of a shadowed pixel of any material. In this case, the empirical line is a segment of a mixing line that includes a fully illuminated endmember and "shade" with a near-zero reflectance (Chapter 4). There are differences between the spectra of shadow and a normally illuminated surface of dark material, however, and neither will have zero reflectance. The choice of a dark reference pixel, therefore, is important when considering how accurate a calibration needs to be.

In-scene relative measurements (normalization)

If the objective is to enhance spectral differences in a scene, rather than to compare image spectra with laboratory spectra, we can normalize an image relative to an arbitrary reference. In-scene, relative spectra have been used for planetary applications for many years. For example, early telescopic spectra of the Moon were calibrated to a standard area in Mare Serenetatis (McCord *et al.*, 1972). An in-scene reference spectrum can occupy only a few pixels, or, it can be an average of an entire scene. A common method for analyzing imaging-spectrometer data is to normalize all spectra to a scene-average value at each wavelength. Relative calibrations of this type sometimes are referred to as "flat field" or "scene-average calibrations." (Actually, they are normalizations, not calibrations.) The resulting relative spectra do not resemble laboratory reflectance spectra. The similarity with the calibration process is that normalization does suppress atmospheric and instrumental effects, and preserves wavelength variations that, with experience, can be interpreted spectroscopically. This is the same process that occurs when one channel value is divided by another to form a ratio image (Section 3.5.1).

Calibration assumptions	
Systems calibration	No change in pre-flight calibration with time.
Atmosphere	Physical processes of scattering and absorption are understood.
	Model parameters are correct and apply at time and place of image acquisition.
	Atmosphere has same properties spatially and vertically over scene.
Solar irradiance	Standard solar output data are applicable.
Topography and photometry	Digital elevation model is correct. Photometric function describes actual surface.

Other factors that may require calibration

Across-image gradients that are caused by atmospheric backscattering (Section 3.1.2) may be suppressed or removed by applying a spatially varying offset function across pixel columns in the scan direction. A simple, empirical correction derived from the image itself is effective if a scene is compositionally uniform and has little topographic relief. However, finding a correction function for a heterogeneous scene may require experimenting with an atmospheric model. The situation is complicated by the fact that an increase in path length is accompanied by more atmospheric absorption as well as scattering. An empirical approach also can be used to remove gradients that are caused by an across-image change in the fraction of sub-pixel shadow (Sections 1.3.3 and 3.1.2). In this case, a varying gain may be the dominant correction needed to suppress the gradient. Atmospheric effects and compositional and topographic irregularities will complicate the problem.

When sunlight is scattered from objects on the ground, some portion of the scattered radiation illuminates other, nearby objects. In principle, to determine the true radiance that is reflected from each pixel footprint one should remove the effects of this "bounce" light from adjacent or nearby pixel footprints. Fortunately, for those parts of terrestrial landscapes that are directly illuminated, the proportion of light that is scattered from elsewhere on the ground comprises only a few percent at most, and can be ignored for remote-sensing purposes. Even in shadowed areas, the light from nearby illuminated objects or adjacent hill slopes rarely exceeds 10% of the total, the remainder being contributed by indirect illumination from the atmosphere.

A special case of indirect illumination from the ground is the light that has been transmitted through green leaves. Green vegetation tends to transmit a relatively high proportion of incident radiation at NIR wavelengths. This effect is ignored when calibrating spectral images; however, it is taken into account in selecting the shade endmember in mixture analysis (Chapter 4), because shadows cast on a soil substrate by green vegetation typically have spectra that mimic dark vegetation.

3.2.2 Calibrating thermal-infrared images

As in the case of VIS-NIR, it is usual to begin calibration of TIR (8–12 μm) images by removing unwanted effects contributed by the sensor system. The main problem is that TIR scanners themselves emit and scatter radiation, and it is unavoidable that some is recorded by the detector elements. The result is an added amount of radiance that varies with scanner temperature. To minimize this problem, engineers typically cool the optics, detectors, and detector housings. In addition, it is common for thermal imaging systems to inspect two different black-bodies, hot and cold, at the end of every image scan, to allow real-time calibration of the data. Correction for these blackbody data takes the form of a linear-transfer equation that relates measured DN to radiance units for each scan line. The corrected data are TOA radiances, because they have not yet been compensated for atmospheric absorption and emission.

Atmospheric calibration in the thermal infrared

There are two general approaches to atmospheric calibration in the TIR. In one method, known areas on the ground are used for an empirical-line correction. In another, radiative-transfer models such as MODTRAN are used to predict atmospheric transmission and absorption. To apply the empirical-line approach, we need to find two areas in an image, one hot and the other cold, of known emittance and temperature. Planck's law tells us that if we know the emittances and temperatures of these reference areas, we can compute their surface-emitted radiances (Section 2.1.1). The reference pixels establish an empirical relationship between DNs and emittance, and define a line for each spectral channel, where the slope of each line represents the gain due to atmospheric transmission, and the axis intercepts are the offsets due to path radiance. Water bodies, if present, make especially convenient reference areas, because the emittance of calm water is the same as the laboratory-measured emissivity, that is, near unity, and water temperature can either be estimated or measured at the time of image acquisition with radiometers or near-surface thermistors. Furthermore, bodies of water often are larger than a few pixels, and tend to be homogeneous. In principle, though, other surfaces such as vegetated areas or sand dunes could be used. For a complete description of the atmospheric effects over an image we would need to have a large number of control sites, because not only must the differences in view angle be captured, but topography also must be considered. Usually, it is infeasible to measure more than a few sites in an image; therefore, the empirical-line approach often is coupled with atmospheric modeling and topographic correction using a DEM.

In the TIR, an atmosphere itself emits and scatters radiation. Atmospheric compensation is complicated, because the amount of path-emitted radiance is a function of the temperature profile, as well as the total amount of atmosphere in the optical path. Upwelling radiance emitted from a warm, low-altitude mass of humid air will be partly absorbed by the air above it; radiance from a high-altitude mass will be less affected. In the modeling approach, measured atmospheric profiles, if available, can be used to constrain the solution. Even if profiles are available, they rarely are numerous enough to define completely the spatial variability in an atmosphere, and the resulting compensations usually are imperfect. Analysts still may want to use the empirical-line approach to make second-order adjustments to the data.

Initial atmospheric modeling using measured or modeled profiles corrects the data to what is known as "surface-*leaving* radiance." Actually, what we want is something known as "surface-*emitted* radiance," which, as the name implies, is the radiance actually emitted by the ground, but that does not include the downward-emitted irradiance from the sky that has been reflected back to the sensor by the ground. To remove the reflected component, we need to know the emittance of the surface (see Kirchhoff's law in Section 2.1.1). However, this is a circular argument, because we cannot estimate emittance until the reflected term has been removed. One way out of the dilemma is to adopt an iterative approach, whereby a best estimate is made of the surface emittance from the surface-*leaving* radiance, and then an estimated (by MODTRAN, for example) amount of reflected sky irradiance is subtracted to give the surface-*emitted* term. The newly estimated surface-emitted radiance is likely to be closer to the actual value, and can be used to refine another estimate. The approach usually works with three to five iterations, but fails to converge unless the sky irradiance is only a small fraction of the surface-emitted term.

Calculating emittance

Once we have calibrated a TIR image to surface-emitted radiance, we would like to extract emittance as a function of wavelength. This is what we are looking for, because we know from laboratory measurements that spectral emittance can be a valuable tool for identifying surface materials (Chapter 2). Given radiance values, we can easily compute emittance, if (a big "if") we know the temperature (Planck's law, again). Surface temperature, though, is not that easy to measure, even on the ground. For example, temperature can change rapidly when even light winds are blowing. It would be much more convenient if we could estimate temperature remotely, but this requires knowing emittance (the circular argument, again). A compromise solution is to select one wavelength channel (reference channel) for which we assume that the emittance of everything on the ground is the same. Based on this reference channel, we then can calculate the temperature (known as a "brightness temperature") for all pixels. The next step is to apply these modeled surface temperatures to all other channels to derive emittance. Although an approximation, this technique can yield satisfactory results. A program for applying this calibration method to TIMS images is included in some commercial image-processing packages.

3.3 Enhancement for photo interpretation

The first digital image-processing techniques were developed to remove defects, enhance contrast, and to combine images into ratios and into red, green, blue (RGB) composites. These methods were designed for spectral images with only a few, broad-wavelength channels. Early emphasis, therefore, was on using spectral variations to assist photo interpretation, rather than trying to extract compositional information. A notable exception was green vegetation, for it had been known since the early days of color aerial photography that chlorophyll could be identified by its spectral properties. Today, standard tools for analyzing spectral images still include multiple ways to assist the analyst in photo interpretation and to make images look better. In this book, we make a distinction between those techniques that are particularly important for facilitating the visual process of image interpretation (Section 3.3) and other techniques that, in addition, involve analysis of spectral data (Section 3.4).

3.3.1 Contrast and color

Contrast enhancement

Everyone who works with digital images quickly learns about the power of contrast enhancement and its critical importance for displaying information. In remote sensing, the data recorded in any particular channel rarely fill the pre-set dynamic range of the sensor system. As a result, the visual quality of virtually all encoded-radiance images can be improved substantially by applying a contrast stretch, and the same is true of many types of images that are derived from the basic data. We refer the reader to remote-sensing texts for details of different contrast-stretch algorithms and examples of what they do to specific images. Contrast stretches, though, are a rather personal matter, because what is useful to one person may not be optimum for another. How we stretch an image depends on the data and the information that we want to retrieve.

Color enhancement

For photo interpretation, visual colors can be an important way to link an image to the ground. Color is a sensation of the human eye–brain system within a narrow wavelength region of about 0.4–0.7 μm. Remote-sensing systems that operate in the VIS-NIR range typically include three channels that approximate the response of the human eye, and afford the opportunity to construct an RGB composite image that resembles natural color. Unfortunately, when displayed, many such images appear to have mostly tones of gray, owing to low spectral contrast that is caused by atmospheric scattering and limited dynamic range. Contrast stretching

Color

There is an extensive literature on color science, and remote-sensing textbooks discuss color transformations. Color transformations are standard tools in image-processing software.

may help, but the colors tend to be far less saturated than we are accustomed to when observing or photographing in the field. Besides trying standard stretches, another way to enhance natural color is to transform RGB images into a color data space such as the one defined by hue, saturation, and intensity (HSI) and selectively stretching the low DNs in the saturation image. After all, it is low saturation that causes colors to appear faint or "washed out," whereas we want to see each color more vividly. Low DNs can be enhanced by applying a logarithmic stretch or by complementing ("inverting") the saturation image and manually adjusting the contrast. The stretched HSI images then are re-displayed as RGB images. Some image-processing packages do this automatically using a "saturation-stretch" algorithm. The advantage of making individual H, S, and I images, though, is that we can control the stretches and control how H, S, and I are assigned to R, G, and B. The results of enhancing images to bring out color are highly dependent on the nature of the landscape and the image measurements. Usually, we have to experiment to adapt the enhancements to our particular needs.

Decorrelation stretch

The decorrelation stretch is another tool to enhance small spectral differences among a few channels. The method is similar to a principal-component analysis (Section 3.4.1), and is best known for its applications to TIR images where channel-to-channel correlation is notoriously high. The same technique, though, can be used to enhance color contrast among three channels in the visible-wavelength range, making it another way to connect image data to familiar visual colors.

In TIR images, emissivity and temperature control the brightness at each wavelength. Emissivity varies little across the spectrum, whereas temperature may vary a lot (e.g., 300–350 K) across a scene, but does not vary with wavelength. To a good approximation, decorrelation stretching subdues temperature variations while exaggerating the emissivity component. With this technique the emissivity information is enhanced and displayed by means of color, whereas the temperature information is expressed by the lightness in the displayed image (Plate 6). Another way to "decorrelate" TIR data is to transform to HSI space and enhance saturation while reducing intensity. This is similar to the method used for visible-color channels, except that intensity needs to be more suppressed in the TIR data.

3.3.2 Topography and texture

Methods for suppressing the variable illumination that is caused by topography are discussed in Section 3.2.1 in the context of calibration. Sometimes, though, we want to enhance topography and texture to facilitate photo interpretation. We can, of course, accentuate topography by overlaying a spectral image on a shaded-relief image calculated from a DEM (Section 1.3.3), or by merging a spectral image with another image having high spatial resolution and high contrast. However, topography and texture also can be enhanced without additional images. One simple technique is to display the complement of (invert) a channel image and apply an interactive contrast stretch. This makes it possible to visually lighten only the low-DN pixels and to darken the highest DN pixels by setting them to zero DN. Keep in mind that by complementing an image, light and dark are reversed, often giving the appearance of changing the Sun azimuth by 180°. Other techniques that can enhance topography and texture include the principal-component transformation (Section 3.4.1), the tasseled-cap transformation (Section 3.5.3), and mixture models (Sections 4.8.2 and 7.2.3).

3.3.3 Cosmetics

We all like images that look good. It is tempting to place less value on images that are not aesthetically pleasing – for example, ones that have ugly colors, or have distracting speckle or stripe patterns. Commonly, and perhaps usually, published images have been "cleaned up." Mathematical filters have replaced "noisy" pixels with new values that enhance the appearance of the image. Classification boundaries have been "smoothed" or redrawn to look better. Boundaries between images in mosaics have been rendered invisible by adjusting the data. Cosmetic changes can greatly improve an image for photo interpretation by making it easier for the eye–brain system to understand certain relationships. When analyzing spectral data, however, it is essential to realize that repairing blemishes may alter the spectral information.

> **Cosmetics**
> Cosmetics are superficial measures to make something appear better, more attractive, or more impressive.

Early Landsat images commonly had various defects that included line dropouts, periodic striping, and line offsets. Techniques evolved to restore these images so that they looked more like the original scene. Restoration was evaluated according to whether an image was visually improved, but less attention was given to how spectral information was affected. Although today's systems have fewer such problems, it is still common to find images that have been processed to remove visually offensive defects. Be aware that the "noise" that someone else removed may have been the "signal" that you were looking for. Cosmetic improvements always can be done as the final step, after we have completed a spectral analysis.

3.4 Data reconnaissance and organization

To explore spectral images and to do thematic mapping, it is essential to have visual and analytical ways to organize pixels that have similar properties. We would like to be able to select a pixel or group of pixels that have meaning to us, and to find all of the other pixels that are similar. In some cases, we also would like to have our computer take the next step, and to classify groups of similar pixels. To do this, we need specific rules that define what we mean by "similarity" and what we mean by "class." At the same time that we are evaluating which pixels are spectrally similar, we would like to know how pixels differ from one another; therefore, we need tools that allow us to assess PP spectral contrast. Classification algorithms are the main tools for measuring spectral similarity. The main tools for assessing and enhancing spectral dissimilarity are principal-component analysis, spectral-mixture analysis and matched filtering. Classification and principal-component analysis are among the standard methods that are widely used. We discuss spectral-mixture analysis in Chapters 4 and 5 and matched filtering in Chapter 6. Ultimately, the methods for data reconnaissance and organization have to be judged in the context of what is on the ground. Experience tells us that mathematically robust measures of spectral similarity or dissimilarity of image data sometimes, but not always, make sense in a field context, and, sometimes even can be misleading. In this section, with the field investigator in mind, we discuss the strengths and weaknesses of principal-component analysis and the main types of spectral classifiers.

3.4.1 Exploring the data space

A synthetic spectral image

To illustrate how spectra and images are related, we created a synthetic image consisting of four spectra sampled by four channels on a DN scale of 0 to 100 (Figure 3.2 and Table 3.1). The spectra of A, B, and C are hypothetical. Their spatial arrangement is such that there are areas of pure pixels and areas of mixed pixels. Mixtures of A and B occupy a ten-pixel-wide zone in the middle of the image. The transition between A and B occurs in increments, such that a pixel of 100% A in column 5 is bordered by 95% A and 5% B in column 6, 85% A and 15% B in column 7, and so on. Spectrum C only mixes with A and B along a narrow zone, part of which is enlarged in the inset in Figure 3.2b. Spectrum S represents shade. Shade was added as horizontal stripes to simulate the topography of asymmetric east–west ridges and valleys with illumination from the top of the image. The repeating shade pattern is: row $2 = 10\%$; row $3 = 30\%$; row $4 = 50\%$; row $5 = 40\%$; row $6 = 20\%$; row $7 = 10\%$.

Table 3.1. *Data number values (0–100) in four channels for A, B, C, and S.*

Spectra	Channel			
	1	2	3	4
A	25	53	59	60
B	68	12	55	64
C	70	76	82	67
S	2	2	2	2

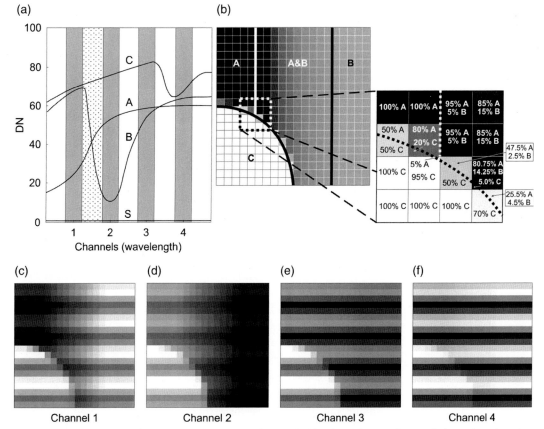

Figure 3.2. Synthetic image. (a) Spectra of A, B, C, and S sampled by channels 1–4. The stippled wavelength region between channels 1 and 2 is discussed in Chapter 6 (Figure 6.2). (b) Distribution of A, B, and C in the image. (c)–(f) Images by channel. Color versions of the synthetic image are included in Plate 9.

Histograms and scatter plots

We recommend taking the time to examine each single-channel image by dissecting its histogram. Interactive programs for contrast-stretching images usually display channel histograms, and most programs allow one to select parts of a histogram to see the corresponding pixels highlighted on the image. At a minimum, we want to be able to find out which pixels on an image correspond to the highest and lowest DNs, and which features in an image account for the main histogram peaks. Simple two-channel scatter plots can be extremely helpful in data reconnaissance and organization.

Standard image-processing tools for exploring spectral data spaces also include tools to make scatter plots of two or three channels (e.g., Figure 3.1). Two-channel scatter plots are used to show where groups of pixels on an image are located in the data space, and how the lightness in different channels co-varies. Three-dimensional plots can be produced by most graphics programs, and when rotated, they help to visualize the data structure. Of course, we are unable to visualize the data from four or more channels in one display, and we must resort to multiple two- or three-dimensional plots (sub-space projections).

Figure 3.3 shows histograms and scatter plots for channels 1 and 2 of the synthetic image. In channel 1, B and C have DNs at the high end of the histogram, whereas A and S are low. In channel 2, C is at the high end of the histogram, and B is near the low end, along with S. The same channel 1 and 2 scatter plot also illustrates some other important features of the data. From the image in Figure 3.2 we know that there are multiple pixels of pure A, B, and C. The locations of these spectra are revealed in the histograms, although the pixels in each group plot on top of one another on the scatter plot. We also see the steps of the transition from A to B, whereas there are few pixels that are mixes of A and C or B and C. The scatter plot further reveals the effects of adding S (shade) to all of the other components. Spectral mixtures, including those with shade as an endmember, are discussed further in Chapter 4.

The structure of real images, although far more complicated than our synthetic one, can be explored using the same basic tools. The biggest difference is that with real images of landscapes we are not sure of the spectra and identities of the surface materials and their mixtures. But we can use our field knowledge and photo interpretation to give meaning to the data plots, and in this way, select channels and algorithms that are best for extracting information. Also, in a photo-interpretive context, histograms and scatter plots can be used to identify pixels that are artifactual or are noisy. An example of histograms and a scatter plot for part of a Landsat TM image of Seattle,

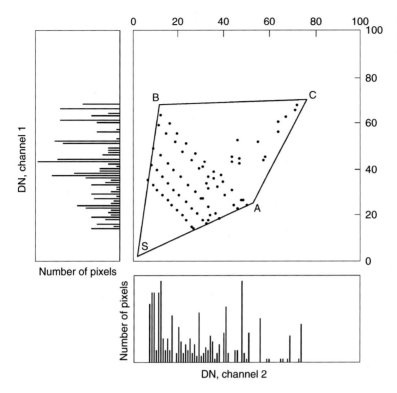

Figure 3.3. Scatter plot and histograms of channels 1 and 2 of the synthetic image.

Washington is shown in Figure 3.4. (See also Figure 3.1 and Web reference figures.)

Scatter plots, such as the one in Figure 3.4, may not be in the optimum orientation with respect to the Cartesian channel axes to show the parts of the data that are most important to us. In Figure 3.4, for example, the shape of the histogram for TM Band 4 is controlled by light-green vegetation at the high-DN end, a large population of dark vegetation overlapping with urban/suburban pixels in the mid range, and water at the low-DN end. In this channel, lighter vegetation is well separated on the DN scale from the other scene components. It is clear that TM Band 4 is a good one to distinguish vegetation from other things. On the other hand, in the histogram for TM Band 3, all green vegetation is compressed into a narrow DN range that overlaps with water and urban/suburban. This is not a good view of the data space to separate green vegetation from other scene components. Thematic mapper Band 3 does separate urban/suburban from water over a small DN range, but the range would be larger if the DN axis were rotated about 45° relative to the TM Band 3 data. Such data-space transformations are common in image processing, and are especially important for exploring detection thresholds (Chapter 6).

Figure 3.4. Histograms and scatter plot for TM Bands 3 and 4 of a subset of the August 1992 Landsat image of Seattle, Washington (Plate 2). The histograms are projections of the data onto the Band axes. The spike in the histograms at low DN values corresponds to the large number of pixels of water. The broad hump in both histograms is the heterogeneous urban, suburban, and rural landscape. Dry grass mimics mixtures of urban and green vegetation in Bands 3 and 4, but it can be distinguished if Band 5 is included. The population of pixels is highest in the light areas in the scatter plot, but it is very low at the high-DN ranges of both channels.

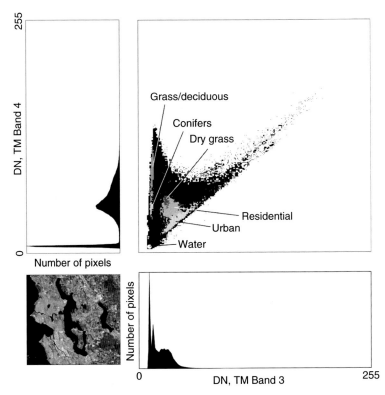

Principal components

There are several methods for projecting the data space onto new coordinate axes to achieve greater PP spectral contrast among scene components (Section 2.2.1). The most commonly used standard tool is the principal-component transformation (PCT). The analytical basis for PCT is discussed in other texts on remote sensing, and will not be covered here. Principal-component transformation is a linear transformation that projects all channel-DN values onto statistically defined orthogonal axes that act as new spectral channels. The first principal component (PCT_1) defines the axis of maximum overall variability in the data set. The second principal component (PCT_2) defines the orthogonal direction in spectral space of the next largest extent of variability, and so on for higher-order components. When the channel-DN values for each pixel are projected onto the principal-component axes, they define new DNs that can be displayed in image form, one image for each axis. The objective is to make images that reveal spectral differences that otherwise would be hard to see with the original channels. The PCT is especially helpful when channel data are highly correlated, meaning that the channel-to-channel variations in DN for each pixel are small,

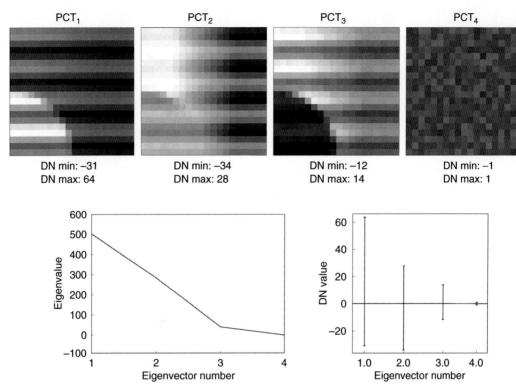

Figure 3.5. Principal-component transformation applied to the synthetic image that was introduced in Figure 3.2. The PCT$_1$ image shows that the main variability in the data is between C and S, the lightest and darkest components; PCT$_2$ shows that the next largest variations are between A and B; PCT$_3$ defines the variation between A and C; and PCT$_4$ shows 5% random noise that was added to the data.

even though the lightness differences from pixel to pixel may be large. In the familiar, two-channel scatter plot, highly correlated data lie near a line, and the spectral contrast is largely restricted to the SL component, where pixels range from dark to light (Section 2.2.1).

Principal-component transformations are applied to numerical data sets, and do not take into account photo-interpretive context; it is up to the analyst to provide the interpretation. It is widely taught that in images of landscapes, PCT$_1$ accentuates topography. This is true when the main variability is between illuminated and shadowed areas, but in the absence of noticeable shading and shadows, other variations in the data will define PCT$_1$. There is no general rule as to the significance of the spectrally controlled components, because the data structure of each image is different. In images of landscapes, higher PCTs may or may not be interpretable in terms of spectrally different materials on the ground. In all images, at some point, the highest components display mostly noise. Figure 3.5 shows PCT images and a

Interpretation of PCT images

Although PCT is a quantitative, mathematical operation, analysts must still use photo interpretation to evaluate PCT images.

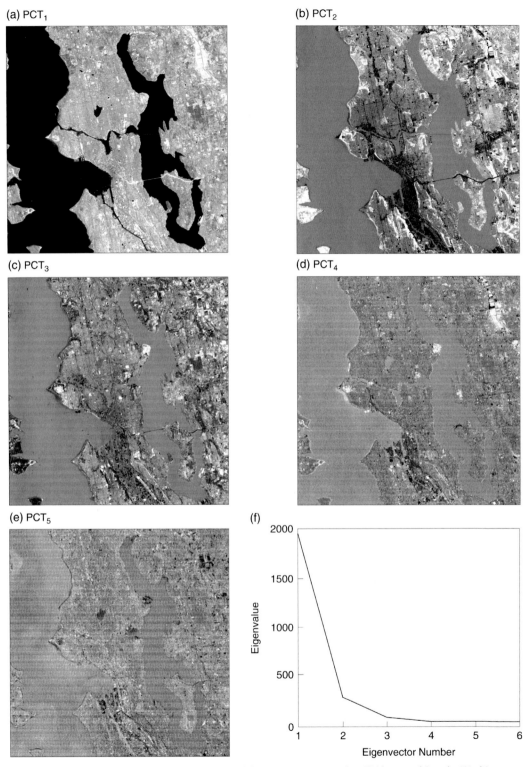

Figure 3.6. Principal-component transform images of the August 1992 Landsat TM image of Seattle, Washington. (a) Image PCT$_1$ is dominated by the lightness difference between dark water (a shade mimic) and lighter land; (b) PCT$_2$: lightest tones are green vegetation, darkest tones are urban areas; (c) PCT$_3$: lightest tones are areas of dry vegetation, darkest tones are urban areas; (d) PCT$_4$: lightest tones are dry vegetation, darkest tones are some agricultural and some urban areas; (e) PCT$_5$: light and dark tones show horizontal banding (sensor noise) and faint patterns in Puget Sound; (f) Plot of eigenvalues. Most of the variability is in PCTs 1 and 2.

plot of the eigenvalues for the synthetic image. These PCT images can be interpreted by referring back to Figures 3.2 and 3.3.

The PCT is a particularly valuable tool for reconnaissance and organization of image data. We like to start by examining a PCT_1 image to get a feeling for the extent to which light/dark variations, including topography, influence the data. Subtle differences in shading and shadowing may become more apparent after applying a logarithmic stretch or by displaying the complemented $(1 - PCT)$ image. By examining images of the higher-rank transforms, we can evaluate the context of the main spectral variability.

In the Seattle image (Figure 3.6), we see that PCT_1 is controlled by the contrast between dark water and lighter land areas and not by topographic shading and shadows. Based on photo interpretation and field work, we interpret PCT_2 as highlighting green vegetation, and PCT_3 as highlighting areas of dry grass; PCT_4 calls attention to soils. We see immediately from these images that the dominant influences, besides lightness/darkness (shade), are the basic spectral components (Section 2.2.3) of the scene – green vegetation (GV), non-photosynthetic vegetation (NPV) and soil. Three of these scene components are displayed in a three-color composite (Figure 3.7 and Web color Figure 3.7) where red $= PCT_3$ (NPV), green $= PCT_2$ (GV), and blue $= PCT_1$ (shade).

> **PCT and scene components**
> Principal-component transformation images are a means of defining the basic spectral components.

In contrast with the Seattle image, a PCT reconnaissance of a Landsat TM image of Death Valley, California, reveals an entirely different organization of the spectral information (Figure 3.8), and emphasizes the requirement for photo interpretation and field knowledge of the area. The first four PCT images of Death Valley emphasize differences among the light-colored salt deposits on the valley floor. This outcome is of interest for study of the composition of the valley floor, but it is not optimum for understanding the rest of the scene. For example, green vegetation is not highlighted until PCT_5. However, if we eliminate the valley floor from the scene by subsetting the image or masking, or we obtain an image of the same scene taken at a different time of day with different shading and shadowing, the PCT results will be different, and they will express the new, spectrally dominant scene components. In the Seattle scene there are fewer spectrally dominant components, and this allows most of the variability to be expressed in a three-color composite (color Figure 3.7; see Web).

It is common to apply standard classifiers to principal-component transforms, because, by using a few PCT images, one can reduce the dimensionality of a multi-channel data set, thereby improving the computational efficiency of the common classification algorithms (Section 3.4.2). But considerable caution is needed when one

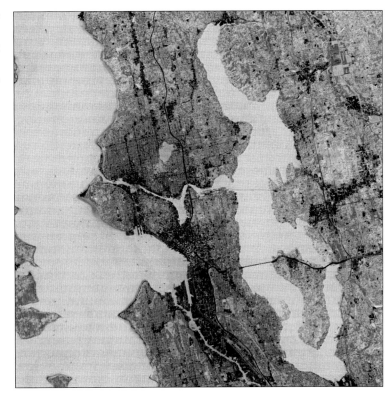

Figure 3.7 and color Figure 3.7 (see Web). Color composite PCT image of the August 1992 Landsat TM sub-image of Seattle, Washington; $R = PCT_3$, $G = PCT_2$, $B = 1 - PCT_1$.

statistical method – a classification algorithm – is applied to the results of another statistical method. Unless there is a strong photo-interpretive context, it is easy to lose track of what the connection is to the ground.

Unsupervised classifications

For some images, unsupervised classifiers can help in reconnaissance and data organization. As the name implies, this form of classification is done without interpretive guidance from an analyst. An algorithm automatically organizes similar pixel values into groups that become the basis for different classes. In one commonly used method, known as K-means, the analyst needs only to specify the number (K) of desired classes. The algorithm begins by assigning each pixel to a group based on an initial selection of mean values. Groups are iteratively re-defined until the means no longer change beyond some threshold. Pixels belonging to the groups are then classified using a minimum-distance-to-means or other method.

Unsupervised classifiers measure the "lumpiness" of the data by locating clusters in the n-dimensional data space, but they cannot

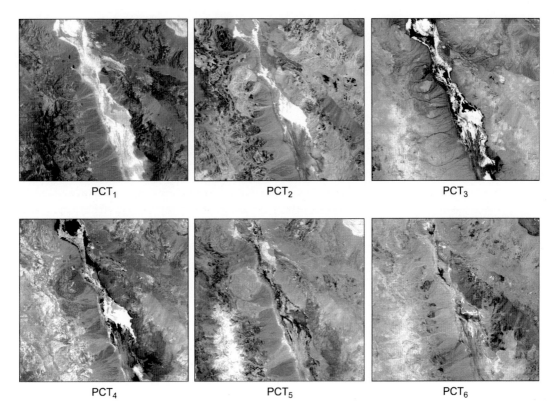

Figure 3.8. Principal-component transforms of a sub-image of the Landsat TM image of Death Valley, California. The first four transforms highlight saline deposits on the valley floor; PCT_5 highlights vegetation. See Plate 1 and reference images on the Web.

determine whether the spectral signatures of the clusters relate to interesting things on the ground. Therefore, the analyst must label and identify the classes that are produced by an unsupervised classifier. The unsupervised approach is of limited use to a field investigator who already has considerable knowledge of the ground. Under some circumstances, though, it can be a starting point for investigating unfamiliar scenes or ones that are imaged at an unfamiliar scale. Thus, an unsupervised classifier sometimes is employed as a precursor to a supervised classification (Section 3.4.2).

The binary encoding algorithm (Mazer *et al.*, 1988) is a special case in which an unsupervised classifier was developed to be applied to imaging-spectrometer data. Ordinarily, classifiers are too unwieldy for images having many channels. The algorithm was designed to match spectral signatures between pixels and reference spectra from a spectral data base, although spectral means from training areas can be used instead of reference spectra. A distance (known as a "Hamming" distance) is measured between vectors, which describes spectral

amplitude and local slope at each wavelength. The distance then is used
to define classes by an iterative, unsupervised clustering method similar
to K-means. A threshold can be set that allows some pixels to be
unclassified. The method is relatively insensitive to shading and albedo
differences, and has been applied successfully to AVIRIS data, especially of geological targets.

Tests of unsupervised classifiers on the synthetic image

K-means and isodata	Unacceptable test results, even varying all parameters. For example, misclassified all shaded pixels as A. Did not distinguish between classes A and C, and between A and B.
Binary encoding	Good results for both training sets and for all threshold settings. Free of striping and noise. Not sensitive to floating-point or byte differences. A and B mixed-pixel zone divided equally, except for highest minimum-encoding threshold (0.7) where some mixtures of A and B were unclassified.

3.4.2 Supervised classification

Standard methods for supervised classification of spectral images originally were developed for use with Landsat Multispectral Scanner
(MSS) and Thematic Mapper (TM) data, and have been applied successfully to a variety of image types having only a few (generally less
than ten) spectral channels. Like unsupervised classifiers, most of the
standard, supervised classifiers do not work well, or at all, with imaging-
spectrometer data that have tens to hundreds of channels, although
they can be applied to fraction images (Chapter 4) when the number
of spectral endmembers is small (generally two to five). A discussion of
how to classify fraction images is deferred to Chapter 5.

Supervised classification, as the name implies, requires human
guidance. The analyst selects a group of contiguous pixels from part
of an image known as a "training area" that defines the DN values in
each channel for a class. A classification algorithm computes certain
properties (data attributes) of the set of training pixels, for example the
mean DN for each channel. Then, the DN values of each pixel in the
image are compared with the attributes of the training set. Classification
is the process by which each pixel in the image is assigned to a class on
the basis of its data attributes. It is important to keep in mind that
classification algorithms only operate on the input data. The algorithms
do not take into account the spatial context of the training pixels or the
intent of the analyst who selected them.

Image classification works best when training pixels have unambiguous data attributes, such as isolated clusters of points in the data

> **Supervised classifiers**
>
> Supervised classifiers do not
> take into account the
> spatial context of the
> training pixels or the intent
> of the analyst who selected
> them.

space, that can be uniquely described by their distance from a mean vector or by their angular position. But, alas, training sets may or may not correspond to discrete clusters in data space; furthermore the data attributes of different training sets often overlap. The challenge in classifying images is to assign pixels to classes in a consistent mathematical way that also is consistent with the image context.

Supervised classifiers employ a two-step process. The first analytical step is to compare the data attributes of each pixel with each training set. Different algorithms measure different data attributes. The second step occurs when each pixel is assigned to a class according to criteria (thresholds) specified by the analyst. In some classification schemes, pixels can belong to more than one class. Four basic types of algorithms are included in most commercial image-processing packages: parallelepiped, minimum distance, maximum likelihood, and spectral angle. More thorough treatment of these and other algorithms can be found in several texts (e.g., Schott, 1997; Schowengerdt, 1997).

Basic types of supervised classifiers

Parallelepiped (also, "level-slice" or "box")

Description	Pixels are classified by whether or not DNs for each channel are within prescribed ranges. Channel ranges can be fixed arbitrarily or described by the standard deviations of the training set. Produces classes for each training set, and one of unclassified pixels.
Options	Set ranges or box limits using standard deviations about means.
Advantages	A simple, computationally inexpensive method. Includes class variance. Does not assume a class statistical distribution.
Disadvantages	Does not adapt well to elongated (high-covariance) clusters. Often produces overlapping classes, requiring a second classification step. Becomes more cumbersome with increasing number of channels.

Minimum distance

Description	Measures Euclidean (or other characterizations of) distance in data space from each unclassified pixel to the multivariate mean value of each training set. Pixels are assigned to the set with the nearest multivariate mean.
Options	Set standard deviation of mean; or set maximum allowable distance.
Advantages	Simple to implement and computationally inexpensive.
Disadvantages	Does not take into account the variability of the training sets. Pixels distant from the means of their correct classes (high variability) may be closer to the means of nearby compact clusters (low-variability classes), leading to misclassification.

Maximum likelihood

Description The distribution of data in each training set is described by a mean vector and a covariance matrix. Pixels are assigned an a-posteriori probability of belonging to a given class and placed in the most "likely" class. This is the only method in this list that takes into account the shape of the training-set distribution.

Options Probability thresholds can be selected.

Advantages Provides a consistent way to separate pixels in overlap zones between classes. Assignment of pixels to classes can be weighted by prior knowledge of the likelihood that a class is correct.

Disadvantages Cluster distributions are assumed to be Gaussian. Algorithm requires enough pixels in each training area to describe a normal population, and assumes class-covariance matrices are similar. Classes not assigned to training sets tend to be misclassified – a particular problem for mixtures. Changes in the training set of any one class can affect the decision boundaries with other classes. Relatively expensive computationally. Not practical with imaging-spectrometer data.

Spectral angle

Description A measure of the angle in the data space between the vector of the mean of a training set and the vector representing each pixel. Smaller angles indicate more similarity. Training spectra can be from the image or from a library.

Options Maximum-angle threshold can be set. Pixels outside the angle as measured from the training spectrum are not classified.

Advantages Simple method; not computationally expensive. Emphasizes spectral differences; insensitive to shading and albedo. Not dependent on the distributions of the populations of training sets.

Disadvantages Misclassification occurs when angle is too large and classes overlap. Ignores differences in albedo. Assumes that offsets (biases) for sensor and atmospheric effects have been removed or are minimal.

Data-attribute ("rule") images

The data attributes that are calculated from basic classification algorithms can be displayed as continuous gray-tone images, one for each class. We refer to these as "data-attribute images." They also are known as "rule images." For example, pixel vectors separated by small spectral angles from the mean of a training area can be assigned lighter (or darker) tones than vectors with larger angles (Figure 3.9). On average, those pixels within a homogeneous training set will have the lightest (or darkest) tones of all. Black-and-white data-attribute images, each based on a different training area, can be combined into color

> **Spectral similarity**
> Data-attribute ("rule") images are powerful tools for locating other pixels that are similar to those in a training set.

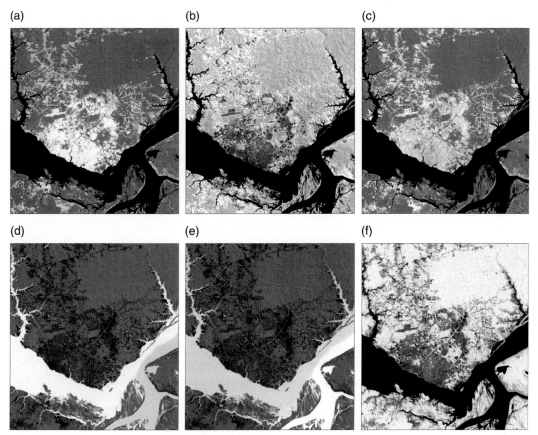

Figure 3.9. Spectral-angle classifier data-attribute ("rule") images of part of a 1991 Landsat TM image of Manaus, Brazil. Images (a)–(f) are based on training areas. Lighter tones indicate spectral angles closer to the means of the training sets. (a) Urban; (b) light-green vegetation; (c) soil; (d) Rio Negro; (e) Rio Solimoes; (f) primary (terra firme) forest. See Plate 3 for locations of features. See also reference images on the Web.

composites. Although data-attribute images are byproducts of standard classification algorithms, few analysts make use of these images, or are aware of them, even though they have been known since the early 1970s. More typically, when a classification algorithm is invoked, analysts are interested only in the final product that shows all classes and class boundaries. This is unfortunate, though, because data-attribute images have a utility that extends beyond being just a first step in classifying. They can be powerful tools for finding the other pixels that are most like (or unlike) those in a designated training area. Another method, matched filtering also can be used to find pixels that are spectrally similar to a training area: however, matched filter images may be ambiguous in cases of spectral mimicking or spectral mixing (Chapter 4).

Class boundaries

Defining boundaries
between classes in the data
space is *not* the same as
defining boundaries
between patches of pixels
in the image.

Bayes classifier

A maximum-likelihood
classifier produces an
a-posteriori probability,
whereas a Bayes classifier
also has an a-priori probability that a pixel belongs
to a given class.

Class boundaries

After pixels have been ranked according to their similarity to the data
attributes of each training set, the final step in the classification process
occurs when boundaries (in the data space) between classes are defined
using pre-selected thresholds. In a "winner-take-all" or "hard" classification, each pixel is assigned to one labeled class, and no pixels
remain that straddle boundaries between classes. We know, however,
that pixel values in the data space often do not occur in isolated clusters
that make it straightforward to assign classes. Instead, it is common for
the vectors defined by training sets to overlap in the data space. In a hard
classification, these ambiguities can be resolved by using a maximum-
likelihood and/or a Bayes classifier to force a decision. Alternatively,
there are "soft" or "fuzzy" classifiers that deal with overlap ambiguity
by allowing pixels to have membership in more than one class, and that
are able to set fuzzy (gradational or mixed) boundary zones between
classes. Whether a hard classification is satisfactory depends on what
information we are looking for in an image. For certain applications we
might be perfectly satisfied to ignore overlap at boundaries, and have a
classifier draw statistically correct divisions between classes. After all,
at some scale the fractal-like structure of boundaries requires making a
best approximation. If a fuzzy boundary is an important feature in itself,
a hard classification will miss information that may be needed for photo
interpretation.

In Figures 3.10 and 3.11 both the minimum-distance classifier and
the spectral-angle classifier partition mixtures of A and B of the synthetic image into one or the other class, or they leave the mixtures as
unclassified. If we go straight to the classified images without inspecting the data-attribute images, we will not be aware that there is a
gradational boundary between A and B. However, the true nature of
the boundary between A and B is clear in the data-attribute images for
both algorithms. The width of the unclassified, mixed zone depends on
the threshold settings for each algorithm. When tightly constrained,
both classifiers are more parsimonious about placing a pixel into either
A or B, and, if the fit is not good enough, mixed pixels are left
unclassified. When more loosely constrained, all mixed pixels are
placed in either A or B. Notice also that the data-attribute images for
the minimum-distance classifier are strongly striped, owing to the effect
of simulated shading. In comparison, the data-attribute images for the
spectral-angle classifier are essentially free of shading effects. The
classification images for both algorithms contain misclassified pixels,
and there is considerable variability from one algorithm to another and
among threshold settings.

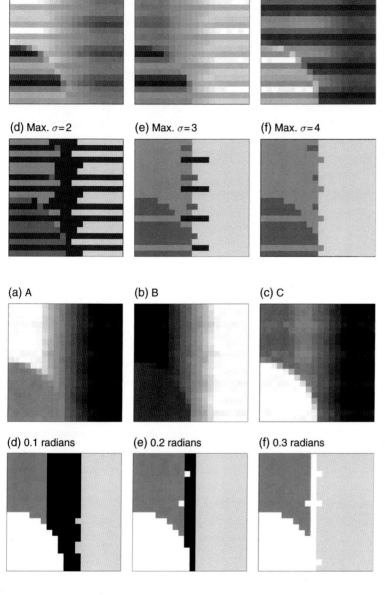

(a) A   (b) B   (c) C

(d) Max. $\sigma=2$ (e) Max. $\sigma=3$ (f) Max. $\sigma=4$

Figure 3.10. Minimum-distance classification of the synthetic image. (a)–(c) Data-attribute images for training areas in A, B, and C, respectively. Lighter tones are closer to the means. Simulated stripes of horizontal shading have anomalously large distances. (d)–(f) Classified images using different standard-deviation thresholds. Black = unclassified, other gray tones designate classes.

(a) A   (b) B   (c) C

(d) 0.1 radians   (e) 0.2 radians   (f) 0.3 radians

Figure 3.11. Spectral-angle classification of the synthetic image. (a)–(c) Data-attribute images for training areas in A, B, and C, respectively. Lighter tones indicate a smaller angle between a pixel and the mean of a training set. (d)–(f) Classified images using different angle thresholds. Black = unclassified; other gray tones designate classes. See Section 2.2.1 and Figure 2.10 for a definition of the spectral angle.

Training areas, classes, and mimics

Training areas, also known as "regions of interest" (ROI), are the starting point for supervised classifications. We use the term "training area" to remind ourselves that they usually consist of multiple, contiguous pixels. The intent, of course, is to select pixels that are understood in terms of image context, and, perhaps, based on outside information.

Problems occur when there are pixels in an image that mimic the spectral attributes of a training area, but that do not have the same meaning. Spectral mimicking often is underestimated because analysts think that the signature of a well-chosen training area is not likely to be duplicated by some other material(s). But in spectral images it is common for the attributes of different training areas to overlap partially or even completely. Particularly insidious is sub-pixel mixing, which can provide ample opportunity for a large number of spectral combinations. In Chapter 6 we discuss how, as a pixel footprint becomes larger, spectral contrast diminishes, and spectra of all pixels start to converge. Spectral mimics are undetectable using spectral information alone, and they can be seriously misleading.

The basic methods of spectral classification often are explained in textbooks by means of two-channel scatter plots that show distinct data clusters (e.g., Figure 3.12). Such diagrams illustrate how basic classifiers assign each pixel to an appropriate cluster. What is missing from these simple illustrations is the concept of spectral mixing. Figure 3.12 illustrates that sometimes pixels that we assign to a cluster intuitively, or that are based on the output of a statistical algorithm, can just as well be interpreted as mixtures. Clustering of the data may be correct, but our interpretation of the meaning may be ambiguous or incorrect.

Variability and size of training areas
On the ground we take for granted the spatial variability of landscapes. There are exceptions, such as monotonous parts of polar regions and some deserts; but, in general, the land surface is variable in lightness due to shading and shadowing and in the degree to which surface materials absorb, reflect and emit radiant energy. If we ignore variability introduced by an atmosphere and by the measurement system itself, the main sources of spectral variability arise from myriad variations in illumination and composition of surface materials at a range of spatial scales. We do not expect one spectrum in a spectral library to describe all spectra of the same material, and we cannot expect a training set to encompass exactly the same variability of a class in all parts of an image.

Training pixels are, necessarily, a sample of a population. Increasing the size of a training area may include more of the desired pixels in a class, but it also will increase the possibility of overlap (and, thus, ambiguity) with other classes. Some algorithms are more sensitive than others to class variability; therefore, it is possible to select classifiers that best fit the nature of the data. For example, maximum-likelihood or Bayes classifiers require training sets that have sufficient variance, whereas minimum-distance-to-means or spectral-angle classifiers can be trained with one or a few pixels. On the other hand, all

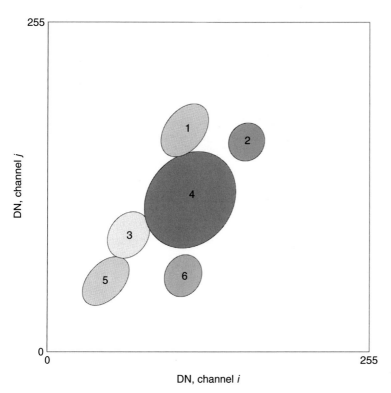

Figure 3.12. Two-channel cluster diagram of the type often used in textbooks to illustrate how various types of classifiers work. Hypothetical clusters 1–6 represent discrete categories of materials on the ground as defined from image training areas. However, cluster 4 contains pixels that mimic mixtures of pixels in clusters 1 and 6, and mixtures of pixels in clusters 2 and 3. Some pixels in cluster 3 mimic mixtures of pixels in clusters 1 and 5. Clusters may have meaning in the spatial context of an image, but do not necessarily reveal their meaning unambiguously in the data space. It is rare to find isolated clusters in real image data. A continuous spread of data points is more typical; for example, see Figure 3.4.

classifiers may perform poorly when a training area is heterogeneous. This point is illustrated in Chapter 7 where we use a minimum-distance attribute image to map the urban core of Seattle, and find that many of the pixels within the training area are far from the spectral mean.

Number of classes

When standard classification algorithms are applied to spectral images it is common to accumulate 20, 30, or more classes – more than a field investigator can interpret easily or validate on the ground. There appear to be two main reasons for so many classes. One reason is that if classification is approached as a statistical problem, multiple classes may be necessary to fully accommodate the spectral variability of a landscape. Although mathematically satisfying, the individual classes actually may just describe spectral variability within one type of land cover, variations in illumination, or gradations from one type of land cover into another. A second reason for excessive numbers of classes is that analysts often do not realize that the data attributes which they isolate from training areas are not always spectrally distinct and separable. We have watched with concern as projects have followed a predictable scenario. A need is defined to have a satellite-based map of the land surface. Memos are circulated about what should be included in the map, and a committee compiles a long list of classes. Training areas are found and a classified image is produced. Those who are not inclined to go into the field display the impressive-looking multicolored classified image on the wall. The field investigators find that some of the classes correspond to what is on the ground, and some do not, and abandon the classified image in frustration and return to using aerial photographs.

In reality, there is no simple answer to the number of useful classes that can be extracted from an image. It depends on the spectral contrast of the surface materials, the spectral variability of the materials, the scale of the pixel footprint, the spectral resolution and range of the detector system, and prior knowledge of the ground surface. However, in supervised classification, it is possible to guide the process by evaluating training areas in terms of their spectral contrast relative to all other classes. From this perspective, each potential class is treated as a spectral target viewed against the background of the other classes and their mixtures. Spectral targets, as defined in Chapter 6, have detection thresholds, and, in the context of classes, any one class must exceed the threshold for detection relative to adjacent classes. In general, detectability of individual classes is better with fewer classes overall. As more classes are defined it becomes increasingly difficult to detect any one class, because sooner or later the class we are most interested in will mimic something else. This is a good reason to start with a shortlist of desired classes, and to prioritize a potential class list according to importance for the application. With knowledge of detection thresholds, field investigators are forewarned that some classes may easily be confused with others. Consider, for example, that an investigator is standing in class L on the ground, and the classified image says it is class M. If the observer knows that classes L and M are difficult to separate spectrally, even though they look quite different on the ground, the misclassification would be of little concern. Without this knowledge the classification would be of limited use.

Too many or too few channels

Many of the standard classifiers were developed for Landsat images in the mid to late 1970s. Recall that early Landsat scanners had four VIS-NIR channels (MSS), later expanded to six (TM). Standard classifiers could be used with a small number of channels, and they led to many applications with Landsat and SPOT images. However, as we have mentioned, the algorithms for some of the standard classifiers do not work when applied to images with too many channels, including

(a)   (b)

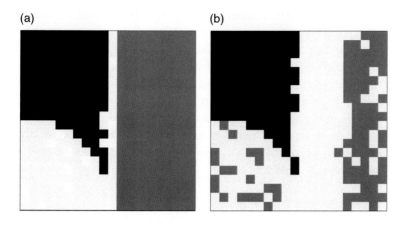

Figure 3.13. (a) Maximum-likelihood classification of the synthetic image using all four channels. (b) Same classifier using only channels 1 and 4. When channels 2 and 3 are deleted it is not possible to separate spectral areas B and C. See Figure 3.2 for spectra and locations of A, B and C.

imaging-spectrometer data having hundreds of spectral channels. A common solution is to delete certain channels, and to use only a selected subset of channels to classify an image. This can be a practical solution, providing that the spectral information in the chosen channels is satisfactory for a particular application. However, a word of caution here: removing one or more channels may not make any difference for one purpose, and at the same time it may be detrimental for another. A few channels that are selected, for example, to distinguish green vegetation from soil may not be adequate to discern soil from dry vegetation. We have to define the information that we want in order to identify irrelevant (often called "redundant") channels that can be deleted. "Redundant" channels do not exist by themselves; they must be defined in an interpretive context.

We can illustrate the general problem of classifying with too few channels by using the synthetic image. In Figure 3.13a we applied a maximum-likelihood classifier to all four channels of the synthetic image. If we select only channels 1 and 4 (Figure 3.13b) we lose the ability to separate areas B and C. The result is consistent with Figure 3.2 where it is evident that channels 1 and 4 show little difference between B and C, even though the same channels separate A and B. Whenever we attempt to classify a large and complex scene we need to be aware that each time we delete channels we may lose the ability to distinguish some of the spectral components. This is consistent with a general rule that extendibility of results (in this case classification) is achieved at the expense of resolution (in this case the ability to distinguish classes.)

Another approach to the problem of too many channels is to transform DNs in all channels to a smaller number of dimensions. Channel ratios (Section 3.5.1) can be used with some standard classifiers, as can combinations of ratios and channel-DN values. Classifications based on

> **"Redundant" channels**
> Deleting one or more channels may not matter for distinguishing classes A and B, but may make it impossible to separate A and C.

ratios achieve a modest compression of the multivariate data and can be effective in enhancing small spectral differences that are important for understanding composition. A principal-component or canonical transformation also can be applied to channel-DN data prior to classification. Known as multiple discriminant analysis, this method also reduces the number of vectors, but it is sensitive to the overall set of classes chosen and can be difficult to interpret. Perhaps the most effective data compression is achieved by transforming channel DNs to fractions of endmembers (Chapter 4). Classification of endmember fraction images is described in Section 5.2.

Topography and illumination

Variations in illumination produced by topography can cause pixels of the same material to range from maximum lightness when fully illuminated to dark when in shadow. This variability produces scatter plots that are elongated and tapered toward the origin (0 DN). This is why PCT_1 images of landscapes often show topography (Section 3.4.1). With increasing amounts of shading/shadowing, scatter plots shrink toward the origin, thereby reducing SL contrast (Section 2.2.1) and making it more difficult to distinguish among classes. Although some supervised classifiers such as spectral-angle and binary-encoding algorithms are relatively insensitive to such topographic effects (Figure 3.11), others, including minimum-distance and maximum-likelihood, require that topographic variability be incorporated into the training sets. It is not always possible, though, to be sure that one has collected pixels for each training set that encompass the full range of illumination. In this case, the result of training only on the less-shaded pixels is that the darker ones may not be correctly classified (Figure 3.10). Other ways to circumvent problems of illumination when classifying include applying an initial topographic correction using DEM data, classifying normalized or ratio images, classifying PCT images, and classifying endmember fractions.

Within a given image, classifiers also respond to variations in illumination that result from changes in view angle across scenes. For this reason, most spectral classifications have been carried out on images made by instruments that have narrow FOV, such as Landsat TM (14.9°). It is more difficult to apply classifiers successfully to aircraft multispectral images and wide-FOV satellite images. Even with Landsat TM scenes, there may be significant changes in classes between the up-Sun and down-Sun sides of an image.

Calibration

One of the advantages of standard classifiers is that they often produce useful results when applied to images with little or no prior calibration.

Canonical transformation

Selected subsets of image data or the means of training areas are used to define the covariance matrix for a PCT, instead of the whole image.

The origin and shade

The origin is approximately the same as the endmember "shade" in spectral-mixture analysis (Chapter 4). Shade is a dark spectrum that potentially mixes with all other spectra.

However, classification results can be adversely affected if there are significant within-scene variations in atmospheric properties or illumination (Section 3.2). Furthermore, unless images can be inter-calibrated, standard classifiers are valid only for the time of acquisition, making comparisons with results from other images difficult or impossible.

Evaluation of standard classifiers

Field investigators commonly are presented with classified images to use for a particular application without the benefit of a clear understanding of what was done to the data. This can be a serious problem when field work reveals that certain classes are incorrect, because it usually will not be evident on the ground what went wrong or how to find a remedy. The field investigator who understands the basic functions of classifiers can be forewarned. For those who wish to do their own classifying, several basic algorithms are included in commercially available image-processing software packages. However, the number of algorithms and the choices of parameters to be set can be daunting, making it unwise just to start clicking on different classifiers to see what happens. Furthermore, the results depend strongly on the nature of the data; and, as usual, the usefulness of the results also depends on what we want to find out in the first place.

> **Judging classifiers**
> "Just tell me which classifier is best. Don't bore me with the details." – "Sorry. It depends on the scene and what you want to find out."

Tests on the synthetic image

One of the advantages of the synthetic image (Section 3.4.1) is that we know exactly how it was constructed, and can use it to judge how well different classifiers perform. We tested classification algorithms from a commercial software package for image processing, using the version of the synthetic image with 5% random noise. Noise was added because several of the classifiers require that the training sample have a normal distribution of the data. We retained the horizontal pattern of shading that produces a maximum shade fraction of 0.5. Tests were run using two different training sets. One was a 5×5 block of pixels that included the lightest and darkest pixels of each of the classes, A, B, and C. The other was a set of 25 pixels consisting only of the lightest pixels for each class. For each classifier we used the default setting and at least three other threshold settings that changed parameters such as the standard deviation from the mean, maximum-distance error, spectral angle, or probability threshold. We also ran all tests in both floating-point and byte form.

The tests with the synthetic image illustrate that there can be significant variability from one algorithm to another, even when the "best"

Table 3.2. *Classification scores for the synthetic image.*

Classification algorithm	% Unclassified	% Class A	% Class B	% Class C	% Misclassified
Parallelepiped (5 × 5)	6.75	9.75	35	48.5	27.75
Minimum distance (5 × 5)	0	34	49	17	5.25
Maximum likelihood (5 × 5)	0	28.5	45	26.5	5.75
Spectral angle (5 × 5)	0	24.5	49	26.5	5.75
Parallelepiped (25 light)	20.5	6.75	47.75	25	13.25
Minimum distance (25 light)	0	47.25	46.25	6.5	12.5
Maximum likelihood (25 light)	0	31.25	57.75	11	8.75
Spectral angle (25 light)	0	23.25	48.5	28.25	7.5
Synthetic image	**0**	**29.5**	**50**	**20.5**	**0**

Each classifier was tested over a range of threshold settings. The best threshold setting was selected based on the lowest number of misclassified pixels. The percent of misclassified pixels for each classifier is tallied in the right-hand column. Pixels were considered misclassified if: C pixels occurred in the A–B mixed zone; A, B, or C occurred in pure areas of the other classes. None of the algorithms correctly modeled mixtures; therefore, mixed pixels were judged correct if classified as any one of the actual components. Training areas for A, B, and C are blocks of 5 × 5 pixels or are 25 light (low-shade) pixels.

settings are known (Figures 3.10 and 3.11). Of course, in real images of landscapes, the correct classification is not known beforehand, which means that there is no a-priori way of knowing which settings are best for an algorithm, or how to judge which algorithm produces the fewest misclassified pixels. If the composition of the synthetic image had not been known, and if we had merely selected one algorithm and tried a few different settings, it is entirely possible that we would have come away with erroneous and misleading results.

We considered a classification satisfactory if it correctly located spectral areas A, B, and C, was not confused by shading, and separated A and B by either drawing a boundary down the middle or not classifying the mixtures at all. We did not expect the algorithms to identify mixtures correctly; thus, for example, mixtures of A and B were considered adequately classified when the pixels were assigned as either A or B (but not C). Although the algorithms are ranked by the lowest number of misclassified pixels, it is important to remember that these results are based on one rather peculiar synthetic image, and this does not predict how any algorithms will act on any other image. Results of the tests are summarized below and in Table 3.2. In general, the classified images were highly variable from algorithm to algorithm and from setting to setting within a given algorithm.

Summary of tests of supervised classifiers on the synthetic image

<u>Parallelepiped</u>	Fair results for the 5×5 training set and standard deviations of 1 and 2. Shade stripes and noise are present. Mixed pixel zone was left unclassified. Light-pixel training set produced erratic results for different settings and between floating-point and byte data.
<u>Minimum-distance-to-means</u>	Acceptable results only for the 5×5 training area. Unacceptable results using the lightest pixels to train. Some misclassification of shaded stripes in area C. A standard-deviation threshold of 0.4 divided the mixed-pixel zone down the middle; whereas a setting of 2 did not classify the mixed pixels and confused C and shade.
<u>Maximum likelihood</u>	Acceptable results only for the 5×5 training area. Classes free of shade-related striping and noise. At low probability settings (0 to 0.3) the mixed-pixel zone was divided equally between A and B, except for a one-pixel-wide interface that was incorrectly classified as C. At a probability threshold of 0.4, the center of the mixed-pixel zone included unclassified pixels.
<u>Spectral angle</u>	Good results at a default setting of 0.1 radians. Classes free of shade-related striping and noise. Central part of the mixed-pixel zone unclassified. Increasing the angle to 0.2 radians narrowed the unclassified zone to 2 pixels, and replaced unclassified pixels by the nearest class A or B. Further increases in the threshold angle to 0.3 and 0.4 radians incorrectly replaced the few unclassified pixels of mixed A and B by C.

Tests using Landsat images

We do not have the luxury of knowing the answers with spectral images of landscapes as we have with the synthetic image. The best we can do is to test algorithms and settings using images that we understand from previous work in the field. Accordingly, we applied the same standard algorithms that we tested on the synthetic image to four Landsat TM sub-images of different types of landscapes that we had studied on the ground: the Seattle, Washington urban area, an area of tropical forest near Manaus, Brazil, an area of conifer forest in western Washington State, and a portion of desert in Death Valley, California. Landsat TM images have been widely used for classification for many years because of their availability and their favorable spectral and spatial resolutions. Indeed, many of the classification algorithms were developed and improved specifically for application to Landsat and similar image data.

Figure 3.14 shows the results of applying supervised classification algorithms to Landsat TM sub-images. Training areas were selected for each image using field experience as a guide. For each algorithm we tested a range of settings, and we selected the settings that produced the best classification, based on our field knowledge. The significant

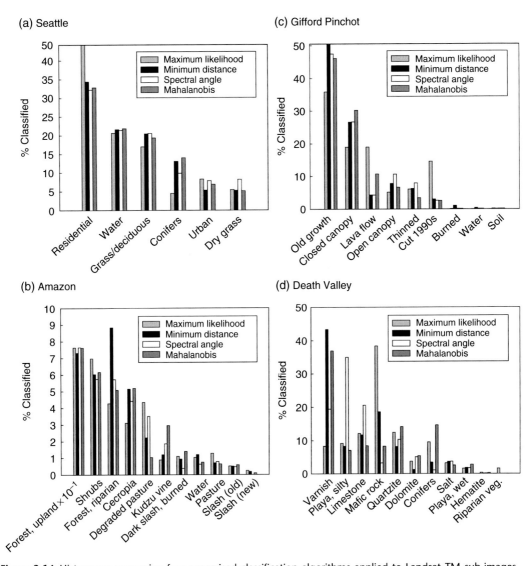

Figure 3.14. Histograms comparing four supervised classification algorithms applied to Landsat TM sub-images. (a) Seattle, Washington; (b) Amazon forest near Manaus, Brazil (Fazendas sub-image); (c) Gifford Pinchot National Forest, southwestern Washington; (d) Death Valley, California. Histograms of the percent of pixels classified are plotted for training areas that were selected based on knowledge of the landscape from ground studies and photo interpretation. Each classifier was tested with several settings, and the best results for each algorithm are shown. The Mahalanobis classifier (not discussed in the text) is a maximum-likelihood classifier that assumes equal class covariances. Classification results vary from one algorithm to another for a given image, and the discrepancy in classification results varies from one image to another. See also Plates 1–5 and reference images on the Web.

conclusion is that our tests with Landsat TM data revealed the same problems that we encountered with the synthetic image, namely that class assignments vary from one algorithm to another and with different settings. Furthermore, the way in which the results vary is different from image to image, and it is not predictable from the tests on the synthetic image. In other words, the results we get when we apply standard classifiers also depend on the image itself. This is reasonable, because the data structure is different from image to image.

Misclassification: prevention and remedies

From the perspective of a field investigator, the value a classified image lies in its ability to predict what is on the ground in areas that have not been visited. In addition to the broader issues of validating results from spectral images (Chapter 1), our tests with standard classifiers illustrate that there are two principal causes of misclassification. One source of error is where vectors from different training areas overlap. The other problem occurs when different algorithms do not agree on where the boundary belongs between adjacent classes. Where do these misclassification errors come from, and what can we do to prevent or rectify them?

Mimicking occurs when spectra of two different materials cannot be distinguished using a given measurement. In some cases, mimics can be resolved simply by keeping the number of classes small and by segmenting images according to context. Mimics are more likely in a large scene with many classes than in a smaller one with only a few classes. The Landsat TM image of part of the Gifford Pinchot National Forest in Washington State provides an example where context and segmentation help resolve mimicking (Plate 5, CD reference images on the Web and Figure 3.15). Standard classification algorithms applied to this image confuse a sparsely vegetated lava flow (right-hand edge of images) with areas of lightly thinned, mature conifer forest. At first it may seem odd that these classes should be confused, but the spectral patterns actually are quite similar, as can be told by plotting the mean spectra of training areas and exploring two-dimensional scatter plots. Thinned forest is characterized by tall, dark conifers with shadows in the gaps formed by the selective removal of trees. The lava flow is only sparsely covered by conifers and shrubs, but it also is deeply shadowed, because it is very rough on scales of meters to millimeters. The lava flow, however, is known from field work and photo interpretation to be confined to one edge of the sub-image, and it can easily be masked to resolve the classification ambiguity in the rest of the image.

Of course, it is not always feasible to mask troublesome mimics. One must know where the classes are on the ground to be able to remove them. In the Landsat TM sub-image of the Amazon Fazendas (Plates 3

(a) lava   (b) 5 σ threshold   (c) 10 σ threshold

(d) thinned   (e) 5 σ threshold   (f) 10 σ threshold

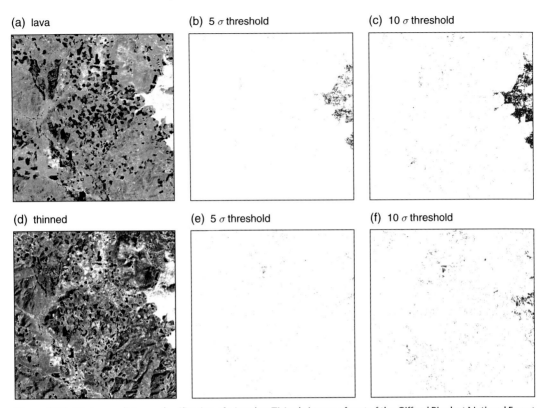

Figure 3.15. Minimum-distance classification of a Landsat TM sub-image of part of the Gifford Pinchot National Forest in western Washington State. (a) Data-attribute image of "lava flow." Lighter-toned pixels are closer to class mean. (b) "Lava flow" class (dark tones) using a 5-σ threshold. (c) "Lava flow" class (dark tones) using 10-σ threshold. (d) Data-attribute image of "thinned forest;" lighter-toned pixels are closer to class mean. (e) "Thinned forest" class (dark tones) using 5-σ threshold. (f) "Thinned forest" class (dark tones) using 10-σ threshold. The two partially overlapping classes can be separated by context, because the lava flow only occurs on the right edge of the image.

Terminology

In this text, "σ" stands for "standard deviation."

---

and 4; reference images on the Web), for example, regeneration of vegetation following clearing of tropical forest has produced areas of "mixed shrubs" and areas of "*cecropia*." These plant communities are easily told apart in the field, but their spectral properties overlap in the TM data space. (See ahead to Chapter 7.) It happens that mixed shrubs and cecropia commonly are interspersed in the field on a fine scale, and there is no simple way to mask either one of the mimics. In this case, one has to be satisfied knowing in advance that these types of land cover cannot be separated reliably by standard classification algorithms. When classes have ambiguous meaning, we naturally try to distinguish them by looking for other information. Photo interpretation and context sometimes solve the problem. In other cases, it may be possible to resolve mimicking classes by applying filters that include non-spectral measurements such as altitude, temperature, or RADAR backscatter.

It has long been recognized that pixels can be misclassified if they straddle a sharp boundary between two units. Gradational boundaries, however, are more difficult to classify, because they may go undetected. As we saw in the example of the synthetic image, none of the classification algorithms tested was able to reveal that there is a transition zone separating areas A and B. In the Landsat images that we tested, part of the variability among algorithms stemmed from differences in the locations of class boundaries. As previously mentioned, some algorithms (fuzzy classifiers) avoid the uncertainties of boundaries between classes by allowing a spectrum to belong to more than one class. The degree to which a spectrum is affiliated with each class is determined by calculating a numerical score, known as a "membership grade." For example, a fuzzy classifier applied to our synthetic image would identify pixels in the zone between areas A and B as belonging to the classes of pure A *and* of pure B, and would specify the weighting of membership grade. Fuzzy classifiers exist for both unsupervised and supervised methods, but are not usually part of standard software packages for image processing. A disadvantage of fuzzy classifiers is that they add another layer of mathematical complexity to the classification process, and they are subject to many of the same problems already discussed. An investigator will want to be satisfied first that a statistically based classification is useful for the field area being studied. If so, it may be worth testing and implementing a fuzzy classifier to improve interpretation of class boundaries.

Fuzzy classification often is described in the context of spectral-mixing. It is important, though, not to confuse mixing at class boundaries with spectral-mixture analysis (Chapter 4). Classes are selected on the basis of data clusters (unsupervised) and training areas (supervised), and there is no requirement that the spectra of classes represent endmembers that mix with one another. Indeed, as we have seen, spectra of classes may be very similar or even mimics. For this reason we do not refer to classes as "endmembers." In some cases the classes make sense as physical mixtures. For example, spectral mixtures of the classes "grass" and "bare ground" might describe a partially vegetated transitional zone between the two. Other classes do not make sense as physical mixtures. For example, spectral mixtures of the classes "pasture" and "corn" are improbable except along a sharp boundary.

From the perspective of a field investigator, applying statistical methods to define boundaries between classes means turning one of the most critical interpretive tasks over to the computer, and hoping that it does a good job of it. A drawback of this approach is that, for a given classifier, each threshold is applied to the whole image, and a setting that works for one boundary is not necessarily best for others. As usual,

Endmembers and classes

In this text, we do not refer to classes as "endmembers." They may or may not be, depending on whether they are basic spectral components of a scene and whether they mix with one another.

thresholds are sensitive to the sampling in the training areas. We can, however, intervene and choose boundaries that satisfy photo-interpretive criteria. For example, we can adjust classification thresholds so that a class boundary is placed where it belongs as judged by looking at the image. An advantage of this approach is that we can explore an image, boundary by boundary, and see the effects of different thresholds without the expense of re-calculating the data-attribute image each time. Ideally, we would like to be able to work through an image boundary by boundary, setting separate thresholds for each one. In practice, however, commercial software packages do not offer this option, and it can be difficult to assemble the final classified image.

Pre-classification and post-classification processing

Standard classifiers also can be applied to principal-component or canonical transformations, tasseled-cap transformations (Section 3.5.2) and mixing models (Chapter 4). Each kind of image that is produced by pre-classification processing defines a new data space that is defined by a few new "channels" that can be operated on by classification algorithms in the usual ways. Having fewer channels significantly improves the computational efficiency of most classification algorithms – an important consideration when working with imaging-spectrometer data. For certain applications where the results of the first analytical operation are well understood, the classification step can be an effective way to organize image information further. However, if we do not fully understand what we produced with the first operation, we may make interpretation even more difficult by superposing a classification algorithm. Post-classification processing includes operations such as combining classes, adjusting boundaries and changing class labels. There are two main reasons that classes are adjusted; one is that the reality of field work or additional information may require modifications. The other is that that there may be important information yet to be extracted from the *spatial* patterns of the spectral classes, and this information can lead to a new level of classification.

3.5 Physical modeling with spectral data

3.5.1 Ratios as spectral probes

As discussed in Chapter 2, reflectance spectra can be compared directly as long as they are measured as a ratio relative to an agreed-upon standard. A ratio also is used to define spectral contrast between wavelengths of a single spectrum (Section 2.1.2). The main advantage of ratios in spectroscopy is that, in the absence of large offset terms, the influences of the illumination source, the illumination geometry, and the detector system tend to cancel, with the result that information on spectral variability is retained and information on the absolute magnitude of the signals is not. In remote sensing, the absolute value of radiance is of less interest for identification and discrimination, because materials can appear highly variable in overall lightness depending on geometric factors such as angle of illumination and particle size. Ratios, therefore, allow us to examine an image where light–dark variations due to topography and roughness at all scales are minimized and where the spectral variations are enhanced. The value of channel-to-channel ratios in images of landscapes was recognized by the pioneers in multispectral image processing at JPL in the 1970s (e.g., Goetz and Billingsley, 1973), and ratio images continue to be a mainstay of remote-sensing analyses (e.g., Rowan *et al.*, 2004).

The choice of spectral channels to ratio depends on the information we are seeking. Usually, channels are selected because of prior knowledge of the predicted spectra of materials on the ground. For example, there are channel ratios for Landsat TM images that are "good for" green vegetation (Band 4 / Band 3), ferric-iron oxides and hydroxides (Band 3 / Band 1), clays and alunite (Band 5 / Band 7) and ulexite (a borate mineral) (Band 4 / Band 7). These and other targeted ratios have been used successfully where field investigators are familiar with the local materials and can rely on photo interpretation (e.g., Sabins, 1996: pp. 367–378). Often, though, image analysts are unfamiliar with the spectroscopic underpinnings of these and other ratios, and as a result are unaware that targeted ratios can be mimicked by other materials. Some analysts dispense with spectroscopy altogether and examine several or all ratio combinations in a search for interesting photo interpretive patterns. In this book, though, we focus on the physical modeling approach and how ratio images can be used as spectroscopic tools.

The nature of ratio images is illustrated by the synthetic image (Figure 3.16) where each of the three reflectance spectra has distinct features, and each can be distinguished using only two channels. For example, spectrum A has relatively strong absorption in channel 1, and differs from other spectra by having the highest ratio of channel 2 to channel 1. On the other hand, in spectrum B, the salient feature is the strong absorption band in the vicinity of channel 2 which produces the highest value of channel 1/channel 2 (Ch_1/Ch_2). From inspection of the spectra we can select one or two ratios that will best measure what we are interested in, and we can predict how any given ratio will appear as an image.

All of the ratios that are derived from the synthetic image suppress the striping that simulates topography. However, even with the

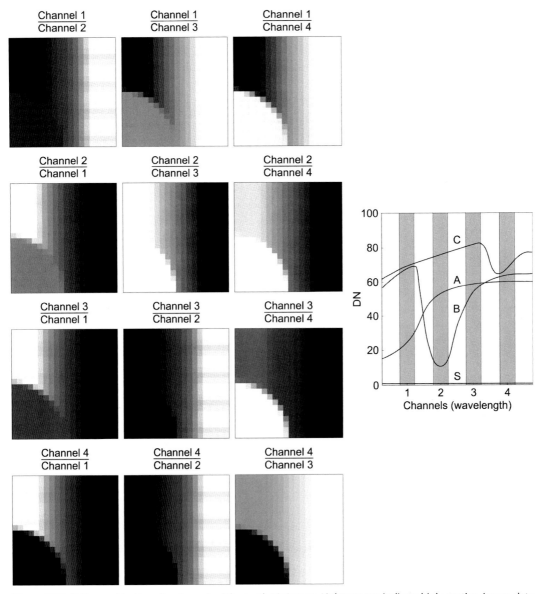

Figure 3.16. Ratio combinations for channels of the synthetic image. Lighter tones indicate higher ratios. Image data are noiseless. Locations of A, B, and C are shown in Figure 3.2.

Gaussian stretch

A nonlinear contrast stretch in which the histogram is made to resemble a Gaussian curve.

"noiseless" version of the synthetic image, vestiges of the simulated shade striping remain in some areas after a linear contrast stretch, and the stripes are prominent after a Gaussian stretch. In this case, residual striping arises from round-off error in the DNs when we superposed the striping on the spectral data. In real, remote-sensing ratio images, "noise" and defects often are apparent in those parts of images such as shadows where the input DNs have low values. This sensitivity of ratios to small numbers sometimes makes it difficult to produce clean images and to remove all traces of topographic shading and shadowing.

Quantitatively, channel ratios will be in error if the measured radiance includes additive effects of the atmosphere or the detector. Either the offset terms need to be subtracted first by calibration or one can minimize errors by ratioing only those wavelength channels where the offset terms are small. However, most people just go ahead and make ratio images without regard for calibration of the offset terms, which suggests that approximate values suffice for most applications.

Defining sample and reference

In the laboratory, reflectance and emittance spectra are ratios in which the numerator is the radiance measured from a sample, and the denominator is the radiance from a reference such as halon, spectralon, or a blackbody. When we make ratios from channels in a multispectral image, it facilitates interpretation also to define the denominator as the reference and to interpret the sample in terms of absorption or emission. We consider three situations:

(1) **One sample, one process**. Consider spectrum B of the synthetic image as a reflectance spectrum (Figure 3.16), and that we are interested in the absorption band centered at channel 2. (If we were dealing with a real spectrum, we might be able to derive information about composition of the material from the wavelength and depth of this band.) Channel 1 samples the shoulder of the absorption band, where it is minimally influenced by the same absorption process. Therefore, we define channel 1 as the denominator-reference for measuring the amount of absorption in channel 2. The ratio Ch_2 / Ch_1 (0.18) does not change with simulated variations in illumination (except for small changes in round-off errors), as can be seen by the absence of horizontal striping in the ratio image. A lower ratio, though, would be interpreted as a deeper absorption band (lower DN in channel 2), and a higher ratio would imply less absorption. Thus, the potentially variable absorption in channel 2 is the "sample," and the ratio is constructed the same way that spectra are measured. In image form, the lower ratios (expressed as darker pixels) establish an intuitive connection to absorption of reflected light. Of course, for purposes of visual display and enhancement, it often is effective to use the inverse ratio (in this case, Ch_1 / Ch_2) that makes the "sample" light against a darker background.

One of the first applications of an image ratio designed specifically to show variations in absorption-band depth is shown in Figure 3.17. The figure displays the ratio of radiance at 0.95 μm to radiance at 0.56 μm of part of the lunar surface that was measured in the mid 1970s

---

Contrast and ratio images

If we complement a ratio image (e.g., $(1 - Ch_1/Ch_2)$), light and dark are reversed, but the contrast range does not change. If we invert the ratio (e.g., Ch_2 / Ch_1) light and dark also are inverted, but the contrast range changes. Ratio values for which $Ch_2 < Ch_1$ will be in the interval 0–1; whereas if $Ch_2 > Ch_1$, they will range from 1 to ∞.

by a multispectral vidicon camera at a large telescope (McCord *et al.*, 1976). Earlier reflectance measurements of lunar samples showed that the common rock and soil types contain Fe^{2+}-bearing pyroxenes that have absorption bands in the 0.9 to 1.0 μm range. The depth of this pyroxene band is greater for rocks and coarse, immature soils that are abundant in and around fresh craters. Those pixels that have low 0.95/0.56 channel-ratio values (darker pixels in Figure 3.17b), therefore, indicate relatively more absorption near 0.95 μm, and they are interpreted as rock and soil that has been disturbed by relatively young impacts. Even though the images cannot spatially resolve the rocks themselves, the ratios provide compositional information and a measure of relative ages of the surface features.

(2) **One sample, multiple processes**. Even at high spectral resolution, resonance bands caused by different processes can overlap. Thus, a single image channel, especially one having a wide wavelength interval, may sample more than one spectroscopic mechanism. Interpretation can be further complicated when the reference channel and the sample channel both vary in response to the same or different processes. An important example (discussed below) is how to interpret red/NIR ratios of green vegetation when one channel responds to chlorophyll absorption and the other varies with scattering by leaf structure.

(3) **Multiple samples, multiple processes**. Ratios become more difficult to interpret when the reference and/or the sample channels are responding to more than one material. With multiple materials, both channels are potential variables, and without additional information there is no way of knowing to what extent each material on the ground is responsible for the recorded signal. The point is illustrated in the Ch_2/Ch_1 image of the synthetic image (Figure 3.16). From the previous discussion we know why area B is dark (low Ch_2/Ch_1 ratio) in this image; however, we cannot apply the same interpretation to explain why area A has a high ratio value (2.12), because the spectroscopic model for B does not apply to A. The Ch_2/Ch_1 ratio has no common reference for both A and B. This example illustrates one of the difficulties in physically interpreting ratio images of most landscapes. One needs to have different spectral models for each material.

Ratios and models

A different ratio may be needed to model each material.

Vegetation ratios and indices: a special case

(See also Section 2.2.3.) Chlorophyll in green plants absorbs strongly at red wavelengths (about 0.66 μm) and the cellular leaf structure is highly reflective (scattering) in the near infrared (from about 0.8 to 1.1 μm). No other common materials on the land surface have such a large spectral contrast in the VIS-NIR. As a result, ratio measurements in these two wavelength regions are used in remote sensing to detect and identify

(a)

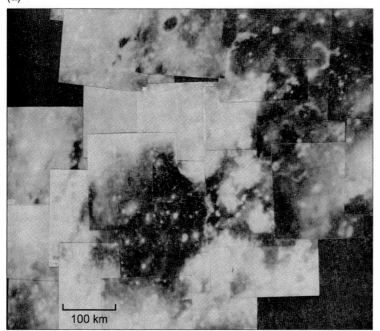

100 km

(b)

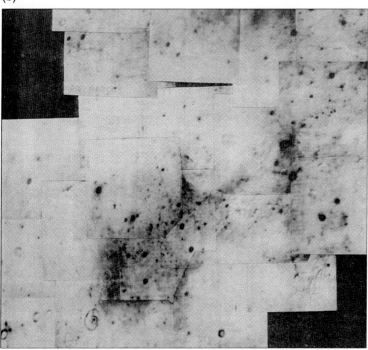

Figure 3.17. Mosaic of telescopic vidicon images of the Mare Humorum region of the Moon. (a) Image at 0.56 µm. Dark areas are visually dark maria (basalts). Light areas are highlands (anorthosite-rich) and relatively immature craters and their impact ejecta in both mare and highland materials. (b) Ratio of channels centered at 0.95 µm and 0.56 µm. Dark areas absorb more solar radiation at 0.95 µm. Absorption is caused by ferrous iron in the mineral pyroxene in immature soils and rocky craters. Adapted from McCord *et al.* (1976). The spectroscopic basis for interpreting the ratio image is further explored in Pieters (1993).

green vegetation. The simple vegetation ratio is usually constructed as NIR/red. A more commonly used version of the ratio is (NIR − red)/(NIR + red), known as the normalized-difference vegetation index or NDVI. The term "index" suggests a key or formula to some property. Indeed, the NDVI has been linked empirically in some field areas with ground measurements of the amount of vegetation cover (e.g., Tucker *et al.*, 1985; Huete *et al.*, 1997) and in other areas with biophysical properties of vegetation such as leaf-area index (LAI) and fraction of photosynthetically active radiation (FPAR). However, we caution field investigators that the simple ratio and the NDVI need to be interpreted carefully from a spectroscopic perspective, and that the same interpretations cannot apply to all landscapes.

Simplistically, the amount of red absorption by green vegetation is governed by the concentration of chlorophyll and the path length of the radiation. Applying the Bouguer–Beer–Lambert law (Section 2.1.2), we could measure chlorophyll concentration if we knew the path length; however path length is uncertain, because radiation is scattered unpredictably through the leaf structure. When we consider multiple leaves, different spacing of leaves, shading and shadowing, and the complex scattering geometry of real vegetation canopies, the problem of measuring chlorophyll concentration becomes intractable. An additional complication is that optical absorption by chlorophyll can reach a maximum (saturate) when either the concentration is high or the path length is large. Indeed, saturation in the red appears to be common for closed vegetation canopies. On the other hand, in the NIR, chlorophyll does not absorb, but instead, scattering and transmission by the cellular structure dominate. The reflectance of vegetation canopies in the NIR varies in response to a number of factors including leaf structure and thickness, arrangement and spacing of leaves, and canopy geometry.

How then should we interpret the NIR/red ratio and the NDVI? It depends on whether we are measuring a single sample or a multiple sample. In the case of an ideal, closed canopy that consists entirely of green leaves, neither wavelength is an obvious choice for a reference, because each one responds to different and largely independent physical processes. For example, if the ratio increases, it could mean that there is more chlorophyll absorption (lower DN in the red channel) or it could mean that there is more leaf scattering (higher DN in the NIR channel). Or both processes could be at work. If the chlorophyll band is saturated, then an increasing ratio would just mean more scattering from the canopy.

In the case of multiple samples, for example green vegetation and dry litter or soil, we adjust our interpretation to fit a predictive model. We know that the NIR/red ratio is much higher for green vegetation than for litters and soils that typically have more gently sloping reflectance spectra in this wavelength region. Accordingly, we attribute the high ratio values to the green vegetation and the low values to the substrate. In this context, a vegetation index can be used empirically to estimate the percentage of vegetation cover. Closed canopies of vegetation also consist of multiple samples when senesced leaves, branches, trunks, flowers, and epiphytes are exposed in addition to green leaves. Again, high NIR/red ratios are attributable to the chlorophyll-bearing component, and lower values suggest the influence of the other spectral components of vegetation canopies (Chapter 2).

These observations remind us that vegetation ratios and indices always need to be interpreted in a field context so that we can sort out which spectral components are influencing reflectance/absorption in the two channels. Because multiple spectral components at the sub-pixel scale produce mixed spectra, we also can extract the same information by applying spectral-mixture analysis (Chapter 4). Images of NDVI and the fraction of the endmember green vegetation (GV) often are approximately equivalent, the main difference being that all channels can be used to calculate GV. In Section 7.2.2 we analyze images of NDVI and GV for the Amazon Fazendas Landsat-TM sub-scene (Plate 4) to illustrate the necessity of applying different interpretations to different areas.

Water bands in vegetation

Although the NIR/red ratio is the best known of all ratios in remote sensing, there is additional information that can be extracted from vegetation using other wavelengths. Water absorbs strongly at 1.4, 1.9, and 2.5 μm, and the broad wings of these bands extend into the adjacent wavelength regions. For example, TM Bands 5 (1.55 to 1.75 μm) and 7 (2.08 to 2.35 μm) echo the amount of absorption in the nearby water bands, and for this reason ratios such as Band 5 / Band 4 or Band 7 / Band 4 can be used to interpret the relative water content of vegetation. Again, as in the case of the Band 4 / Band 3 ratio, it is essential to understand the image context to interpret the ratio images.

Ratio-composite and hybrid images

With standard image-processing software, three ratio images can be combined into a color-composite image. The example shown in Figure 3.18 and color Figure 3.18 (see Web) is from an AVIRIS image of Cuprite, Nevada. The AVIRIS channels were combined to synthesize a Landsat TM image. Ratio images also can be combined with channel images or other derived products to make color composites. There is an obvious advantage in merging different kinds of data for display when the product shows the extracted information in a form that is easily interpreted. But we know that even a single ratio can be ambiguous,

**Figure 3.18 and color Figure
3.18** (see Web). Color-ratio
composite from an AVIRIS
image of the Cuprite, Nevada
area. The AVIRIS channels were
combined to synthesize a
Landsat TM image. Image by
R. Clark, USGS. From
http:/speclab.cr.usgs.gov/.
Compare this image with
mineral-identification images
made using full AVIRIS spectral
resolution in Figure 3.20 and
color Figure 3.20 (see Web).

Synthesized TM Bands

R = TM Band 5 / TM Band 7
(1.67 μm / 2.22 μm)

G =TM Band 5 / TM Band 4
(1.67 μm / 0.84 μm)

B = TM Band 3 / TM Band 1
(0.67 μm / 0.48 μm)

2 km

N

especially when the local context is uncertain. Not surprisingly, ratio
composites and hybrids can be difficult or impossible to understand
unless there is a key that specifies what each color represents spectroscopically and what are the inherent assumptions and ambiguities.

3.5.2   A fixed model: the tasseled cap

A fixed spectroscopic model is one that is designed to extract a
specific kind of information each time it is applied to an image, and
that specifies the type of scene it can be applied to and the system
channels and their matrix-transformation coefficients. The NDVI
could be considered a fixed model for Landsat, SPOT or AVHRR
whenever we specify whether the scene (or part of a scene) is closed-
or open-vegetation canopy. The tasseled-cap transformation (TCT) is
a fixed spectral model for measuring the state of vegetation in agricultural areas. It was designed by Kauth and Thomas (1976) for
Landsat MSS images, and later was adapted to Landsat TM images.
"Tasseled cap" gets its whimsical name from the shape of the

two-dimensional scatter plot that includes the axes that define soil and green vegetation. The TCT is superficially like the PCT in that it adjusts the coefficients (weightings) of the values in each of the spectral channels to produce new (orthogonal) coordinate axes; however it differs from the PCT in that the TCT involves a physical model that weights channels the same way each time and is not just a best fit to the data. The coefficients for the TCT need to be specified for the specific channels of each remote-sensing system, which means that we cannot just click a TCT button in our image-processing program and apply the transformation to any spectral image. For example, the coefficients differ from one Landsat TM system to another. Furthermore, one needs to be sure that the image-processing software actually uses the correct coefficients. (We have found one widely used program that contains erroneous coefficients.)

As originally defined for Landsat MSS, $TCT_1 =$ "soil brightness," $TCT_2 =$ "greenness," $TCT_3 =$ "yellow stuff" and $TCT_4 =$ "none such." Based on further studies using Landsat TM, TCT_3 evolved into "wetness," and TCT_4 into "haze." Higher-order coefficients for Landsat TM are published, but because these transforms do not have consistent interpretive labels, they are not fixed spectroscopic models. The basic idea of the TCT is that agricultural areas should change over time from bare soil (TCT_1) to green vegetation (TCT_2) to senescent vegetation (TCT_3). Pixels having high values (lighter in a TCT image) are interpreted as having more of the specified component.

Many analysts today consider the TCT to be an old tool that only applies to crop fields in Landsat images. We have a more generous view. The TCT is perhaps the first, and certainly the best known, example of a model that predicts how landscape components are expected to change spectrally over time in response to physical processes. We see process models as a neglected but fruitful subject in spectral remote sensing (Chapter 8). Once we understand the spectroscopic basis for the TCT model, there are other interesting applications, particularly in forest studies (Chapter 7).

The spectroscopic basis for the TCT is closely connected to the discussion of vegetation ratios and indices. When we examine the TCT coefficients (Figure 3.19), we see that the two largest coefficients for green vegetation (TCT_2) are in Band 4 ($+0.72$) and Band 3 (-0.55). These are the same channels that are used for the NDVI and that have the largest coefficients for the endmember GV in mixing models. In effect, TCT_2 simply "views" the data space from the same perspective that NDVI does, and that is used for various forms of GV. We already have established the spectroscopic basis for interpreting TM Bands 3 and 4, and we have pointed out that NDVI and GV images are closely similar.

Figure 3.19. Landsat 5 tasseled-cap transformation (TCT) coefficients for soil brightness (TCT$_1$), greenness (TCT$_2$) and wetness (TCT$_3$) from data tabulated in Crist *et al.* (1986). Also shown is the complement of wetness (1 − wetness) that in most contexts can be interpreted as "dryness." The plots of the coefficients resemble the reflectance spectra of the scene components.

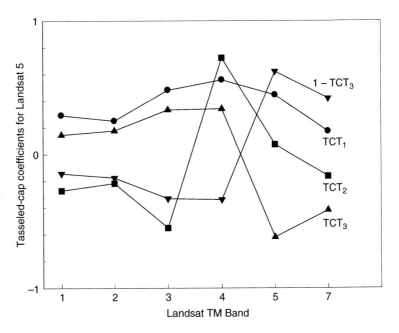

The coefficients for TCT$_1$ (soil brightness) are positive in all channels which is consistent with relatively flat soil spectra. (Keep in mind, though, that if the context does not include soil, TCT$_1$ will show "brightness" of other scene components such as rocks or urban surfaces.) In contrast, TCT$_3$ (wetness) has large negative coefficients for TM Bands 5 and 7 where the wings of the main water absorption bands overlap. Not surprisingly, a TCT$_3$ image will portray water and wet soil as light tones, and dry soil (or other substrata) will have dark tones. In the context of a closed vegetation canopy such as a crop field, light tones in a TCT$_3$ image indicate relatively strong absorption by water in (or on) green foliage, whereas darker tones are consistent with senesced vegetation and woody materials – the basic spectral component that we call NPV or non-photosynthetic vegetation (Chapter 2). Ironically, TCT$_3$ for Landsat TM images of closed-canopy crops actually does respond to the "yellow stuff" (senescent crops) that was the focus of TCT$_3$ for early Landsat MSS. (Landsat MSS with fewer channels relied on visible channels to detect "yellow stuff.") Thus, it turns out that for Landsat TM images, TCT$_3$ can be used to highlight "wetness" and/or "yellow stuff." The interpretation depends on field knowledge of the landscape in the same way that we saw for NDVI. When we consider the tasseled-cap transformation as a spectral tool that distinguishes basic spectral components (Section 2.2.3), we can broaden the applications of TCT beyond agriculture. In Section 7.2.2 we apply TCT to forested landscapes and compare NDVI, TCT, and spectral-mixture analysis results to extract information about basic spectral components.

3.5.3 Mixture models

Spectral-mixture analysis (SMA) has been used as a technique for applying physical models to images of landscapes since the mid 1980s, although the idea of mixed pixels and their effects on classifying Landsat MSS agricultural scenes dates back to the early 1970s. The objective of SMA is to model an image as linear mixtures of a few basic spectral components, usually three to five, that are termed "endmembers." The output of SMA consists of images for each endmember and an image for the RMS residual which is the difference between the model and the data. The DNs of the endmember images are scaled to represent fractions from 0 to 1 and are called "fraction images."

Like the PCT and the TCT, SMA results in a reduction in the dimensionality of the data from the starting spectral channels to a few, newly defined axes in the data space. The endmember axes in SMA are not necessarily orthogonal, and they are not defined by the structure of the data space as is the case for PCT. In contrast with the TCT, the SMA axes are not fixed according to pre-set scene components, and, therefore, the matrix coefficients are not channel specific or system specific. Thus, SMA can arrive at nearly the same results as the TCT by selecting as endmembers a spectrum of "shade" (the equivalent of the complement of TCT_1), GV (equivalent to TCT_2) and NPV (the equivalent of $1 - TCT_3$) (see Section 3.5.2). With SMA, however, one has the flexibility to select different varieties of these endmembers or to choose entirely different scene components to model. A further, significant advantage of SMA is that by examining RMS-residual images and fraction-overflow images, one can evaluate how well a model fits the data, and, if necessary, adjust a model until the results are more satisfactory. Because SMA is a relatively new method that is still unfamiliar to some image analysts, and because it is powerful for testing physical models, the topic is explored in more depth in Chapters 4 and 5.

3.5.4 Resolved resonance bands

Technological advances in remote-sensing systems in the 1980s made it feasible to acquire "hyperspectral" images in hundreds of channels. The first images available to the public were from JPL's AIS, followed by AVIRIS. The AIS and AVIRIS images had spectral resolutions equivalent to many laboratory spectrometers, and geologists seized the opportunity to apply physical models based on mineral absorption

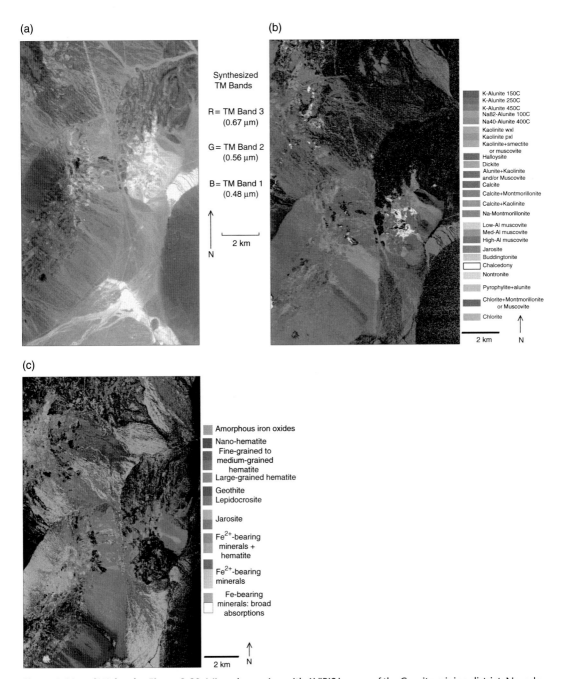

Figure 3.20 and Web color Figure 3.20. Mineral mapping with AVIRIS images of the Cuprite mining district, Nevada. (a) Three-color composite image (three AVIRIS bands out of 224). (b) Interpretative mineralogical map, from spectral matching with library reflectance spectra. (c) Map of iron minerals. Images by NASA and R. Clark and G. Swayze, USGS. From http://speclab.cr.usgs.gov/.

bands to remote sensing (Chapter 2). Since AVIRIS became operational, several other experimental imaging-spectrometer systems have acquired a variety of images, but a field investigator probably will find that AVIRIS images are still the most accessible. Furthermore, there is an extensive literature on AVIRIS results and an annual conference at JPL devoted to hyperspectral remote sensing.

Roger Clark and his group at the US Geological Survey have developed many excellent examples of how AVIRIS images can be used to identify minerals and other materials using resolved resonance bands. This work can be accessed at: http://speclab.cr.usgs.gov/, along with a tutorial on how to work with imaging-spectrometer data. Furthermore, the USGS group has compiled a comprehensive and well documented library of reflectance spectra, and has created an algorithm ("Tetracorder") that uses the spectral library as a reference to identify and map the distributions of minerals from AVIRIS data (Clark *et al.*, 2003); see also http://speclab.cr.usgs.gov/PAPERS/tetracorder. Examples are given in Figure 3.20 and color Figure 3.20 (see Web).

Many of the standard image-processing tools that were developed for broad-band images cannot be used with imaging-spectrometer data. Some of the standard classification algorithms and the PCT, for example, turn out to be computationally expensive when applied to hundreds of channels. There are two notable exceptions among the classifiers – spectral angle and binary encoding (Section 3.4.1). With hyperspectral data there are many more opportunities to use channel ratios (inexpensive spectral transformations) to extract compositionally specific information (Section 3.5.1). But when only two out of hundreds of hyperspectral channels are selected, it is important to be testing spectral models, rather than just trying combinations to see if anything interesting turns up. The tasseled-cap fixed model presently is restricted to Landsat data (Section 3.5.2). However, SMA (Section 3.5.3 and Chapter 4) is well suited to hyperspectral images.

Next

In the next chapter we discuss how SMA actually works. Questions addressed are the following. What are the main strengths of SMA in terms of making a connection between images and materials on the ground? What are the limitations of the method?

Chapter 4
Spectral-mixture analysis

Copernicus Crater, the Moon. Photograph by NASA

In this chapter

This chapter is a tutorial on the basics of spectral-mixture analysis; SMA is an especially powerful tool for testing spectroscopic models, and for defining the basic spectral components of landscapes in a photo-interpretive context. Only under special circumstances can SMA be used as an accurate measure of the proportions of materials in a scene.

Simple models for complex surfaces

Much of the compositional and structural information that we are seeking in spectral images lurks at a scale below that of a pixel footprint. At the pixel scale, measurements integrate the radiant flux from all of the spatially unresolved materials in the field of view, regardless of whether or not we know or suspect their identities. Virtually all pixels of landscapes incorporate spectral mixtures.

We use the term "spectral-mixture analysis" (SMA) to describe a method for analyzing spectral images that entails physical modeling (Chapter 1), and that is designed primarily to enhance photo interpretation. It is *not* always the main goal of SMA to make accurate measurements of the proportions of mixture constituents, although that is possible under some circumstances. Nor is the method simply an algorithm or a formula for classifying images, although the results can be used to construct classes for thematic mapping. Perhaps the main advantage of SMA for someone working on a field problem is that it is easier to interpret an image in terms of the approximate proportions of certain materials on the ground than it is to interpret pixel DN values that represent radiance, reflectance, or emittance.

A mixing model converts image DNs in all spectral channels to numerical fractions of a few endmembers. Thus, there is a transformation from "spectral space" to "fraction space."

Transformation from "spectral space" to "fraction space"

Pixel *x,y* as a spectral measurement   Pixel *x,y* as a fraction measurement

Channel 1 50.8 DN		
Channel 2 28.8 DN	Data Transformation	Endmember A 0.4
Channel 3 56.8 DN		Endmember B 0.6
Channel 4 62.8 DN		

Endmembers are spectra that are proxies for materials on the ground. In a perfect model, endmembers represent all of the basic components of a scene; thus the fractions of the endmembers sum to 1 for the same reason that the fractions of the surface components sum to 1.

Typically in remote sensing, the models that are applied to spectral images are constructed to match the complexity of a scene; however, our approach to mixture analysis concedes that most landscape surfaces are far too complicated to let us interpret spectral data uniquely. Furthermore, only a few materials can be distinguished unambiguously by their spectra, and there are a number of complicating factors in making remote spectral measurements. For these reasons, SMA is designed primarily to work with the few spectrally dominant components (Chapter 2) that also represent important materials on the ground. Thus, the basic spectral components of landscapes are the best candidates for endmembers in SMA. The spectral complexity of landscapes, among other factors, also constrains our ability to select endmembers automatically, to have a large number of endmembers, and to measure endmember fractions with high accuracy. Fortunately, for photo interpretation and for most field investigations, it is not necessary to know the identities or the proportions of all of the spectral constituents. Many constituents simply are not important. And we rarely need to know the proportions of the spectrally dominant materials to better than, say, 10 %, or even 20 %. For example, if we were trying to measure the percent cover of vegetation, we might be well satisfied with an estimate of 48 % if the true, field value turned out to be 44 %. As usual, it depends on what we are trying to find out.

Basic spectral components

Basic spectral components are the best candidates for endmembers (see Section 2.2.3.).

Spectral mixing

Interest in spectral mixtures has been around since at least the 1970s, when scientists first attempted to compare laboratory reflectance spectra with telescopic spectra of thousands of square kilometers on the Moon and Mars. At about the same time, analysts classifying Landsat (then ERTS) images, addressed the problem that mixed pixels cause at the boundaries of classes, such as different types of crop fields. The subject received little further attention in the remote-sensing community until the late 1980s when images with increased spectral and spatial resolution became widely available, and SMA began to be developed as an analytical technique.

4.1 Endmembers, fractions, and residuals

Albee's law

Albee's law was named by us after Professor Arden Albee of Caltech who for many years has taught principles of mixing in the context of metamorphic petrology and geochemistry. Interestingly, the basic concepts and the mathematics of mixing in those fields carry over directly to spectral mixing as applied to remote sensing.

The fundamental idea of mixture modeling can be summarized somewhat whimsically by Albee's law: "All of something equals the identified parts of it plus what is left over."

In equation form, Albee's law can be written as in Equation (4.1),

$$V = F_a V_a + F_b V_b + \ldots F_n V_n + r \qquad (4.1)$$

where V is the measured value of some property of a mixed sample; V_a, V_b and V_n are measured values of some or all of the components of the mixture; F_a, F_b and F_n are the fractions of the components; and r is a residual that includes measurement error and contributions from any components that have been left out. When all of the components are known, r simply describes measurement error; however, Albee's law and Equation (4.1) include the possibility that not all of the components have been found, in which case the missing ones are expressed in the residual. It is useful to state the mixing problem this way, because when we attempt to model mixed pixels we typically do not know what all of the components are. Equation (4.1) describes linear mixing if the total measured signal is the sum of the signals from the components, weighted by their fractions. Linear mixing, because of its relative computational simplicity, is a convenient formulation for testing models, and works well for many remote-sensing applications. But spectroscopic mixing is not constrained to be linear, as we discussed in Chapter 2.

Terminology

Member: a constituent part of any structural or composite whole.
End member: (mineralogy) either of two pure compounds occurring in various proportions in a series of solid solutions that comprise a mineral group.
Spectral endmember: a spectrum that is a constituent part of a spectral mixture; a conceptual construct that is used to model spectral mixtures.

4.1.1 Spectral endmembers

To describe spectral mixing we need to be clear about what gets mixed. Our definition of "endmember" is not quite the same as in other fields. In mineralogy, for example, "end members" (two words) are pure compounds that occur in various proportions in a series of solid solutions that comprise a mineral group; thus, albite and anorthite are end members in the plagioclase solid-solution series, and forsterite and fayalite are end members in the olivine solid-solution series. Chemical compounds, minerals, and groups of minerals are used to describe mixing in geochemical, mineralogical, and petrological applications. For the purposes of SMA we define "spectral endmembers" simply as spectra that are constituent parts of a spectral mixture. We use "endmember" as one word to distinguish our definition.

Endmembers are spectra, not materials, although spectra may represent materials. Endmember spectra are not necessarily spectral signatures of "classes," however, proportions of endmembers can be used to classify pixels (Section 5.2). Mixing, by its nature, is scale dependent; therefore, endmembers for the same scene may vary depending on spatial resolution and the particular application. Endmembers may be spectra of "pure" materials, or they may be spectra that themselves are mixtures of materials at some more fundamental scale. For example, the spectrum of a granular mixture of plagioclase feldspar and olivine could be modeled by two endmember spectra, plagioclase and olivine, even though each of the corresponding mineral end members is itself a mixture of two other minerals having distinct spectra. At remote-sensing scales, we expect virtually all pixel spectra to consist of mixtures of the spectra of materials that we could measure on the ground. Thus, endmembers that are defined from pixels of training areas are not "pure" when viewed from a field perspective, but they may be relatively pure at image scales. For example, spectra of green vegetation and of soil might be useful to model mixing at scales of tens of meters to kilometers; however, these same endmember spectra comprise mixtures of varied spectra at a scale of centimeters. All spectra are potential endmembers, but, in an image context, once we find that a spectrum does not mix with anything, it is not useful to designate it as an endmember in a model. In fact, inclusion of unnecessary spectra as "endmembers" may confound unmixing (inversion of a model).

> **Candidate endmembers**
> If a spectrum does not mix with anything, it is not an endmember.

4.1.2 Unmixing spectral images

The mixing equation for spectral images is given in Equation (4.2),

$$DN_i = \sum_j F_j DN_{i,j} + r_i \quad \text{and} \quad \sum_j F_j = 1 \qquad (4.2)$$

where DN_i is the measured value (vector) of a mixed pixel in spectral channel i; DN_j is the measured value (vector) of each endmember (the jth endmember); F_j is the fraction of each endmember; r is the residual that accounts for the difference between the observed and modeled values. We impose the constraint that the fractions sum

> **Endmember fractions**
> Fractions should sum to 1, but need not have values between 0 and 1.

Table 4.1. *Data numbers in four channels of the synthetic image for spectrum 0.4A, 0.6B; $V_{A,B}$ = the DN value of spectra A and B, respectively; $F_{A,B}$ = the fractions of A and B, respectively.*

Channel	V_A	F_A	V_B	F_B	$V_A * F_A$	$V_B * F_B$	$\Sigma V * F$
1	25	0.4	68	0.6	10.0	40.8	50.8
2	53	0.4	12	0.6	21.2	7.2	28.4
3	59	0.4	55	0.6	23.6	33.0	56.6
4	60	0.4	64	0.6	24.0	38.4	62.4

to one; however, we do *not* constrain the fractions to be between 0 and 1. A set of selected endmembers defines a mixing model. The objective is to solve the mixing equation for the fractions of the selected endmembers while minimizing the residual, a process sometimes referred to as "unmixing." Unmixing a pixel into fractions of endmembers is an inversion problem involving a linear vector-matrix equation.

To illustrate how mixture models work, it is convenient to use the synthetic image that was introduced in Chapter 3 (Section 3.4.1). In constructing the synthetic image (Figure 3.2) we specified the fractions of each endmember for each pixel. When we know the fractions of each endmember for each pixel, and we know the DN values in each channel for each pure endmember, we can solve for the DNs of mixed pixels. This is just the inverse of unmixing, where we solve for the fraction of each endmember, Equation (4.2). In the case of the synthetic image, the residual is zero, because the endmembers completely describe the image. An example of a mixed spectrum from the synthetic image that consists of 0.4 A and 0.6 B is shown in Table 4.1 and Figure 4.1.

4.1.3 Residuals

The residual, r, is measured for each channel i as shown in Equation (4.3).

$$r_i = \mathrm{DN}_i - \sum_j F_j \mathrm{DN}_{i,j} \qquad (4.3)$$

To solve for the residual, the number of channels N must be equal to or greater than the number of endmembers n. We will see later that it generally is not practical to use more than three or four endmembers at a time, regardless of the number of channels. When we have chosen endmembers appropriately the value of the residual will be small.

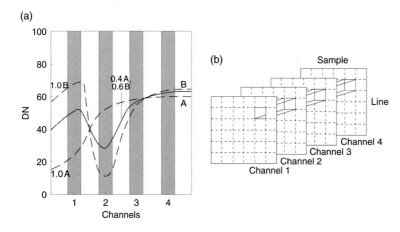

Figure 4.1. Mixed spectrum 0.4 A, 0.6 B. (a) Linear mixing of spectra A and B from Figure 3.2. See Table 4.1 for calculations. (b) Spatial setting of a pixel that consists of 0.4 A and 0.6 B in a four-channel image.

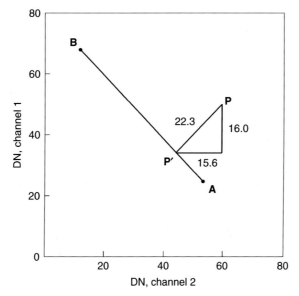

Figure 4.2. Geometric illustration of channel residuals and RMS residual; DN is in arbitrary units. Point P does not fit on mixing line AB. The residual in channel 1 is 16.0 DN; in channel 2, 15.6 DN. The RMS residual is 22.3 DN, and is the minimum distance from P to point P′ on the line AB, which is the square root of the sum of the squares of the residuals in each channel. Residuals similarly can be measured from a point to mixing planes and volumes. Spectra A and B are from Table 3.1 and Figure 3.2.

Residuals for all channels can be combined into a root-mean squared (RMS) residual, Equation (4.4). The RMS residual is a convenient measure

$$\text{RMS} = \left[\frac{1}{n} \sum_i r_i^2 \right]^{1/2} \tag{4.4}$$

of the overall fit of the set of endmembers to the image (Figure 4.2). Both the RMS residual and the channel residuals can be displayed as images. High residual values call attention to pixels that need to be modeled by different sets of endmember spectra, and for this reason are essential for evaluating mixing models, as will be discussed below.

4.1.4 Fraction overflow

A corollary of Albee's law is that all of the parts of something add up to the whole thing, Equation (4.1). Accordingly, when we solve Equation (4.2) we impose the constraint that the fractions sum to 1. Even though it may seem counter-intuitive, we choose not to impose the additional constraint that the fractions must lie between 0 and 1. We apply the term "fraction overflow" to describe fractions that are < 0 or > 1. It turns out that fraction overflow provides an additional way to evaluate the set of endmembers used in a mixture model. (See ahead to Figure 4.12 and Figure 7.13.) We lose this evaluative tool if we constrain the fractions to be between 0 and 1 when solving Equation (4.2).

We need to be careful to distinguish fraction overflow in a spectral model from the actual proportions of materials. It is mathematically correct to have 1.1 of one spectral endmember and -0.1 of another; but sober field observers never find negative amounts of materials on the ground, or, for that matter, amounts of things that exceed 100 %. But, adding the mathematical constraint that the fractions of endmembers must be between zero and one just eliminates some endmembers from consideration without signaling what to do next to improve the model.

Figure 4.3 illustrates a simple example of overflow fractions, using a case in which the residual, $r = 0$. Mixing occurs between spectra A and B, the same ones that were defined as endmembers in Figure 4.1 and Table 4.1. We imagine A and B to be connected by mixing line AB. The line AB is fully defined by two or more channels; in this case we use channels 1 and 2 for purposes of illustration. In terms of fractions, point A represents 1.0 A and 0.0 B, whereas point B represents 1.0 B and 0.0 A. Points along the mixing line AB represent various proportions of A and B, such as 0.4 A and 0.6 B which was illustrated in Figure 4.1. If mixing is linear, the fractions of A and B will be linearly proportional to the distance along the line AB.

Points along the projection of line AB have no residuals, but they can be described by overflow fractions of A and B. For example, D specifies 1.3 A and -0.3 B. One way to interpret an overflow spectrum such as D is that it represents a more "pure" endmember than A, and that A actually is just a mixture of D and B. If we re-defined our model as the mixture [D, B], A would become 0.91 D and 0.09 B. Similarly, all other fractions would be redefined in terms of D and B. This suggests a way of searching for "pure" endmembers among overflow spectra,

Overflow fractions

Overflow fractions are a tool for evaluating mixture models. If endmember fractions are constrained to be between 0 and 1 this tool is lost.

Notation for models

We enclose the symbols for endmember spectra in square brackets – e.g., [D, B] – to designate a mixture model.

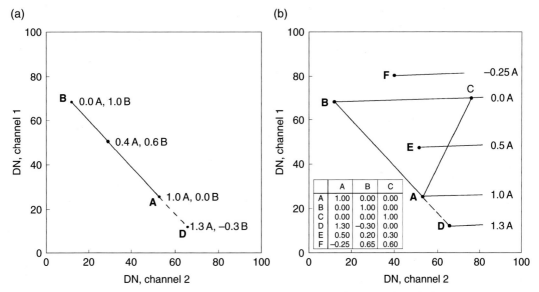

Figure 4.3. Fraction overflow. (a) Spectrum D lies on the projection of mixing line AB, and has the fractions 1.3 A, − 0.3 B. The point labeled 0.4 A, 0.6 B corresponds to the mixed spectrum in Figure 4.1. (b) Spectra D, E, and F are co-planar with ABC, but D and F are outside the mixing polygon ABC. Fractions for D, E, and F are shown in the insert table.

because when mixing is described by a plane or volume, rather than a line, the overflow spectra always occur outside a "convex hull" defined by the endmember spectra (vectors). Various methods have been devised to find endmembers by expanding the convex hull to include the maximum number of pixels in a shape of minimum volume.

Another interpretation, of course, is that D just happens to fall on the extension of AB, but is unrelated in composition. Or, for that matter, that A just happens to fall on the mixing line between B and D. Mimicking among spectra and their mixtures is common. Mimicking is strongly influenced by the choice of channels: the fewer channels, the greater the chance for mimicking. We select channels to separate materials from one another and to avoid mimicking. But the channels that nicely distinguish two materials will not necessarily separate a third from the first two or their mixtures. For this reason, as a general rule, it is advantageous to have as many spectral channels as possible when investigating mixtures.

> **How many channels?**
> As a general rule, it is advantageous to have as many spectral channels as possible when investigating mixtures. More channels mean less mimicking.

4.1.5 Endmember variability

In the examples so far in this chapter, each endmember occupies a point in the spectral hyperspace that is defined by the spectral channels. There is an implicit assumption that the spectral properties of endmembers do not vary; only their fractions do. We know from experience, of course,

Figure 4.4. Variable
endmember having a mean
spectrum of A. Here, all of the
points surrounding A are part
of a general class of A-like
spectra. Variations in A might
be caused by instrumental
noise, variable measurement
conditions, or different
expressions of the same
material. Spectrum A_1 lies on
the AB mixing line; therefore,
the identity of A_1 is ambiguous.
A_1 can be interpreted as pure A
(1.00 A) plus noise, or a
mixture of A and B (0.75 A,
0.25 B). Spectrum D, discussed
in Figure 4.3, is at the edge of
the field of A-like spectra, and,
therefore, is a mimic.

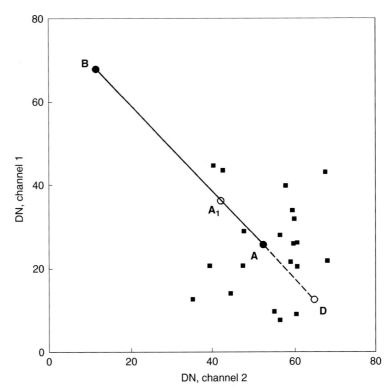

Figure 4.4. Variable endmember having a mean spectrum of A. Here, all of the points surrounding A are part of a general class of A-like spectra. Variations in A might be caused by instrumental noise, variable measurement conditions, or different expressions of the same material. Spectrum A_1 lies on the AB mixing line; therefore, the identity of A_1 is ambiguous. A_1 can be interpreted as pure A (1.00 A) plus noise, or a mixture of A and B (0.75 A, 0.25 B). Spectrum D, discussed in Figure 4.3, is at the edge of the field of A-like spectra, and, therefore, is a mimic.

that there can be considerable natural variability within a given class of materials and their spectra. It simplifies a mixing model to select an average spectrum or a representative spectrum for an endmember, but we know that we are departing from reality by doing so. Furthermore, the variability of endmembers is scale-dependent.

The effect of endmember variability on estimates of fractions is illustrated in Figure 4.4, which uses the endmembers A and B, previously defined in Figure 3.2. However, in this case we consider that A is the mean spectrum for a cluster of points that describes the variability of A. The mixing line AB passes through point A_1 which can be interpreted either as a natural variant of A (1.00 A), or as a mixture of A and B (0.75 A, 0.25 B). Thus, a pure endmember (in this case A_1) can mimic a mixture, or a mixture can masquerade as a pure endmember, and, in either case, the fractions are ambiguous. Other variations of A do not fall on the AB mixing line; however, when projected onto the AB line they also can be interpreted as mixtures of A and B. Endmember variability can be an important consideration when the objective is to estimate fractions quantitatively, and, as we see in Chapter 6, when we want to assess the detection threshold of a spectrum.

4.2 Shade

It has long been recognized that shadows plot at or near the origin in the data space that is defined by spectral channels. For example, Kauth and Thomas (1976) referred to the "point of all shadows," and cautioned that its location varied with atmospheric effects. A conceptual breakthrough for us in SMA was the realization that a spectrum having values of zero (or near-zero) in all channels could be used as an endmember that mixed with other spectra to model variations in illumination (Adams *et al.*, 1986). Dictionary definitions of shadow, shading and shade illustrate that these terms are hopelessly muddled in common usage. For our purposes we define the terms more narrowly (see margin), and in mixture analysis we restrict the term "shade" to a dark spectral endmember that represents shadows *and* the effects of shading. We noted in Chapter 1 that shading and shadows are important or dominant components of most images of natural surfaces. Because surface roughness occurs at all scales, including sub-pixel scales, shading and shadows do, also. In the realm of reflected light, shading and shadows lower the spectral values in all channels. Although useful for modeling illumination effects, shade can be ambiguous, because dark materials may mimic the spectrum of shade.

As discussed in Section 3.4.1, shade was added to the synthetic image by creating a horizontal pattern of light and dark that resembles shading produced by east–west trending ridges and valleys in a real image. The shade pattern is superposed on the distribution of A, B, and C shown in Figure 3.2. In scatter plots, shade is at or near the origin (Figure 4.5). Recall, however, that shadows typically have small, non-zero values on Earth, primarily due to skylight (Chapter 3).

Spectra A, B, C, and S are plotted in two- and three-channel combinations in Figure 4.5. Endmember S can mix with any of the other endmembers or their mixtures. With the addition of a fourth endmember we have defined a mixing volume, where one corner of the volume is anchored at S, near the origin. Fractions of mixtures of S and the other endmembers can be calculated from Equation (4.2) in the usual way. For example, a pixel that consists half of B and half of shade plots half way along the mixing line SB in each combination of channels.

By adding shade to Figure 4.5 we have created a more realistic model of a remote-sensing image. We also have introduced some interesting problems. Recall the general rule that each time we add an endmember we tend to introduce ambiguities caused by mimicking. For example, when we view the data space in channels 2 and 3 (Figure 4.5d), endmember A mimics a mixture of C and S; A can be interpreted as a darker version of C. Alternatively, C can be interpreted

---

Shadows and shading
Shadow: a dark image cast on a surface by a body intercepting light.
Shading: a reduction in the amount of light reflected from a surface that accompanies a departure from normal (90 degree) illumination.
Shade: an endmember spectrum having zero or low values in all channels.

Modeling shade
Although shading and shadow can be modeled using a shade endmember, many materials that have dark spectra in reflected light can mimic shade. Shade look-alikes often can be revealed by spatial context or by other information such as thermal properties.

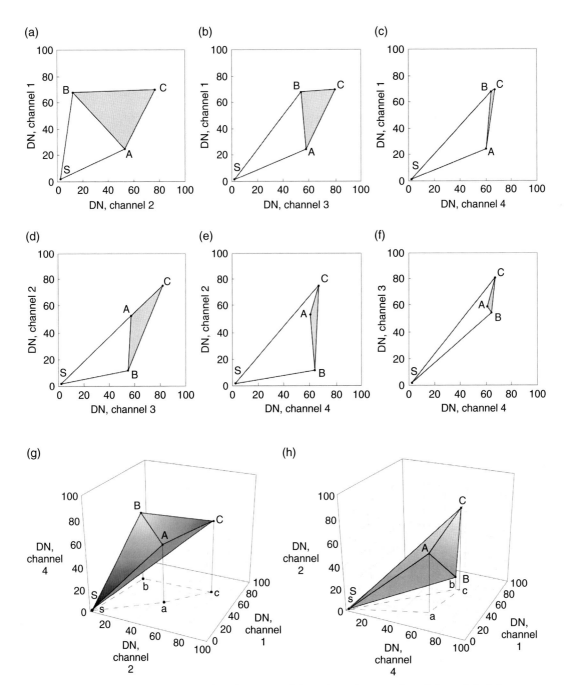

Figure 4.5. Endmembers A, B, C, and S plotted in two- and three-channel scatter plots of channels 1–4. Spectra correspond to those in Figure 3.2. The lines that connect two endmembers describe two-endmember mixtures. Lines connecting three endmembers describe mixing planes, for example ABC (gray tone). All four endmembers define a mixing volume (tetrahedron). The mixing tetrahedron has six different projections onto the two-channel data planes (a)–(f). The mixing plane ABC has the largest projection area in channels 1 and 2, and the highest spectral contrast between A and B. By comparison, in channels 3 and 4 the spectral contrast between A and B is lower. In (g) and (h) the tetrahedron ABCS is shown in the volume defined by Channels 1, 2, and 4.

as a more fully illuminated version of A (fraction overflow). In our synthetic image these ambiguities can be resolved by using other channels, such as 1 and 2. But for images in general, it is easy to imagine combinations of channels and endmembers where ambiguities persist. Again, we press the point that more channels usually reduce mimicking and are advantageous for interpreting the spectral data.

4.3 Fraction images

4.3.1 Interpretation

The fraction of each endmember can be displayed for each pixel to create an image for photo interpretation. Figure 4.6 shows the fractions of the endmembers A, B, C, and S expressed as gray tones. We call such images "fraction images." Each fraction image incorporates the information from all of the individual channels. In this case we have limited the example to four channels just to make it simpler to build the synthetic image. If the data had consisted of 200 channels all of the information still would be compressed into the four fraction images. The transformation from spectral measurements to fractions can accomplish a significant reduction in the dimensionality of the image data with little loss of information, provided that the residuals are small compared to that expected from measurement imprecision. Notice that there is no system noise in this version of our synthetic image.

> **Encoding endmember fractions in images**
>
> In Figure 4.6, fractions are displayed as DNs on an eight-bit scale by equating a fraction of zero to a DN of 100, and a fraction of one to a DN of 200. This is an arbitrary conversion, but an easy one to remember. It also leaves DNs 0 to 99, and 201 to 255, to display overflow fractions.

The gray tones in the fraction images are lighter with higher fractions. The same convention is used for the image of the RMS residual. The choice is arbitrary; sometimes it aids visualization of patterns to invert the image tones, which is easily accomplished with most image-processing software. Shade images, for example, usually are more intuitive when inverted (complemented) so that darker tones indicate less illumination; however, we did not complement shade in Figure 4.6. Gray-tone fraction images also can be "contoured" into fraction ranges as an aid to visualizing gradations, and using three endmembers we can assign each to one of the RGB colors to display a color composite. Contrast stretching of black-and-white or color-composite fraction images can assist visualization of patterns; however, stretching distorts fraction values for quantitative use.

To draw attention to just one endmember, it is effective to display higher fractions as lighter tones; however, when a mixing model consists of three or more endmembers, the darker pixels are unidentified, and simply indicate less of the displayed endmember. The reverse occurs when we display the complement of shade $(1 - \text{shade})$ in order to make a more intuitive image where shading and shadows are dark; then, the lighter tones just mean high fractions of unidentified

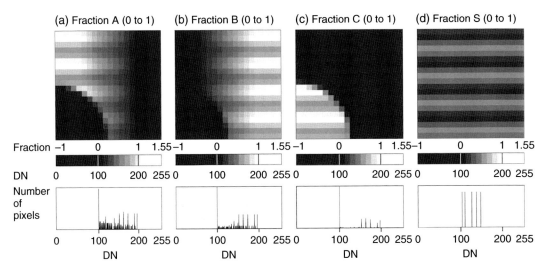

Figure 4.6. Fraction images. (a) Endmember A; (b) endmember B; (c) endmember C; (d) endmember S. Lighter tones indicate higher fractions of endmembers. In contrast with Figure 3.2b, shade (S) modulates the fractions of the other three endmembers. Because the image is "perfectly" modeled, the residuals are zero for all pixels in all channels, and fractions of all pixels are between 0 and 1.

non-shade endmembers. A complemented shade image typically reveals topography or roughness to the eye, and is an important qualitative check that the mixture model has indeed separated the shade component from the other endmembers.

4.3.2 Normalization of fractions

One of the most important advantages of working with fractions of endmembers, as compared with channel values, is that we can select combinations of endmembers to display without losing spectral information contributed from any of the channels. Although the number of endmembers displayed simultaneously by a color monitor is limited to three, the fractions of each endmember always express the contributions from all of the spectral channels used in the model. When we select some endmembers to display, while leaving out others, we do not omit information from any of the channels, but we do change how the channels are weighted, which follows from Equation (4.2).

We can decrease the number of endmembers displayed by deleting one or more of them and recalculating the fractions of the remaining ones so that they sum to one. For example, consider a pixel having fractions 0.33 A, 0.33 B and 0.33 C. If we do not want to display the contribution of endmember C in the image we can delete it and re-sum the fractions of A and B. The new fractions would be 0.5 A and 0.5 B.

The normalized fractions can be displayed as gray tones or color contours to show the relative proportions of just the two endmembers. This is an effective way to remove unwanted endmembers from an image, thereby focusing attention for photo interpretation on the remaining ones.

Examples of normalized fractions from the synthetic image are shown in Figure 4.7 where S has been deleted, thereby removing the horizontal striping that is present in Figure 4.6. The proportions of A, B, and C are more easily seen when the confusing pattern of S is removed. Furthermore, the fractions and the image pattern now are the same as the map of A, B, and C in Figure 3.2b. The effects of removing endmembers S and C are shown in Figures 4.7d,e: here, the visual information includes only the relative proportions of A and B, and does not measure the "true" non-shade fractions of either A or B. When a spectral image is fully described by only two endmembers the fraction images complement one another. Lighter tones indicate more of one of the endmembers, and darker tones mean more of the other.

We also can select three endmembers from a model, normalize their fractions to sum to 1, and display them as a three-color composite. Here, mixtures of colors, rather than gray tones or color contours, express the relative fractions of the endmembers. Color-contour (density slice) images made from two normalized endmembers, and color composites made from three normalized endmembers, that do not include shade, can be "modulated" by the complemented shade image. By modulation we mean that the lightness of each pixel in the color-contour or color-composite image is decreased by an amount that is proportional to the shade fraction. This has the visual effect of draping a (topographically) flat color image over topography. Because the fractions of the non-shade endmembers are normalized, their proportions are consistent with how surface components are mapped in the field, where shade is not counted. Thus, the proportions of the non-shade endmembers are maintained, and the important visual clues of shade for photo interpretation are incorporated in the image. The main ways to display fraction images are summarized in Table 4.2. The principles discussed in this section are applied to remote-sensing images in the next chapter (Section 5.1).

4.3.3 Testing models

Figure 4.6 showed the results of applying the "perfect" endmember model to the synthetic image. The residuals for every pixel in every channel are zero, and no pixels have overflow fractions. Although we

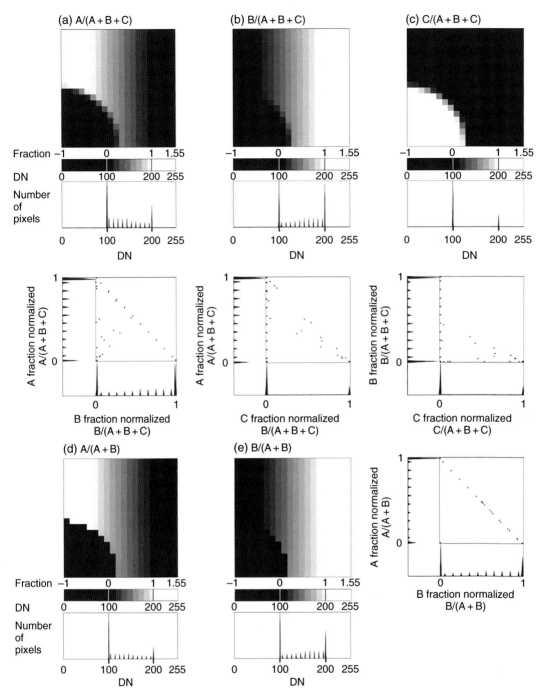

Figure 4.7. Normalized fraction images and scatter plots of A, B, and C. Endmember S has been deleted from all images. Lighter tones are higher fractions: (a) A fractions: A/(A + B + C); (b) B fractions: B/(A + B + C); (c) C fractions: C/(A + B + C); (d) A fractions: A/(A + B); (e) B fractions: B/(A + B). The data for scatter plots for a–c fall within right triangles where the endmember that is not plotted lies at the 90-degree angle. The data for scatter plot d lie along a line, and the missing endmember (C) is a single point at zero fractions of A and B. When only shade is deleted, normalized fraction images correctly show the distributions of A, B, and C in the original synthetic data (Figure 3.2). When endmembers other than shade are deleted, the images show relative proportions, but not true fractions.

Table 4.2. *Displays of fraction images.*

Number of endmembers displayed per image	Normalization	Display	Interpretation
1	No	Gray tones or color contours.	Other fractions ambiguous if model has $\geq$ 3 endmembers.
2	Yes	Gray tones or color contours.	Fractions complement.
3	Yes	Two endmembers, contoured and modulated by shade.	Fractions + topography.
3	Yes	Color composite, shade not included.	Fractions; image appears topographically "flat."
4	Yes	Three-endmember color composite modulated by shade.	Fractions + topography.

cannot expect to produce perfect models for "real" remote-sensing images, we can evaluate and improve models of real images by interpreting residuals and fraction overflow. A limitation of SMA is that we have no analytical means to tell whether the fractions that fall between 0 and 1 are correct. Evaluating such fractions depends on knowledge of the image context and other "outside" information. In this section we use the synthetic image to illustrate how "errors" in mixture models are expressed by residuals and fraction overflow, and how this information can lead to improved models.

A limitation of SMA

There is no analytical way to tell whether the fractions that fall between 0 and 1 are correct.

Let us use the synthetic image to examine what happens when an endmember that should be present is left out of a mixing model. Figure 4.8 shows a scatter plot of the data in the plane of channels 1 and 2. (A more complete "view" of the data space can be gained from Figure 4.5.) All vectors lie within the convex hull defined by endmembers A, B, C, and S. When endmember C is deleted from the model, a new convex hull is defined by A, B and S that does not include those pixels having spectrum C and its mixtures with other endmembers. When modeled by [ABS], C is defined by a negative fraction of S. When endmember B is deleted, the model becomes [ACS], and B is defined by a negative fraction of A.

Figure 4.9 shows the images that result from applying the [ABS] model. The fraction images do not agree with the same fraction images in Figure 4.6, which result from applying the perfect, complete [ABCS] model. Instead, when C is missing from the model, the pixels that contain spectrum C (lower left quadrant of the image) are incorrectly modeled as consisting of A, B, and S. The pixel spectra that do not fit the

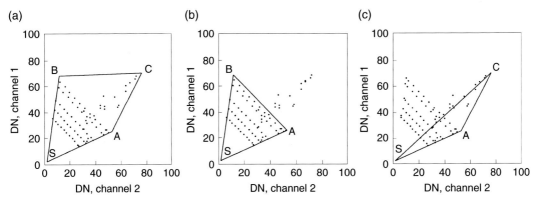

Figure 4.8. Scatter plots of fraction overflow for channels 1 and 2 of the synthetic image. Data points are DN values. (a) All pixel spectra fall within the convex hull defined by [ABCS]; (b) endmember C is deleted from the mixing model, leaving some data points outside the convex hull [ABS]; (c) endmember B is deleted from the mixing model, leaving some data points outside the convex hull [ACS]. Points that lie outside the polygons defined by the mixing models [ABS] and [ACS] have overflow fractions. For example, in scatter plot b, S has a fraction of 1 at S, and S has a fraction of 0 at the AB line, and the points to the upper right of the AB line have negative fractions of S.

[ABS] model well are revealed as having high RMS residuals and fraction overflow. Notice that the vector C has a negative overflow of S, which is consistent with Figure 4.8. Even if we did not know how the image was composed, it is easy to see that something is quite wrong with the [ABS] model.

Residuals

The resulting fraction value for each endmember is a best fit – in effect, a compromise. When a "compromise" fraction is multiplied by the DN of each endmember, and these products are combined (as in Table 4.1), the resulting DN values in each channel may not agree with the image DN value. The difference is the residual, r. In our examples, where we leave out endmembers that should be present to model the image correctly, we cannot invert Equation (4.2) to model correctly the channel DNs for those pixels that contain the spectra of the missing endmember. Images of residuals call attention to those pixels that for one reason or another are not well modeled, and spatial patterns of residuals provide important clues that can lead to improvement of a model.

It can be difficult to interpret channel residuals individually, especially when there are only a few channels. Although Equation (4.2) produces a best fit over all channels, it does not emphasize the role of any one channel, even if that channel is the sole source of residual error. An important exception occurs in the case of hyperspectral data. When fraction fit is measured over many channels the weighting of any one channel is relatively small, and residual error in any one channel can stand out as being anomalous. In hyperspectral data, anomalous single-channel residuals usually signal a defective channel. However, when anomalous residuals occur over a few adjacent channels they may reveal resonance bands in surface materials.

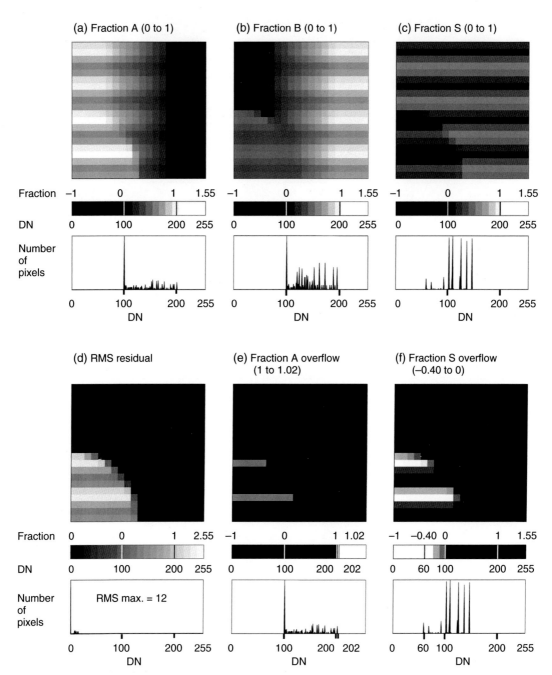

Figure 4.9. Synthetic-image mixing model [ABS] and RMS residual after deliberately leaving out endmember C. (a) Fraction image of A; (b) fraction image of B; (c) fraction image of S; (d) RMS residual image. Residuals are high in the area occupied by pixels of spectrum C; (e) A-fraction positive overflow (light tones); (f) S-fraction negative overflow (light tones). Fraction overflow occurs for pixels having spectrum C. Least-shaded pixels of C have the following fractions: A = 1.02; B = 0.38; S = −0.40.

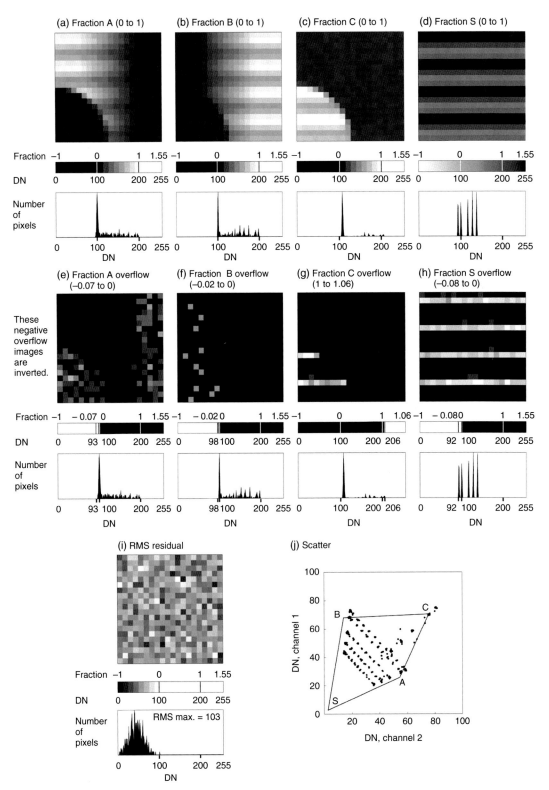

Figure 4.10. Synthetic image with 5% random noise modeled by [ABCS]. (a)–(d) Fraction images; (e)–(h) overflow fractions; (i) RMS residual; (j) scatter plot for channels 1 and 2. Random pattern of noise is expressed in fraction overflow and in residuals. Pixel DNs extend outside convex hull ABCS. In the fraction-overflow images, both the positive and negative fractions are displayed as light tones.

The critique of the [ABS] model leads directly to a remedy that allows us to model all of the image correctly. (Recall, though, that if part of an image is not of interest to us we can ignore it or mask it off, and concentrate on the rest of the image that is well modeled.) To improve the model we look for another endmember in the image having the spectrum of the non-shaded part of the lower left area. This, of course, gives spectrum C. Alternatively, we could examine the scatter plot for the pixel spectrum that has the highest negative value of S and that produces a new convex hull that encompasses the pixels in the lower left of the image. Again, we are led to the missing endmember, C.

In the above example, pixels having residuals and fraction overflow have spatial patterns that are determined by the choice of endmembers. Noise or artifacts in an image also can produce characteristic patterns of residuals and fraction overflow. Figure 4.10 illustrates what happens to the [ABCS] model when 5 % random noise is added to the synthetic image. The fraction images for each endmember resemble those in the "perfect" model of Figure 4.6, but images of residuals and fraction overflow reveal a random pattern that looks, well, like noise. The scatter plot for channels 1 and 2 shows that the introduction of noise causes the DNs of some pixels to spill outside the convex hull ABCS, thereby creating overflow fractions. We made a difference image (not shown) between the fractions measured with and without noise, which confirms that noise also has affected fractions within the convex hull.

4.4 Finding endmembers

4.4.1 Where endmembers come from

Strictly defined, an endmember is just a constituent part of a spectral mixture. But this definition is not complete for our purposes in remote sensing. Endmembers also must help us to interpret spectral images. We are especially interested in the spatial relationships among endmembers that represent materials on the ground that we understand intuitively from field observations. This means that we need to select endmembers that make sense for image interpretation, in addition to their meeting the criteria of fraction and fit for the mixing equation. Sometimes we can easily select endmembers, because we already know what materials are on the ground. This is especially true for scenes that we study over periods of time, for example when monitoring environmental change. We need to experiment to find useful endmembers whenever scenes are unfamiliar, or, when we need to extract special information from images.

There are two main places to find endmembers: spectral images themselves, and collections of laboratory and/or field spectra.

> **Endmembers**
> *Image endmembers* do not need to be calibrated to the image.
> *Reference endmembers* must be calibrated to the image.

Endmembers derived from images are called "image endmembers," whereas endmembers selected from laboratory and field spectra are called "reference endmembers." By far the most convenient place to select endmembers is within the image being studied. The reason for the convenience is simple. Endmembers taken from image spectra can be used without calibration (Section 3.2). To use laboratory and field spectra as endmembers, each image must be calibrated and converted to reflectance.

Examples from the synthetic image

One way to illustrate how to find and use image and reference end-members is to imagine how we would analyze the synthetic image if we did not know the spectra in Figure 3.2 or how the image was constructed. We start by displaying images of the individual channels and making color composites using combinations of three channels. Using photo interpretation we soon discover that areas A, B, and C are relatively homogeneous spectrally, whereas the areas separating them are not. When we display pixel spectra from image areas A, B, and C we see that spectral homogeneity is modulated by a horizontal pattern of light–dark variability, suggesting a pattern of shading and shadow. We would like to sample "pure" spectra from areas A, B, and C, but, without knowledge of the construction of the synthetic image, we have no way of knowing whether or not some fraction of shade is present. Contamination by shade can be minimized, however, by sampling the lightest pixels. We designate the sampled spectra A_I, B_I, and C_I as a reminder that they are derived from the image.

Defining an image spectrum for shade presents a special problem, because examination of the darkest pixels reveals that they all are dark versions of A_I, B_I, and C_I. There are no pixels where the DNs in all channels are near zero, such as in a recognizable shadow. So we suspect that shade is present, but we have no "pure" pixels of dark shadows to define it. If we select a mixed spectrum as an endmember we know that this will produce incorrect fractions. One approach is to use the scatter plots such as the ones in Figure 4.8 to define mixing lines between shade and other endmembers. For example, lines drawn from A, B, and C through data clusters that are aligned with the origin, intersect at values of 2 DN. Although approximate, this method yields a realistic spectrum for S in the synthetic image. Another approach is to make a good guess at the spectrum of shade. We know that a laboratory spectrum of pure shadow is zero DNs in all channels, and that values for shade in terrestrial images typically follow wavelength patterns that mirror atmospheric scattering (Chapter 3). For the example below we will introduce a "reference" endmember for shade (S_R) having a DN value of 0 in each channel.

We now have four candidate endmembers: image endmembers A_I, B_I, and C_I and a reference endmember, S_R. Although all are possible endmembers this does not mean we should use all of them immediately in a mixing model; in fact, there are good reasons not to. Recall that we want to keep the number of endmembers to a minimum, because the solutions to Equation (4.2) become more unstable as the number of endmembers increases, even when there are many channels.

Sequential modeling

We start by testing models that use just two endmembers at a time. From the horizontal stripes in the single-channel images, we suspect that shade occurs throughout the image. Applying Equation (4.2), we find that the two-endmember models, $[A_I S_R]$, $[B_I S_R]$, and $[C_I S_R]$, have small RMS residuals and fraction-overflow in the image areas of A, B, and C, respectively. By testing the three models sequentially we can find all of the potential image endmembers. As an example, Figure 4.11 shows the results of the model $[A_I S_R]$. We also can find all endmembers by applying either of the three-endmember models, $[A_I B_I S_R]$ or $[A_I C_I S_R]$ as was shown with [ABS] in Figure 4.9.

When we subtract the fraction images that are produced by [ABCS] from the counterparts produced by $[A_I B_I C_I S_R]$ we can measure the fraction error for each pixel spectrum that is attributable to our approximating the endmembers rather than using the "perfect" ones. For the least-shaded pixels of A, B, and C the difference for each endmember is 0.03; thus, by using image endmembers and guessing at the spectrum of shade we have overestimated the fractions of A, B, and C respectively by 3%. In this example, when the fractions of A, B, and C are re-normalized, leaving out shade, the relative proportions of A, B, and C are the same as in the "perfect" model. Unfortunately, when working with real, remote-sensing images there is no perfect model for comparison. But there also is no free lunch.

Endmembers depend on spatial scale

Image endmembers, although convenient because they do not require calibration, do not always work in mixing models. For image endmembers to work well, there must be a good match between the scale of a pixel footprint on the ground and the scale at which meaningful materials occur in relatively pure form on the ground. In the best case, we can use an image endmember if an image has at least a few pixels entirely occupied by "pure" surface materials that we understand. Just selecting a spectrally distinct part of an image (such as C in the synthetic image) does not help unless the objective is to model a few mixed

Spatial scale of pure pixels

The best image endmembers are the spectra of pixels of pure, rather than mixed, materials. Homogeneous areas such as lakes or agricultural fields are *not* good candidates for endmembers, because they only mix with other spectra at sharp boundaries.

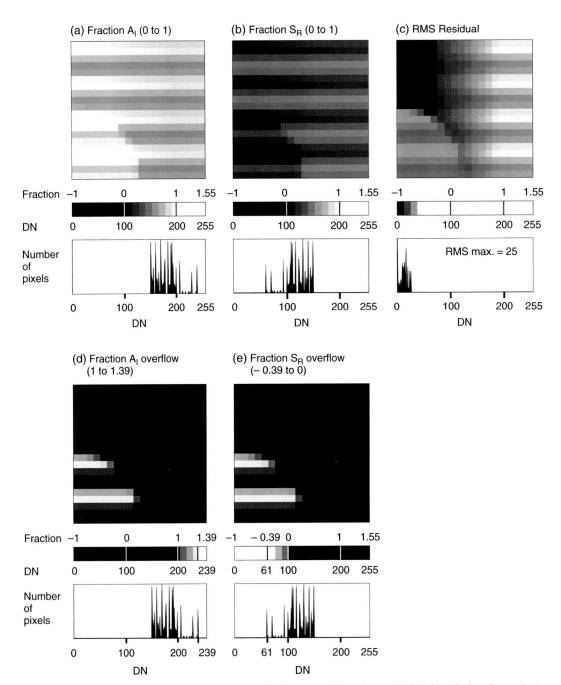

Figure 4.11. Fraction images and RMS residual image for the two-endmember model [$A_I S_R$] applied to the synthetic image: A_I is an endmember derived from the lightest pixels in area A in the image; S_R is a reference endmember, where DN = 0 in all channels. (a) Fraction image of A_I. (b) Fraction image of S_R. (c) RMS residual image. Dark pixels have low residuals. Only the upper left quadrant (area A) is well-modeled. The largest RMS residuals are in area B. (d) Image of fraction overflow of A_I. (e) Image of fraction overflow of S_R. The largest fraction overflow is in area C. New endmembers need to be defined in areas of B and C.

boundary pixels. We also would like to avoid selecting pixel spectra that we do not understand, or that are mixtures of more fundamental spectra that we do understand.

The construction of the synthetic image itself provides an example of how scaling affects image endmembers. The synthetic image originally was made as a 200×200 pixel array in which some pixels of A, B, and C were "pure," that is, uncontaminated by shade. When we re-sampled the original array to make a 20×20 image, all pixels incorporated some shade, which is the main reason why the model $[A_I B_I C_I S_R]$ is not "perfect." If we had continued to re-sample pixels, for example by making a 10×10 pixel image, each pixel would become even more mixed. The extreme example, of course, is when further re-sampling leaves us with one pixel that is a mixture of all components, and that yields no useful image endmembers.

How many endmembers?

A common mistake made by image analysts is to collect a large number of candidate "endmembers" from an image, and to use them all at once in a mixing model. There are serious problems with this approach. If we try to introduce too many endmembers, usually more than four or five, we may end up paying a price in fraction accuracy and model usefulness, regardless of the number of channels available. Mathematically, the mixing equation is a sensitive beast. It is easily upset. We cannot just keep adding endmembers or else the solutions become unstable. Spatial patterns in the fraction images lose their coherence, because fraction values fluctuate sensitively in response to small changes in channel DNs. As the number of endmembers grows, there is increasing likelihood that any one endmember will mimic some mixture of the others. In effect, it becomes increasingly difficult to detect each endmember against the spectral background of the other endmembers, which, of course, contradicts the definition of "endmember" in the first place.

One way to minimize the number of endmembers is to avoid selecting spectra that do not mix with any other endmembers in a given image. The purpose of mixture analysis is to model mixtures, not necessarily to model everything in an image; therefore, we can let spectra that do not mix with the selected endmembers to be expressed in an RMS residual image (Figure 4.9). We also want to avoid spectra that already are mixtures – that is, they are better described as mixtures of other, more interesting spectra. Mixtures of mixtures are difficult to interpret in an image context. We recommend starting with combinations of two endmembers, as we did with the synthetic-image example, to see if there are any parts of the image that are well modeled as judged by RMS residuals and fraction overflow. Build from there to three or

> **A common mistake**
> Analysts often try to use too many endmembers in spectral mixture modeling, perhaps because they are acutely aware of the complexity of many physical landscapes and try to match it in their model. But in many cases a few endmembers describe almost all of the spectral content, and more endmembers only destabilize the model.

maybe four endmembers if there are enough channels. Visually inspect the fraction images to make sure that they make sense.

It may seem counter-intuitive that parsimony in the number of endmembers is essential for successful mixing models. After all, we know that the real world on the ground can be spectrally complex. One reason that images of natural scenes often can be modeled well by only a few spectrally dominant endmembers is that many of the other potential endmembers have small fractions; therefore, the fractions of the dominant endmembers are minimally affected by the other components. Of course, there are natural scenes where there is too much spectral complexity to be usefully modeled by a few endmembers at any scale. Expanding the field of view on any scene eventually will encompass more and more variety (Section 2.2.1) with the apparent need for more endmembers. Closer inspection, though, may reveal that not all endmembers mix everywhere in the image, as is seen with endmember C in the synthetic image. Indeed, some of the most interesting small targets may mix spectrally only very locally (Chapter 6).

> **Fewest possible endmembers**
>
> The optimum spectral models are parsimonious. They use the fewest possible endmembers to explain as much of the spectral data as possible.

Number of endmembers

We have shown in this chapter how to determine the number of endmembers by starting with a minimum number (two or three) and adding endmembers one by one after evaluating RMS residuals and overflow fractions until a good model is achieved. It also is possible to estimate the number of possible endmembers at the start using the data-reconnaissance tools discussed in Chapter 3. For example, by applying a PCT and inspecting the eigenvalues (variances of the PC images) we can estimate the intrinsic dimensionality of the data and the potential number of endmembers. The number of endmembers can be adjusted up or down, guided by the RMS and fraction images, and candidate endmember vectors can be tested in mixing models by iteration. Although PCT can help guide endmember selection, the reconnaissance approach is inherently empirical – not physical – and the final test will be whether the selected endmembers comprise a good model that makes sense in the image context.

Image segmentation

The alternative to using many endmembers at once is to organize images into discrete areas (segments), each of which is modeled as mixtures of a few endmembers. Not all images lend themselves readily to this approach. Some images contain so much spectral complexity that many segments involving many endmembers or variations of endmembers would be needed to construct good mixing models. A general guideline for segmenting an image for mixture analysis is to begin by setting aside parts of an image that are not of interest. This can be done by working on subsets of an image, masking unwanted areas or simply ignoring areas where high RMS residuals or fraction overflow are unacceptable. Even within a selected portion of a larger image there may be certain sub-areas that warrant the most attention for further analysis.

When we use different sets of endmembers to model individual image segments, we are potentially able to apply a larger number of endmembers to the overall image than could be accommodated in any single model. However, we then need to integrate the different mixture models to create a mosaic for display and photo interpretation. It turns out that the segmentation step is relatively easy to accomplish, like taking apart a watch. Putting the pieces back together again can be more of a challenge. Let us consider how segments can differ from one another.

Case 1. Different segments of an image are modeled by a few, but not all, of the identified endmembers. This avoids the problem of using all of the endmembers to model the whole image, and having too many endmembers to solve Equation (4.2). It also minimizes the errors that arise when we apply any endmember to parts of an image where it is not actually present. If we proceed area-by-area, and apply combinations of two, three, or possibly more endmembers, the result is a collection of different models. These models have to be reconciled for display and photo interpretation.

Case 2. Each endmember is allowed to vary slightly in its DN values from one image segment to another. (Endmember variability is discussed in Section 4.1.5.) For example, we could select one spectrum as a "primary" endmember, because it is broadly representative of that type of material. We might know, however, that this primary endmember is not the best one everywhere in the scene, so we define two other "flavors" of the endmember, making a total of three to test on each image segment. We can do the same for each endmember until we have defined subsets consisting of a primary endmember and its flavors. The objective is to find combinations from among the subsets that improve the fit of the mixing models, using an iterative analysis. This approach can easily get out of hand computationally if too many endmember flavors are involved; however, consideration of context often can help us to decide on a small number of appropriate spectra.

We can integrate the results from these multiple mixing models by a bookkeeping step that considers the fractions of flavors as fractions of the primary endmember. In this way, endmember variability is taken into account for the purposes of modeling, but it is ignored in the display of the fraction image that is labeled according to the name of the primary endmember. However, the information on the flavors is not lost, because we can make an image of any individual flavor and examine its spatial distribution.

Case 3. One or more endmembers differ fundamentally from segment to segment. In this case we have little choice except to display separate sets of fraction images for each segment. It can be relatively easy to assemble a few modeled segments into a larger image, providing that it is clear what the endmembers are in each segment. It is more difficult to combine many modeled segments. Having a well-modeled image that consists of many segments presents an organizational problem that is akin to doing field work. How the data are displayed will depend on what type of thematic information is of interest (Chapter 1).

Case 4. Endmembers are allowed to vary from pixel to pixel. This case includes the above three, but allows the size of the segments to be as small as individual pixels (Roberts *et al.*, 1998a; 1998c). Although computationally expensive, this approach has the advantage of applying the best-fitting combination of many types of endmembers throughout an image. As in the previous cases, the multiple different models need to be assembled in a way that facilitates photo interpretation.

4.4.2 Finding image endmembers

So far, the methods that we have discussed for selecting endmembers have relied heavily on image context. It would be useful if there were ways to find endmembers without having to do photo interpretation, because this would open the door to automated SMA using just the DN values for each pixel. Considerable work has gone into the quest for analytical methods for endmember selection that bypass image interpretation, with some, but by no means universal, success. Recall that to be a useful endmember a spectrum must mix with one or more other spectra and must help us to interpret a spectral image. Even endmembers chosen analytically must conform to this definition, and, therefore, eventually must pass the test of relevance in a photo-interpretive context.

One approach that has been extensively explored seeks to find image endmembers by defining the convex hull of the data volume (e.g., Boardman, 1993). This method can be a useful tool for screening some images or segments of images for candidate image endmembers. The basic idea can be illustrated using the synthetic image. For example, the data volume sampled in two and three dimensions in Figure 4.5 is entirely encompassed by the convex hull ABCS. All of the other data points can be described as mixtures of various combinations of A, B, C, and S, without making any reference to the spatial arrangement of the pixels in the image. When these spectra are used as endmembers in a mixing model, and the results are checked against the image, we see that, in fact, these endmembers produce no RMS residual and no overflow fractions, and they make sense for photo interpretation.

Real images, of course, typically have much messier data volumes than our synthetic one, and herein lie some of the potential problems with deriving endmembers from the data volume alone. Because a convex hull is defined by data points without regard for their meaning in the image, it often consists of noisy pixels or defects that are located outside the main data population. Such "endmembers" clearly fail the test of relevance for photo interpretation. Even when defective pixels are eliminated, spectra on the convex hull often consist of unusual scene components that do not mix with other scene components.

In the two-channel scatter plot of the Seattle Landsat TM image (Figure 3.4), a strictly mathematical selection of extreme pixels to define a convex hull produces at least six "endmembers" just in the Band 3, Band 4 plane alone. Most of these candidate endmembers consist of one or a few, isolated pixels that may be difficult to interpret. For example, in the scatter plot of Figure 3.4 there is a high-DN "tail" that extends well beyond the main population, and it consists mainly of unusually reflective buildings and other isolated urban features. The spectra of these extremely light urban pixels may be "pure" and they may be on the convex hull, but they are not good choices to represent "urban" (hereafter referred to as URB) as a basic spectral component, because they do not provide an intuitive basis for understanding the proportions of urban materials elsewhere in the image.

A more intuitive endmember would be an average spectrum of the downtown center of Seattle. Let us examine Figure 4.12 to understand this important point. The figure is derived from the data of the scatter plot in Figure 3.4; URB is the average spectrum of the urban core, URB_1 is the average spectrum of a lighter urban area, and URB_2 is an extreme pixel of the Sun-facing side of the roof of the Kingdome, the former sports stadium. All three points fall on the mixing line between shade (S) and URB. A similar situation exists for potential endmembers for green vegetation (GV); GV is an average spectrum for deciduous trees, GV_1 is an average for a golf course, and GV_2 is a single, extreme pixel of unknown vegetation composition. Mixtures of GV and URB fall along the GV–URB line, and are easy to interpret, because we can visualize the basic spectral components that occur together at a sub-pixel scale to form tree-lined streets or suburban yards. Most of the pixels in the Seattle image lie within the GV–S–URB triangle. (The other endmember, non-photosynthetic vegetation (NPV), is not distinguished in the TM Band 3, Band 4 plane, but it becomes clear if the data for Band 5 are plotted.)

In Section 2.2.1 we pointed out that the contrast between spectra consists of two components, the spectral angle (SA) and the normalized difference in spectral length (SL). (The spectral angle also is used for

> **The best endmembers**
> Well-chosen endmembers not only represent materials found in the scene, but provide an intuitive basis for understanding and describing the information in the image.

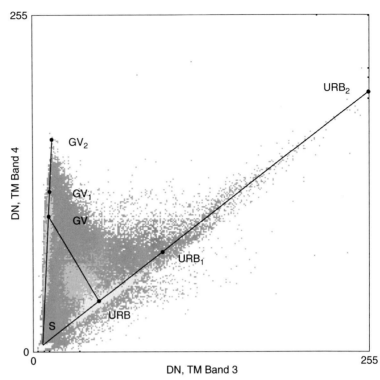

Figure 4.12. Scatter plot of image endmembers, green vegetation (GV), urban (URB) and shade (S) in the plane of Landsat TM Bands 3 and 4 of the 1992 Landsat image of Seattle, Washington; GV is derived from closed-canopy deciduous trees. GV_1 is an average of lighter-tone green vegetation on a golf course; GV_2 is an isolated pixel of unknown green vegetation or of noise; URB is an average spectrum from downtown Seattle; URB_1 is an average of light-tone urban material; URB_2 is a single pixel of the Sun-facing side of the roof of the Kingdome (a former sports stadium). Shade is an average spectrum from deep water of Puget Sound. Mixtures of "pure" GV and shade lie along the line S–GV, and mixtures of "pure" URB and shade lie along the line S–URB. Mixtures of GV and URB are along the line GV–URB. The maximum spectral contrast between GV and URB in these channels is defined by the spectral angle GV–S–URB. See also Figure 3.4 for an explanation of the scatter plot. Fraction images for these endmembers are in Figures 5.1 and 5.3.

Endmembers and spectral angle

Spectral angle contrast is the most important single characteristic of useful end-members besides image context.

the spectral-angle classifier (Section 3.4.2)). In Figure 4.12 the most important factor in selecting endmembers is the angle GV–S–URB, and not the spectral length of GV or URB from S. For the Seattle sub-image, the angle GV–S–URB in the TM Band 3, Band 4 plane encompasses essentially all of the data. Any smaller angle would not include the most "pure" endmembers of both vegetation and urban pixels. Candidate endmembers along the mixing lines with S differ only in SL. However, SL varies with illumination of the scene, and the SL component of contrast is lost when we remove shade by normalizing in order to measure the "true" fractions of the other endmembers (Section 4.3.2). Therefore, it is not the convex hull per se that is important in selecting endmembers; instead, it is the spectral angle in the data space between potential non-shade endmembers.

If we select URB and GV as the endmembers in Figure 4.12, even though they are not on the convex hull, we have the huge advantage of establishing an intuitive framework for interpreting mixing in the image. Then, pixels that contain the other candidate endmembers such as URB_1 and URB_2 and GV_1 and GV_2 have overflow fractions > 1 of the selected endmembers, and overflow fractions < 0 of shade. Overflow fractions are easily interpreted as lighter "flavors" of the selected endmembers, and the overflow is eliminated in the process of normalization to remove shade (Sections 4.3.2 and 5.1.2).

There also is a problem with selecting endmembers from the convex hull if the spatial scale and the context are such that all of the pixels in the data volume actually are mixtures of more fundamental endmembers. We can see this in scatter plots having compact data volumes and where many data points are required to define convex hulls. (Visualize, for example, nearly spherical data volumes in three-dimensional space.) It may be analytically correct to define ten, twenty, or more spectra that define a compact convex hull, but this many spectra will produce an unstable solution, and they are unlikely to make sense in the image context. It is useful to keep in mind that there is no reason, a priori, why an analytically selected spectrum should make a good endmember. If it is difficult or impossible to find useful image endmembers, we may need to expand the search to include virtual endmembers or reference endmembers.

> **Mathematically correct endmembers**
>
> Not all analytically correct endmembers make sense for interpreting an image.

Virtual endmembers

Sometimes, endmembers are implied by the image data, but do not actually exist as image spectra. We call such spectra "virtual endmembers." A simplistic example of this idea is illustrated in Figure 4.3. Consider that spectra A and B in Figure 4.3a are measured image spectra. The line AB defines a mixing line that is populated by other spectra that are potential mixtures of A and B. The angular separation of A and B defines the spectral contrast, where the angle is measured from the origin (Section 2.1.2 and Figure 4.12). Spectrum D, actually might be a better endmember to describe the image (greater spectral contrast); however, because it does not occur in the image, we would be unaware of its existence. If we project the AB line we define the locus of spectra, including D, that we can test as virtual endmembers using a trial-and-error approach. Alternatively, using a calibrated image, we could use the same analysis to look for laboratory or field spectra that fit the extension of the AB line and make contextual sense. Finding virtual endmembers in a multi-channel data set is far more complicated than the simple two-channel, two-endmember case in Figure 4.3. Analytical methods exist for finding convex hulls and exploring for virtual endmembers, and are among the several tools available for selecting endmembers (e.g., Boardman 1993; Tompkins *et al.*, 1997).

4.4.3 Finding reference endmembers

For many images, no pixels can be found that make optimal image endmembers. We already have seen an example of this when modeling the synthetic image, where we used shade as a reference endmember. We know intuitively that the larger the pixel footprint on the ground the more difficult it is to find pixels of "pure" materials: 1-km pixels are less likely to yield useful image endmembers than 10-m pixels. The problem is not simply one of pixel dimensions, however, because we know that pixel footprints of any size encompass mixtures of more fundamental spectral components, all the way down to the scale of a few molecules. In keeping with our objective of using spectral images for photo interpretation, we are looking for endmembers that occur on a spatial scale that we understand. The spectra of natural materials are

Reference endmembers and spatial scale

How do we characterize a landscape using a few spectra that were measured at the centimeter scale?

best understood at the scales of laboratory measurements, typically in the range of centimeters to millimeters. Laboratory spectra of samples that have been well characterized physically and chemically are the foundation of our understanding of reflectance and emittance spectra (Chapter 2). It is laboratory spectra, and to a lesser extent, field spectra, that we turn to when we need reference spectra for mixture models. Reference endmembers have received much less attention in SMA than image endmembers. One obvious barrier to using reference endmembers is the need to calibrate images to reflectance or emittance. A more subtle problem is that of learning how to describe the land surface at remote-sensing scales using spectra measured at centimeter to millimeter scales.

Calibration

Many image analysts do not even attempt calibration. The process is widely viewed as too formidable, and the results can be difficult to validate. Our focus here, though, is not with the calibration process itself, but with the effects of calibration error, or failure to calibrate, on mixture models. Because the output of a mixture model is the fraction of each endmember for each pixel, errors in the input data, including calibration errors, will be expressed as fraction errors. If we have made a good choice of endmembers and the fit of the model is good, we arrive at the question of how much fraction error we are willing to tolerate for a given image or segment of an image. For many applications it is sufficient to know the relative proportions of materials on the ground. If we do not require high fraction accuracy we can apply an approximate calibration, use reference endmembers, and test the results in an image context, employing the usual iterative approach.

Representative endmembers

Experienced field observers are keenly aware that a few centimeter-sized samples of materials usually cannot fully describe the natural complexity seen on the ground in a small area, to say nothing of a large region such as that covered by a satellite image. Segmentation of an image can narrow the spectral variability, but, even then, for reference endmembers to work in a mixing model, reference spectra must represent broader suites of materials.

Because we tend to think spatially and not spectrally, it often is difficult for us as field observers to visualize reference endmembers that are representative of the spectrally dominant materials on the ground. For example, everyone knows intuitively what a forest looks like spatially, but it is not immediately obvious what the spectrally dominant components are. Although forests have reflectance spectra that superficially resemble laboratory spectra of green leaves and chlorophyll, reference spectra of "forests" do not exist, for the reason that we cannot measure such large areas under repeatable, controlled conditions as we do in the laboratory for small samples (Chapter 2). From experience it has been found that many types of forests can be modeled as mixtures of the spectra of green leaves, bark, shade and, for open canopies, substrate such as litter and soil. Each of these spectra represents a category of spectra that can be quite variable, because each forest component, such as green leaves or bark, consists of rich spectral detail on the scale of laboratory measurements. We cannot model such fine detail at remote-sensing scales; nor is this detail always needed for image interpretation. For example, even if we could un-mix seven different types of bark from a forest spectrum it might not be clear how to use this information. We can, however, use the relative proportions of, say, green leaves and bark to give us information on annual variations in deciduous forests and on composition.

Collections of laboratory (or field) spectra are the sources for endmembers that represent categories of materials. In some cases it may be advantageous to average the spectra in a category; otherwise, individual spectra can be selected. The main objective of modeling images or segments of images with representative endmembers is to describe the principal causes of spectral variation; therefore, these endmembers must comprise a significant fraction of pixels of interest. To facilitate photo interpretation, these endmembers should represent basic spectral components as defined in Chapter 2 and Table 2.1. By characterizing this spectral "background" we set the stage for finding other spectra that may only occur in just a few pixels or in small amounts in many pixels. In Chapter 6 we discuss strategies for finding spectral "targets" in a spectrally cluttered background, and why it is important to define the main sources of spectral variation in an image as a first step.

Endmembers vary with the spectral window
Because endmembers are spectrally distinctive, the sets of endmembers used to describe a scene may differ depending on the spectral window through which the image is acquired. Green leaves and bark are distinctive in the VIS-NIR, but much less so in the TIR.

Reference endmembers improve with time
As field experience accumulates in a particular area, it is possible to refine the choices of reference endmembers. Thus, over time, we can

improve the reference endmembers that are applied to any one segment of an image. This can happen because at field or laboratory scales we can continue to make the measurements that define spectrally dominant materials. We also can see which materials change spectrally over time, how they change, and what processes are involved. As an example, spectral changes in an image of a deciduous forest can be understood by ground measurements of seasonal changes from green to yellow to brown leaves. Each time a new image is made of the same field area we can use some or all of the reference endmembers that we have used before, because the endmembers themselves tend to remain constant even though their proportions may vary in response to changing conditions. Reference endmembers, therefore, are an important tool for monitoring changes on the ground surface. We discuss this topic further in Chapter 8 (Section 8.2).

Universal endmembers and standard models

Not only do some reference endmembers stay the same over time, some are the same, or similar, from place to place. We use the term "universal endmembers" for those that can be exported from one image to another. Universal endmembers are derived from basic spectral components (Table 2.1). Sets of universal endmembers can be assembled to create "standard" models for certain types of scenes. For example, a standard model for the Moon consists of the following set of universal endmembers: highlands, highland light craters, maria, and maria bright craters. A standard VIS-NIR model for forests consists of green vegetation, non-photosynthetic vegetation, and shade.

We tested a standard model on the TM images of Amazon and Washington State forests (reference images on the Web), and found that the same reference endmembers (GV, NPV, and shade) worked surprisingly well for both scenes, in spite of the entirely different compositions of the forests. The explanation for this non-intuitive result is that at the scale of 30-m pixels, the largest spectral variability is in the proportions of the basic spectral components, rather than in the spectral "flavors" within the components. When viewing forests in the field, though, our attention is drawn to the fine-scale spectral variability of leaves, bark, flowers, etc., but these details are lost at coarser scales because of spectral mixing.

A standard model also turns out to be a good starting point for analysis of a spectral image. We can start by making fraction images in the usual way, along with images of RMS residuals and fraction overflows. If a model fits fairly well, it may suffice for photo interpretation. However, the most interesting information may be found in the ways that a standard model does *not* fit. From this perspective, a standard model serves as a bench mark against which to measure unusual spectral features or to

Constant endmembers

Reference endmembers tend to remain constant among images of similar landscapes. It is their proportions that vary within an image and from image to image over time.

Standard models

Standard models are bench marks. We can use them to detect unusual materials or changes in a scene.

compare images in a time series. Another potential use is to allow different analysts working on different images to compare results.

4.5 Calibration feedback

So far we have focused on what happens when a mixing model is incorrectly formulated. However, residuals and fraction overflow also can be highly sensitive to those characteristics of an image that do not contain information about the ground surface and that are not related to the choices of endmembers. We already showed in Figure 4.10 that the introduction of random noise to the synthetic image results in a random pattern of RMS residuals. The image pattern of noise is diagnostic in the sense that we can use it to infer the presence of random DN variability usually associated with instrumental/systems response. The image patterns caused by other measurement defects such as dropped lines, striping, and misregistration usually can be easily recognized, particularly in RMS-residual images. Image defects will affect the accuracy of endmember fractions. In some cases, such as misregistration, it may be possible to correct the problem and thereby improve fraction accuracy. We at least want to be aware of such problems, even if we cannot correct them.

Endmember fractions will be distorted by channel offsets and gains that are produced by various atmospheric, instrumental, and other factors. These problems can be circumvented by using image endmembers, but calibration becomes essential when using reference endmembers. Interestingly, reference endmembers, because they are free of the usual atmospheric and instrumental calibration problems, can be used to calibrate an image, providing that there are equivalent image endmembers. Calibration consists of solving for the combined gains and the combined offsets that are needed to adjust the fit between the image and reference endmembers. The gains and offsets then can be applied to the image (Adams *et al.*, 1986).

Consider a case where a landscape is well understood, and a set of reference endmembers has been defined that models spectral images well. Pure pixels of the endmembers need not be available as long as the endmembers are known to be correct at the sub-pixel scale. In each subsequent image of the same scene we expect the same set of reference endmembers to apply, although their proportions can change, and not all endmembers need to be present in every image.

Atmospheric correction factors, however, are expected to vary over time, and even the sensor calibration factors can change. If each subsequent image is perfectly calibrated, the reference-endmember set will produce minimal fraction overflow and residuals. However, any changes in image calibration factors will be expressed as fraction overflow and residuals. In the examples below we show how channel offsets and gains that are applied to the synthetic image influence endmember fractions and residuals, and how to spot calibration problems. We find that in some instances a correction can be applied directly to the fraction data; otherwise, we need to return to the image data and attempt to improve the calibration using standard methods (Chapter 3).

Effects of offsets

Channel offsets can arise from a variety of factors, but the most common in spectral images is atmospheric path radiance. The standard empirical method to correct path radiance is based on the premise that dark pixels, such as those associated with shadows or dark materials, should have DNs near zero in each channel (Chapter 3). We simulated the influence of path radiance on the synthetic image in Figure 4.13 by adding 20 DNs to all pixels in channel 1, 5 DNs to channel 2, 1 DN to channel 3 and 0 DN to channel 4. When unmixed using [ABCS] there is a 2.23-DN RMS residual for every pixel, which produces a uniform RMS image. In this example, the residuals differ from channel to channel, but they are uniform within each channel-residual image. The fraction images correctly represent the spatial distributions of the endmembers; however, there is significant fraction overflow, which, along with the RMS residual, alerts us that the raw endmember fractions are not correct. The lack of any spatial pattern to the RMS-residual image indicates that the problem is evenly distributed over the image, which is consistent with the simulation of a uniform, scattering atmosphere.

Mixture-analysis provides an interesting alternative way to correct for path radiance by adjusting *fractions* to zero. A path-radiance correction can be made by finding pixels that are known to have a zero fraction of one or more of the reference endmembers. In the synthetic image, for example, pixels in area C do not include A or B. However, in Figure 4.13, where path radiance is simulated, the histogram for the C-fraction image has a spike at 143 DN that is produced by the pixels of A and B and their mixtures. Because we know that these pixels have zero fractions of A and B, we know that their correct fraction in the C-fraction image is 0. (Remember that in our examples a fraction of 0 is set at 100 DN and a fraction of 1 is set at 200 DN.) When we subtract 0.43 (43 DN) from all pixels in the C-fraction image we reset the histogram to fractions between 0 and 1, and produce a corrected fraction image. The histograms for the other fraction images can be adjusted similarly. We need to add 0.56 (56 DN) to A-fraction pixels, and subtract 0.04 (4 DN) from B-fraction pixels. After correction, the least-shaded pixels

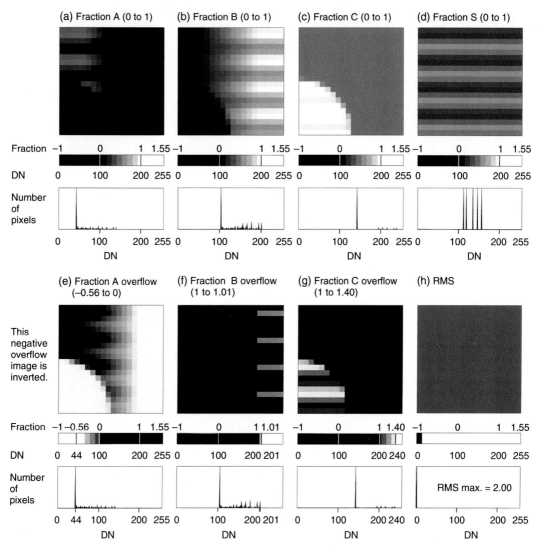

Figure 4.13. Mixing model [ABCS] applied to the synthetic image modified to simulate an optically scattering atmosphere. Each pixel in the image was changed as follows: + 20 DN in channel 1; + 5 DN in channel 2; + 1 DN in channel 3; (no change in channel 4). (a)–(d) Fraction images. (e)–(g) Fraction-overflow images for A, B, and C. (h) Root-mean-square residual image. Residuals are spatially uniform, although they differ from channel to channel. The uniform pattern of the residuals is consistent with a uniform "atmosphere" affecting all pixels. The fraction images can be corrected by applying fraction offsets: $A = + 0.56$; $B = - 0.04$; $C = - 0.43$. The offsets for each endmember are defined by selecting pixels that are known to have zero fraction of an endmember (for example, area C has zero fractions of A and B), but nonetheless show positive or negative fractions. The corrected fraction images are identical to those in Figure 4.6.

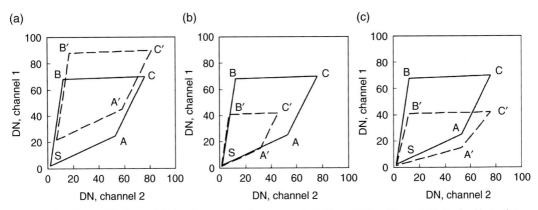

Figure 4.14. Channel 1, channel 2 plots for endmember spectra A, B, C, and S (see Figure 4.5). Dashed lines show positions of ABCS polygon after applying offset or gain (a) adding offsets of 20 DN to channel 1, and 5 DN to channel 2; (b) multiplying channel 1 and 2 DNs by 0.6; and (c) multiplying only channel 1 DN by 0.6. Offsets translate the ABCS polygon to a new position, but retain its size and shape. Equal gains for both channels shrink (or enlarge) the polygon, but retain its shape. Differential gains shift the position of the polygon and change its shape.

of A, B, and C each have respective fractions of 0.967, which is correct for the synthetic image at the scale that is determined by the 20×20 pixel format. The remaining 0.033 fraction is shade. When we normalize A + B + C without shade, the fractions converge with those in Figure 4.7, and we have removed both shade and the effects of path radiance. We also could adjust the histogram of the shade-fraction image by subtracting 0.122 (12 DN) which would set the lowest-shade pixels to 0 (100 DN). But this correction presumes that the lightest pixels have zero shade, which cannot be correct, because the lowest-shade pixels of A, B, and C have fractions of 0.0967. In fact, we need to set the lowest-shade pixels in the histogram not to 0, but instead to a fraction of 0.033 (103 DN), an offset of 0.090 (9 DN).

The effects of channel offsets can be visualized using Figure 4.14, where the data for channels 1 and 2 of the synthetic image are plotted. Channel offsets do not change the distances between A, B, C, and S, but the offsets do translate the ABCS polygon relative to its original position (Figure 4.14a).

Effects of gains
Figure 4.14b illustrates how gains affect the data of the synthetic image. By applying the same gain (0.6) to both channels, the ABCS polygon shrinks by 40% in the direction of the origin (perfect shade.) Because Equation (4.2) describes linear mixing, we also shrink all endmember fractions by 40%, and the residuals are zero. This example is not very realistic as an example of what might occur in an image of a landscape, because we would not generally expect the same gain in every channel.

However, the example illustrates that the endmember, shade, is the same as a gain that applies to all channels. For any given pixel, then, we can correct the fractions of the other endmembers if we know the uniform gain or the fraction of shade.

It is another story if we apply gains differentially to the channels. For example, Figure 4.14c shows that if only the DN of channel 1 is multiplied by 0.6, the ABCS polygon changes shape, because we have changed the identities of the spectra. There is no reason to expect that the reference endmembers will fit these new spectra, and, indeed, the result of applying the mixing model is overflow fractions and residuals. Furthermore, the presence of an uncorrected gain factor compromises our ability to identify and remove the effects of channel offsets. For example, if we apply a gain of 0.6 to channel 1 and then apply the simulated path-radiance offsets (Figure 4.13) we get large residuals and fraction overflow. If we did not have independent evidence that the reference endmembers were correct, we would not be able to tell whether there was a calibration problem or whether we had selected the wrong set of endmembers. However, if we are sure of our reference endmembers, fraction overflow and residuals would alert us to the need to re-examine our calibration. Unfortunately, the pattern of channel residuals is not a reliable indicator of which channel needs to be re-calibrated.

> **Shade as a gain factor**
> The fraction of shade in a pixel affects all other endmembers in the same way that a single calibration gain does.

Interpreting channel residuals

Calibration problems that are confined to individual channels may be evident in channel-residual images whenever there are many channels, as in hyperspectral images. However, single-channel calibration problems may not be apparent in channel residuals when there are only a few channels. For example, channel-residual values do not follow the pattern of the changes that were made to the channels to simulate a path-radiance effect. The largest residual in Figure 4.13 is in channel 3, not in channel 1, where all pixels were increased by 20 DN. When there are only a few channels, the channel residuals and overflow fractions do not necessarily tell us which channels of which endmembers are incorrect, but merely signal that something is wrong. In hyperspectral images, however, individual channels have proportionally less weighting, Equation (4.2), and channel-specific errors are more likely to be revealed by channel residuals. In a hyperspectral image, for example, a faulty detector in one channel can cause a high residual in that channel, but will have minimal expression in the RMS residual and in the fraction overflow.

Tweaking endmembers

Sometimes a standard model can be adjusted by fine-tuning the spectra of the endmembers for the whole image or the endmembers for selected image segments. When analyzing hyperspectral images, it is possible to target individual channel values of the endmembers for adjustment by examining channel-residual images. Altering universal endmembers needs to be done with considerable care, however, because they represent a group of reference spectra. If adjustment of a spectrum in one or more channels results in a new spectrum that lies outside the range of the original group, then we may simply have invented an imaginary spectrum. Furthermore, tweaking reference spectra can have unexpected results, because each change in one endmember will propagate to changes in fractions of the other endmembers in ways that are not intuitive.

Before spending much time trying to fine-tune a standard mixing model it is useful to consider how accurately we really need to know the fractions of endmembers that are representatives of variable groups of materials. A significant limitation of the mixing-model approach is that we only can test for fraction overflow. To evaluate fractions between 0 and 1 we need additional information from image context and ground validation. In reality, it often is difficult or impossible to measure the proportions of representative endmembers on the ground. On the other hand, if our goal is to produce fraction images that are useful for photo interpretation, relative proportions of representative endmembers, when seen in a spatial context, can reveal materials and processes that otherwise might not be evident (Chapter 1).

4.6 Nonlinear mixing

Nonlinear mixing in absorption bands
The Bouguer–Beer–Lambert law, Equation (2.2), reminds us that nonlinear effects are most important at wavelengths that are strongly absorbed.

The discussion in this chapter has been about linear spectral mixing as defined by Equation (4.2). We know from laboratory experiments, though, that spectral mixing can be both linear and nonlinear. As discussed in Chapter 2, nonlinear mixing occurs when radiance is modified by one material before interacting with another one. Theory and experiment demonstrate that we will get the fractions of endmembers wrong by using a linear model when spectral mixing actually is nonlinear (Mustard and Sunshine, 1999). Fraction error using linear mixing is greatest in strong absorption bands and when strongly absorbing materials are intimately mixed with translucent materials (Chapter 2). For example, Mustard et al. (1998) showed that light and dark particulate materials from lunar highlands and maria, respectively, mix in a nonlinear way. At a mare–highland boundary they found that a nonlinear model yielded ~50 % mare, whereas a linear model gave ~60 % mare material. The nonlinear model was considered more accurate based on independent lines of evidence. In the example, spectra were converted to single-scattering albedo that mixes linearly (Chapter 2). The conversion requires ancillary information (or assumptions) on parameters such as particle size, shape, and spacing. This approach, however, has been used only rarely for remote-sensing applications, and it is not a tool familiar to most image analysts. Whether a nonlinear analysis is warranted depends on how accurately we need to know fractions. For the study of lunar soils across compositional boundaries, it was considered important to measure fractions of materials with high accuracy. As a practical approach to remote sensing, a general guideline is to develop the best possible linear model first, and then modify the model by introducing nonlinear mixing if the situation warrants.

4.7 Thermal-infrared images

Spectral mixing in TIR images is more apt to be inherently nonlinear, because the various scene components can have different temperatures, and radiance is exponentially related to surface temperature by Planck's law. In VIS-NIR analysis, there is just one unknown parameter per component in a pixel–the reflectance, and we can solve the mixing equations for the same number of fractions as there are channels, allowing one degree of freedom for a measure of a residual. In the TIR, there are two unknowns, because each fraction has a temperature associated with it. Although this makes it more difficult to unmix TIR data satisfactorily, simplifications can be made to linearize thermal mixing that to lead to useful results.

Planck's law

$$R(\lambda, T) = \frac{c_1}{\pi \lambda^5} \frac{1}{\exp\left(\frac{c_2}{\lambda T}\right) - 1}$$

Planck's law predicts the TIR radiance (W m^{-2}sr^{-1}μm^{-1}) emitted at wavelength λ from an ideal surface at temperature T K (c_1 and c_2 are constants). For real surfaces, the Planck radiance is scaled by the emittance, which ranges from zero to one, like reflectance. Although radiance rises exponentially with temperature, over the relatively small temperature ranges characteristic of terrestrial landscapes, the curve is approximately linear. For example, for scanners with a noise level (NEΔT) of 0.3 K, the temperature range must exceed ~25 K for the nonlinearity to be detected.

One simplification is to assume that there are only two significant endmembers, and that they are at different temperatures. If we have two or more measurements at different wavelengths, and we assume that we know the emissivity, and perhaps even the temperature, of one material (such as snow) and the emissivity of another material (such as forest or rock), we can calculate the mixing ratio of the two endmembers. In the

Figure 4.15. Mixing lines and synthetic radiance data pairs for three endmembers (squares, triangles, and diamonds) at various temperatures from 300 to 340 K, measured with random "noise" in TIR channels i and j. The projections of lines fitted to the data intersect at a low value analogous to the "shade" endmember in VIS-NIR data, and called "virtual cold."

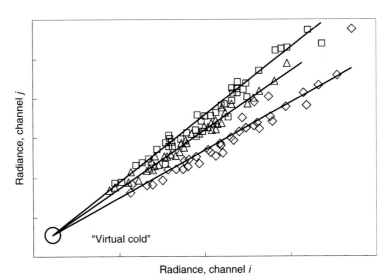

second approach, it is assumed that the scene within a single pixel is isothermal, and that the scene components are described by linear mixing (Gillespie, 1992). In one variation of this approach, only the mixing fractions of the scene components are sought. The effective surface temperature is estimated and used to calculate the mixed "emittance" of the pixel. From here, the approach is just like unmixing of VIS-NIR data, with the exception that there is no "shade" endmember – all the endmembers correspond to physical scene components. In another variation, it is assumed that the scene components at different temperatures define a straight line in DN or radiance space (Figure 4.15). Over a large enough temperature range, of course, this line actually is curved, because Planck's law is nonlinear. However, the line is nearly straight over the range of temperatures of many landscape surfaces. The straight line segments for the different endmembers can be projected geometrically to an intersection point corresponding to a very low temperature, a temperature that does not actually occur anywhere in the scene. This point in radiance space acts just like shade in the VIS-NIR problem – and can be treated in the same way. We call this point "virtual cold," as a reminder that it really is not shade. The fraction image for "virtual cold" is proportional to the pixel temperature.

As in the VIS-NIR, fractions of endmembers in the TIR are most useful for photo interpretation. However, it is more difficult to achieve high fraction accuracy with TIR images, because of the unavoidable uncertainties about temperature distribution at the sub-pixel scale. If a surface is not isothermal, and one scene component is darker, absorbs more sunlight, and is hotter than the others, a disproportionate amount of radiance will come from the hot components, and this will be expressed in higher apparent fractions.

Next

Once we have made fraction images, what do we do with them? How can we use information on fractions of scene components to see landscapes in new ways? What does shade tell us about sub-pixel texture?

Chapter 5
Fraction images of landscapes

Melas Chasma, Mars. Photograph by European Space Agency, Mars Express

---

In this chapter

This chapter is a tutorial on how to make and interpret fraction images of landscapes. Fraction images can be further processed and displayed in many different ways to link image data with the properties of the ground.

5.1 What to do with fraction images

The main power of spectral-mixture analysis lies in facilitating photo interpretation and thematic mapping. Although this may come as a surprise to some readers who are focused on obtaining accurate numerical fractions of endmembers, recall our basic premise in Section 1.1 that spectral images are proxies for being on the ground. When we make thematic maps in the field the first thing we do is determine the qualitative differences between materials and their *relative* proportions in a spatial context (Chapter 7). Once we know what materials are there, and how they are arranged, we can try to make quantitative measurements of amounts (Clark's law, Section 2.1). Accurate, quantitative measurements of endmember fractions are difficult to achieve, because SMA, by its very nature, entails testing simple spectral models on images of spectrally complex natural landscapes. Earlier we discussed factors that affect fraction accuracy, such as selecting the right endmembers and the right number of endmembers, the natural spectral variability of endmember categories on the ground, measurement distortions introduced by atmospheric effects and imaging systems, and nonlinear mixing. Even when we think we have modeled an image to low residuals, it may be difficult to verify fraction measurements on the ground (Section 1.5). When an image is well-modeled and spectral contrast is high, fraction accuracy can approach or even exceed the accuracy of field measurements. Although our emphasis is on mapping, many applications do require quantitative and accurate estimates of scene constituents, often as input for process models that predict physical, but unmeasured, aspects of a scene such as roughness or soil moisture. In these cases, the extra work may be justified.

5.1.1 Displays of fraction images

Basic displays

In Section 4.3, when we applied SMA to the synthetic image, the product was one fraction image for each endmember, including shade (Figure 4.6). Similarly, the "raw" (not corrected for shade) SMA output from remote-sensing data consists of fraction images for each endmember, plus an image for the RMS residual. An example is shown in Figure 5.1 for the Landsat TM sub-scene of Seattle, Washington. As was discussed in Section 4.3.3, such images are useful for evaluating our choice of endmembers, and they can be used for photo interpretation when the shade image is complemented to make shading and shadows appear dark. Raw fraction images, however, do not provide correct fractions of the materials on the ground. The shade fraction first must be removed by normalization.

Contrast stretching

Apparent endmember fractions are changed by contrast-stretching the fraction images. A first step is to use the histogram sliders to set fractions <0 to 0 DN and to saturate fractions >1 at 255 DN. Do not be satisfied with a default contrast stretch that your image-processing program applies to a fraction image. Experiment with each image, keeping in mind that the objective is to interpret the scene in terms of the selected endmembers and not just to display all of the data. If we contrast-enhance the low fractions, we invariably reveal the limits of a mixing model and the noise threshold of the image. (Recall that fractions of the selected endmembers will be assigned to all pixels, including pixels for which the endmembers are not

Contrast stretches of fraction images
Adjust the stretch of a fraction image so that endmembers do not show up in places where you know they are absent on the ground.

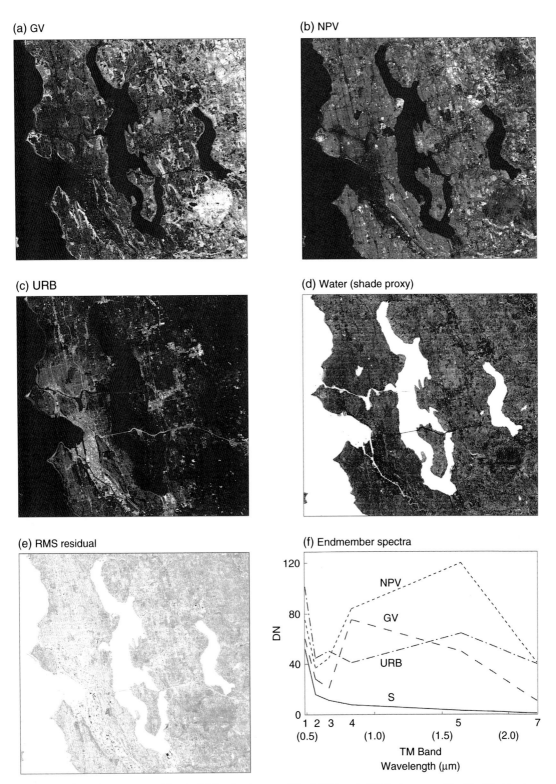

(a) GV

(b) NPV

(c) URB

(d) Water (shade proxy)

(e) RMS residual

(f) Endmember spectra

Figure 5.1. "Raw" (not corrected for shade) fraction images of the 1992 Landsat TM sub-scene of Seattle, Washington. (See also Plate 2.) (a)–(d) Fraction images (lighter tones are higher fractions); (e) complemented RMS-residual image (darker is higher RMS residual); (f) spectra of the four image endmembers: green vegetation (GV) from deciduous trees; non-photosynthetic vegetation (NPV) from a field of dry grass; urban (URB) from downtown Seattle; shade (S) from water in Puget Sound.

right or the data are noisy.) Too strong a stretch will force an endmember to appear to be everywhere in an image, even in areas where it obviously does not belong. For example, vegetation may appear to be present in bodies of clear, dark water, or soil may appear to be a component of a closed-canopy forest. If we know that an endmember is *not* present in a particular area, we need to adjust the threshold of the display of low fractions to show that. Whenever possible, thresholds should be determined for each endmember as a way of evaluating the validity of the mixing model.

Mapping and interpreting overflow fractions

We can find pixels that have fractions <0 and >1 in a fraction image by displaying the corresponding histogram using the contrast-stretch tool and making a black-and-white image (Figure 5.2). We also use a program that allows us to toggle displays of pixels having fractions >1 in red and pixels having fractions <0 in blue. As with the RMS-residual image, it is important to know the image context of pixels that have overflow fractions to be able to evaluate where and why a simple mixing model departs from the reality of the image (see also Sections 4.1.4 and 4.3.3). In some parts of an image, it may not matter whether the fractions are too high or too low, but elsewhere, overflow fractions can guide needed improvements in the endmembers.

In the Seattle image, all of the endmembers have overflow fractions in some pixels. This is an expectable consequence of applying a simplistic model to this complex scene. The endmembers GV, NPV and URB were selected to represent low-shade examples of the basic scene components and to provide an intuitive framework for interpretation. The URB endmember, for example, is an average spectrum from the spectrally variable downtown area. Pixels having lighter buildings and roads have fractions of URB >1 and fractions of shade <0, and we interpret them as being lighter versions of the URB endmember. As pointed out in Section 4.4.3, this does not mean that the lighter pixels would make a "better" URB endmember. Indeed, the real test of how well we choose endmembers requires that we remove shade and evaluate the proportions of the tangible endmembers in view of what is on the ground.

5.1.2 Removing shade and other endmembers

Removing shade by normalizing

Thematic maps that are made in the field (Chapter 7) never include shading and shadows, whereas images do. Shading must be removed from raw fraction images to be able to measure the proportions of the other endmembers. For the synthetic image, normalization of endmember fractions after deleting shade recovers the correct fractions of

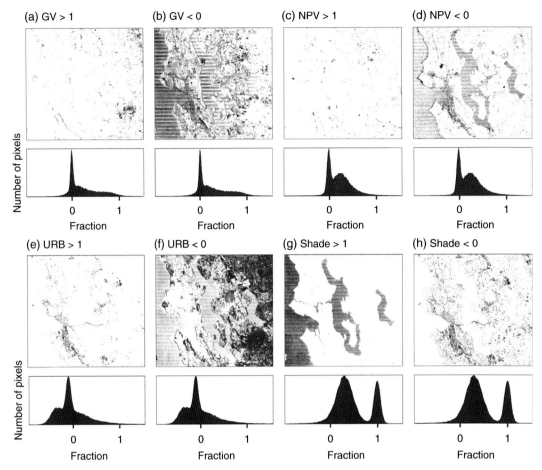

Figure 5.2. Overflow fractions for the raw fraction images in Figure 5.1. Darker pixels indicate larger overflow fractions, either positive or negative. The four image endmembers are: (a, b) GV from deciduous trees; (c, d) NPV from a field of dry grass; (e, f) URB from downtown Seattle; (g, h) S from water in Puget Sound. Images have been contrast-stretched to emphasize the spatial context of the overflow fractions.

A, B, and C for each pixel (Figure 4.7). Of course, in that special case, we used endmembers that perfectly modeled the image, and this means that all fractions were >0 and <1. However, as was discussed above, it is common to have some overflow fractions with remote-sensing images, owing to natural spectral variability in the scene (Section 4.1.5). Mathematically, however, overflow fractions present a problem when normalizing. We adopt a practical solution to this dilemma. Prior to normalizing, we set the fractions that are <0 to 0, and we set the fractions that are >1 to 1. The rationale for this truncation is that, from an interpretive perspective, fractions <0 mean that none of an endmember is present, and fractions >1 mean that an endmember is the only one present. Truncation of overflow fractions does not cause problems for those pixels that are reasonably well-modeled, and when this is the case, the proportionality of the tangible endmembers relative to that in the raw fractions is maintained. As might be expected, though, proportionality will be lost when the amount of overflow is large. However, large values of fraction overflow are always a signal that we are applying an inappropriate model for the affected pixels. Normalized fractions for the Seattle image are shown in Figure 5.3.

Removing multiple endmembers by normalizing

When there are four or more endmembers, we can remove more than one of them and normalize the remaining fractions so that they sum to 1. The process is illustrated for the synthetic image in Figure 4.7 where we calculated $A/(A + B)$ and $B/(A + B)$. Depending on the scene and the information that we are trying to extract, it can be useful to remove not only shade, but also one of the tangible endmembers. For example, by deleting shade and GV, an image can be modified to show only the distributions of types of substrate (e.g., Smith *et al.*, 1990). Although removing the distracting clutter of vegetation can be effective for photo interpretation, caution is needed, because fraction accuracy may be reduced significantly if the remaining endmembers occupy only a small proportion of a pixel or when a pixel is not modeled well.

Some drawbacks of normalizing

Normalization enhances noise. In some images, such as Figure 5.3 where water is used as a proxy for shade, normalization amplifies very small fractions of non-shade endmembers that occur at the level of the system noise. From an interpretive view, we know that GV, NPV, and URB are not present in the water areas (especially as stripes!) and we can set the water areas to zero fractions for all non-shade end-members. We did this in Figure 5.3 by applying a mask that makes the water areas black for each of the normalized fraction images. Noise patterns are less apparent in images having only small patches of shadowed and other dark areas. Masking may not be warranted for such images as long as we remember that normalized fraction values may not be reliable.

We can make better estimates of "true" fractions by normalizing without shade; however, we need to be aware of the limitations of this method. By removing shade, we compensate for *shading* (Figure 1.9) at all scales including spatially unresolved surfaces. The aggregate of sub-pixel *shadows* and dark materials, however, mimics the darkening of a pixel that would be contributed only by shading. Thus, by normalizing without shade, we effectively are assigning the fractions that are derived

> **Normalizing without shade**
>
> Normalizing the fractions of the non-shade endmembers so that their fractions sum to 1 has the effect of taking the information from the measured part of a pixel and extending the same proportions into the unknown (shadowed) parts.

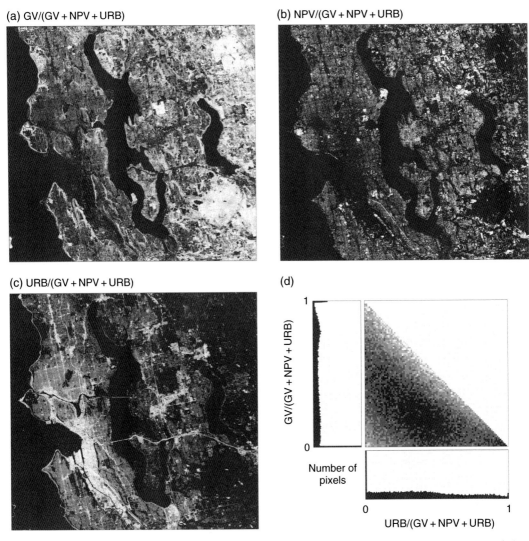

Figure 5.3. Normalized fraction images of the Seattle Landsat TM sub-image. (a)–(c) The fraction images are derived from the raw fractions in Figure 5.1. Shade has been deleted, and the remaining fractions are scaled such that the sum is 1. (d) Histograms and scatter plot of GV/(GV + NPV + URB) vs. URB/(GV + NPV + URB). In the scatter plot, NPV$_{norm}$ has a fraction of 1 at the locations where GV$_{norm}$ and URB$_{norm}$ have fractions of 0. Fractions < 0 were set to 0 and fractions > 1 were set to 1 prior to normalizing. Urban and GV histograms each have large spikes at 0, and small spikes at 1. Striping noise in water areas was suppressed by a mask derived from TM Band 4, DNs > 10.

from elsewhere in a pixel to the unresolved shadows and dark areas. We expect, therefore, that the accuracy of normalized fractions will be lower for darker pixels. Although normalizing for shade is straight-forward, when we remove tangible endmembers such as vegetation and normalize, the fractions of the remaining endmembers are relative only, and their absolute values may be misleading.

5.1.3 Sub-pixel shading

Spectral images contain information on shading and shadows at all spatial scales from mountains to sand grains; however, topographic-scale shading and shadows dominate the appearance of images of rough terrain. In well-modeled images, the apparent topography is partitioned preferentially into the shade image, and the images of other endmembers appear relatively "flat." If we remove the topographic component of shading from a shade image using data from a digital elevation model (Section 3.3.2), we can produce a remainder image that contains the component of shade that occurs at sub-DEM resolution. By isolating sub-DEM shade, we gain information about fine-scale topography (roughness) and texture. For example, in the Gifford Pinchot TM image, sub-DEM shade is a key indicator of forest structure and composition (Figure 5.4). This topographically rough area contains a mosaic of clear cuts, and regeneration forest stands of various ages and compositions, along with a few areas of old conifers. Mature conifer stands have more shading and shadows and lower albedo than younger ones, and conifers are darker than deciduous species; however in raw fraction images, these distinctions are locally obscured by topographic-scale shade. This ambiguity also is evident in classified images that include training areas of dark and light closed-canopy forest (references images on the Web). Sabol *et al.* (2002) mapped old conifers in the Gifford Pinchot scene by subtracting DEM-scale shading from the original bands, prior to applying a mixing model. This is equivalent to subtracting DEM-scale shading from the shade-fraction image. In Figure 5.4f the old conifers stand out as the dark areas in the DEM-corrected (1-shade) image.

Adding sub-pixel shade to a color fraction composite
The normalized fraction images of GV, NPV, and soil in Figure 5.4 no longer contain shade at any scale. We now have the "true" fractions of these scene components that can be displayed as colors. Figure 5.5 displays them as: R = NPV, G = GV and B = soil. Because the sub-pixel shade contains critical information about forest structure and composition, we would like to combine this information with the color fraction image. In Figure 5.5 we added the DEM-corrected shade image to the color image. Now, the darker tones of the shade image modulate the colors, and dark green (GV) indicates closed-canopy forest having relatively high amounts of sub-DEM-scale shade. Field observations confirm that this model correctly predicts the areas of old conifers. This example illustrates the important point that fraction images can be combined in various ways to show those aspects of the data that are most important for photo interpretation.

Figure 5.4. Sub-pixel shade in a Landsat TM sub-scene of the Gifford Pinchot National Forest, Washington. See also Plate 5. (a)–(c) Normalized fraction images: (a) GV/(GV + NPV + soil); (b) NPV/(GV + NPV + soil); (c) soil/(GV + NPV + soil); lighter tones indicate higher fractions; (d) (1−shade) image, where darker tones indicate higher fractions; topography is evident in the (1 − shade) image; (e) digital-elevation model (DEM) of the same area. The model uses the same sun elevation (52°) and azimuth (128°) as the image, and assumes a cosine function for shading; (f) (1 − shade) image corrected for topographic shading using the DEM data. Notice that the image appears "flat," because the visual impression of DEM-scale topography has been removed. Rectangular patches are where coniferous forest has been clear cut and is in various stages of regeneration. Dark areas in the topographically corrected (1 − shade) image are the few remaining patches of mature forest. Adapted in part from Sabol *et al.* (2002). By permission of Elsevier.

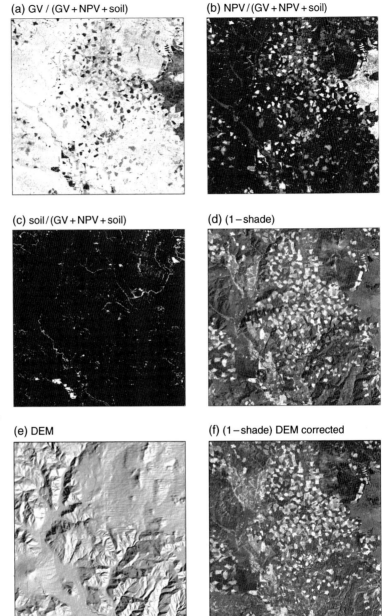

(a) GV / (GV + NPV + soil)

(b) NPV / (GV + NPV + soil)

(c) soil / (GV + NPV + soil)

(d) (1 − shade)

(e) DEM

(f) (1 − shade) DEM corrected

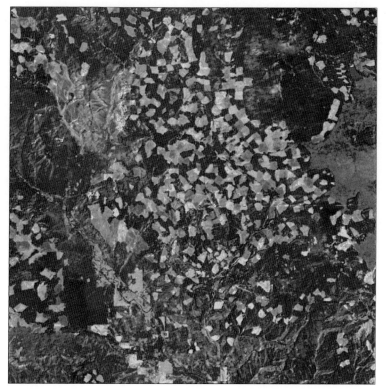

Figure 5.5 and color Figure 5.5 (see Web). Gifford Pinchot National Forest. Color composite of normalized fraction images, modulated by gray tones of a DEM-corrected (1 − shade) image; $R = NPV_{norm}$; $G = GV_{norm}$; $B = soil_{norm}$. See Figure 5.4. Suppression of DEM-scale topography renders the image "flat," and emphasizes the sub-DEM-scale shade that is associated with old conifer stands. Field observations confirm that the dark green areas are old conifers.

5.1.4 Fraction intervals

Standard image-processing tools for making density slices and masking can be applied to fraction images to show desired fraction intervals. To display and evaluate fraction intervals, it is convenient to use the contrast-enhancement tool on the fraction histograms. It is particularly helpful if we can set the slider bars that control the contrast stretch to bracket any fraction interval, say, 10 or 20 DN. This allows us to illuminate only those fractions within the set interval. To do this we must be able to set the DNs outside the slider bars to 0. It also is helpful if we can lock the slider bars together at any DN interval and move them back and forth across a histogram in tandem. This produces a dynamic display of a fraction interval on an image as we sweep the sliders over a histogram. Using this technique, we can easily locate the highest and lowest concentrations of endmembers, and, by sweeping the locked sliders, we can see the spatial context of transitions from high to low fractions. This is an especially important tool for examining gradational contacts between materials on the ground and for deciding on class boundaries (Section 5.2.3). One of the first applications of this method was a

Density slices

A digital image is said to be divided into density slices when colors are assigned to specific DN intervals.

Dynamic displays

A visual impression of adding or removing a component from a scene can be produced by stepping through successive fraction intervals. Examples include "removing" vegetation or soil from an image.

(a)   (c)   (d)

(b)

Figure 5.6 and color Figure 5.6 (see Web). Six-channel Viking Lander 1 images of Mars showing apparent removal of dust/soil from rocks and the spacecraft. Images were modeled by mixtures of three endmembers: rock, soil, and shade. Fractions of rock and soil were normalized to sum to 1 and recombined with a (1 − shade) image to recover the visual-impression texture and topography. The normalized images were density sliced to show pixels having different relative fractions of rock (gray) and soil (brown). In the first fraction interval, (a) and (c), brown = pixels having > 0.50 soil. In the second fraction interval, (b) and (d), brown = pixels having > 0.75 soil. On a computer screen, the full range of rock/soil fraction intervals can be sampled using a dynamic (mouse-driven) display to give the appearance of removing or adding dust/soil to the scene. The object at the left side of a and b is part of the Lander spacecraft. The object in the center of c and d is a detached piece of the spacecraft, and a footpad of the Lander is at the bottom of the images. The normalized spectra of the spacecraft parts mimic the spectrum of the rock endmember; therefore, dust disturbed and re-deposited during landing appears to be removed from the spacecraft as well as the rocks.

dynamic display of normalized fractions of dust and rock in Viking Lander images of Mars (Adams *et al.*, 1986) that produced the visual impression of removing dust from the rocks (Figure 5.6 and color Figure 5.6; see Web). Similar dynamic displays have been made of normalized vegetation and soil/rock fractions in Landsat TM images (e.g., Adams and Adams, 1984; Smith *et al.*, 1990), where moving the sliders in tandem appears to strip away the vegetation cover – a process that Andy Green of CSIRO (Australia) dubbed "spectral defoliation."

(a)
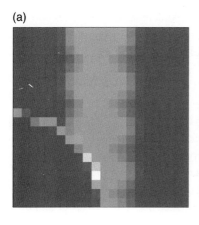

(b)
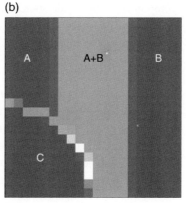

Figure 5.7 and color Figure 5.7 (see Web). Color composite of the synthetic image. Red = endmember A; green = endmember B; blue = endmember C. (a) Raw fractions. See also Plate 9 for a different contrast stretch. (b) Normalized fractions without shade. Yellow pixels are mixtures of A and B. Magenta pixels are mixtures of A and C. White pixels are mixtures of A, B, and C.

5.1.5 Color fraction images

Color composites

As we illustrated in Figure 5.5 and color Figure 5.5 (see Web), three fraction images can be combined in a color composite to express the relative proportions of endmembers. Figure 5.7 and color Figure 5.7 (see Web) show color composites of the synthetic image in which endmember A = red, endmember B = green, and endmember C = blue. The mixed pixels of A and B are yellow (red + green) and there are a few white pixels at the edge of C where A, B, and C mix. Contrast stretching changes the fraction displays and can force colors to mix even when fractions are very low. Given that fractions sum to one, and that whenever DNs for one endmember are high the DNs for the other ones must be low, the colors that represent "pure" endmembers are complementary. Mixtures of red, green, and blue occur only if the endmembers mix. As a result, fraction composites often do not exhibit mixed colors to the extent that we are accustomed to when we display channel data that typically are more highly correlated.

Color images of fractions may include or omit the shade endmember. When three, non-shade endmembers are displayed in a color composite, the visual effect of topography is suppressed, and the compositional information is enhanced; however, as previously discussed, the fractions are not correct unless the endmembers have been normalized by omitting shade (Figure 5.7b). If there are only three endmembers, one of which is shade, a color composite may or may not help photo interpretation. It depends on the image.

In Section 4.3.3 and Figure 4.9 we saw how an RMS-residual image defines the parts of an image that are not consistent with the applied endmember model. Sometimes a residual image shows an interesting

Color RMS residual

An RMS-residual image can be used in a color composite.

component of the scene such as an endmember that has been left out. When that is the case, the RMS-residual image, even though it is not a fraction image, can be used in a color composite for purposes of photo interpretation.

Shade-modulated fraction intervals

As we saw in the Viking Lander images of Mars in Figure 5.6, a visually effective way to display true (normalized) fractions of one endmember is to assign colors to pixels in a certain fraction interval (density slice), and then combine the fraction-interval image with the complemented-shade image. Modulating the intensity of the colors by the gray tones of the complemented-shade image produces the visual effect of draping a fraction image over the topography. Within each fraction interval, the least-shaded areas will appear as saturated colors, whereas the more shaded and shadowed regions will have dark colors or black. An added benefit of such a display is that the assigned colors are most evident in well-illuminated parts of a landscape, and they are less distinct in deeply shaded or shadowed areas – a pattern that also happens to express the lower degree of confidence in the fraction accuracy in high-shade areas. Keep in mind, though, that a drawback of relying on the complemented shade image to visualize topography is that dark materials on the surface may mimic shading and shadow.

5.1.6 Color-enhanced shade

We have developed and tested a new way visually to enhance the topography and texture of landscapes by using color to highlight the fraction of shade. The technique is especially effective when applied to images of vegetated surfaces, although it can be applied to other types of images as well (see Plates 7 and 8). The method is illustrated in Figure 5.8 and color Figure 5.8 (see Web) using the Gifford Pinchot, Landsat TM image to show how the visual impression of topography and/or texture can be enhanced.

To make color Figure 5.8 (see Web), we assigned red to (1 − shade), green to GV and blue to NPV, and applied contrast stretches to the fraction images individually. It usually is necessary to apply a strong contrast stretch to the low-shade end of the (1 − shade) histogram (Figure 5.9). To interpret the result, we need to think of red as "non-shade." This means that red is activated in the computer monitor only for pixels of *light* surfaces. Thus, GV that has a low shade fraction will be yellow, because green (GV) and red (1 − shade) are both activated. Whenever GV is dark (high shade), red is *not* activated and the result is green. Similarly, light NPV will be magenta, because blue (NPV) and red (1 − shade) are both

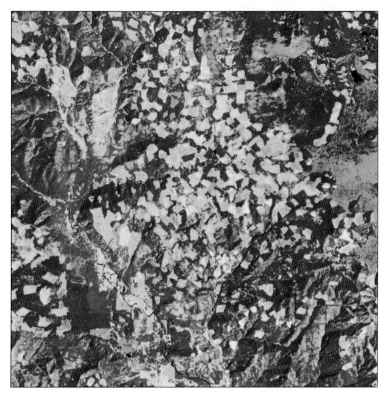

Figure 5.8 and color Figure 5.8 (see Web). Topography and sub-pixel texture rendered in color for the 1991 Landsat TM, Gifford Pinchot sub-image. Red = (1 − shade); green = GV; blue = NPV. Interpretation of colors is given in Table 5.1. Green is dark (high-shade) vegetation on slopes with high incident angles and/or in mature-conifer forest. Yellow is light (low-shade) vegetation on Sun-facing slopes, clear cuts, and riparian areas. Notice how color emphasizes the shapes of the small, vegetation-covered cinder cones at the right-center margin of the image.

activated. Whenever NPV is dark, red is absent, and the result is blue. Root-mean-square residuals, and endmembers other than those assigned to green and blue, do not have a color assignment; therefore, pixels of "other" are black whenever the shade fraction is high, and they are red whenever the shade fraction is low. In the Gifford Pinchot image, although a soil endmember is part of the mixing model, it was not assigned a color for color Figure 5.8 (see Web). Areas of soil appear red, because they have low shade but no other assigned color. Bodies of water are black. When the same color assignments are applied to the Seattle, Landsat TM sub-image, the urban endmember has no color assignment, and pixels with low shade and a high URB fraction appear red.

A drawback of this technique is that color (1 − shade) images are somewhat counter-intuitive, because we are not used to seeing light–dark patterns of shading portrayed in color. It may help to refer to Table 5.1 and Figure 5.9 to decode the colors. Software for image processing usually includes a simple toggle function that allows one to manipulate red, green, and blue color combinations, and this is an excellent way to visualize what happens when we stretch the histograms of color (1 − shade) images.

In color Figure 5.8 (see Web) the information on forest texture is difficult to distinguish from topographic effects. However, we showed

Table 5.1. *Interpretation of (1 − shade) images of a vegetated landscape, modeled by GV, NPV, and shade.*

Color assignment
Red = (1 − shade)
Green = GV
Blue = NPV

Image colors	Interpretation
Red	Light "other" (not GV and not NPV)
Green	Dark GV
Blue	Dark NPV
Yellow (red + green)	Light GV
Magenta (red + blue)	Light NPV
Cyan (green + blue)	Dark (GV = NPV)
White (red + green + blue)	Light (GV = NPV)
Orange/brown (red > green)	Light (GV + NPV and/or "other")
Black	Shadow, dark surfaces, or dark "other"

RMS residuals and endmembers other than those assigned to green and blue do not have a color assignment. Pixels of "other" are dark when the shade fraction is high. Pixels of "other" are red when the shade fraction is low, as is shown for the synthetic image in Plate 9.

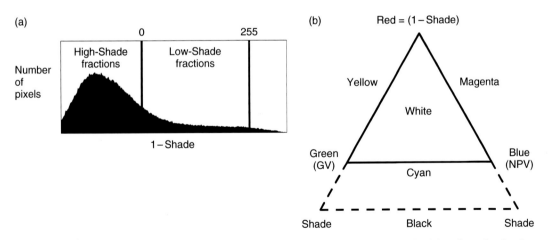

Figure 5.9. Explanation for making a color (1 − shade) image. (a) Histogram of the (1 − shade) endmember for the 1991 Landsat TM, Gifford Pinchot sub-image. Because the shade image is complemented, low fractions have high DNs, and the right side of the histogram has low fractions of shade. In the illustration, the 0 DN and 255 DN slider bars have been placed to contrast-stretch the low-shade fractions. (b) Color triangle for (1 − shade) interpretation. Red is assigned to (1 − shade). Green and blue are assigned to GV and NPV, respectively. The combination of high GV (green) and low shade (red) produces yellow, and the combination of high NPV (blue) and low shade produces magenta. Other color combinations are shown in the triangle and explained in Table 5.1. See also Plate 9.

Figure 5.10 and Web color Figure 5.10. Sub-pixel texture rendered in color for the 1991 Landsat TM, Gifford Pinchot sub-image: red = (1 − shade); green = GV$_{norm}$; blue = NPV$_{norm}$. Interpretation of colors is given in Table 5.1. Topography has been removed from the shade image by a DEM correction (see Figure 5.4). Green is (high-shade) mature-conifer forest. Yellow is (low-shade) vegetation (deciduous species and young conifers) in clear cuts and riparian areas. Orange areas have low shade and a mix of green vegetation and non-photosynthetic vegetation and/or soil. By subduing topographic shading, and color-enhancing sub-DEM shade, it is easier to distinguish and map areas of mature forest. For example, the irregular patches of thinned and clear-cut forest in the upper right corner of the image are clearly visible in yellow and red; the same areas are less apparent in color Figure 5.8 (see Web).

previously (Section 5.1.3, Figure 5.4 and color Figure 5.5; see Web) that removing the main visual representation of topography makes it possible to interpret differences in forest texture and to distinguish areas of mature forest. We achieved the same goal in color Figure 5.10 (see Web) by constructing a color (1 − shade) image using DEM-corrected shade and the normalized fractions of GV and NPV. The color (1 − shade) image exaggerates very small differences in shade fractions, and provides a sensitive measure of surface texture.

5.2 Classification using endmember fractions

5.2.1 Organizing fractions

In some textbooks and image-processing software, mixture analysis is treated as a classification tool. Standard classification algorithms, such as distance to means of a training set, organize data into labeled groups that have defined boundaries in a data space, which is defined by the spectral channels (Section 3.4.2). Unmixing algorithms, however, do not organize endmember fractions into groups of vectors in fraction space. Nevertheless, standard classification algorithms can be applied directly to fraction images, because fractions are just a transformation from channel DNs. Why would one go to the trouble of doing a mixture analysis and then classifying the result with a standard classifier? Why not just apply a classifier to the channel-DN data? One reason, already discussed, is that the transformation of channel DNs to endmember fractions can accomplish significant data compression. This opens the possibility of applying standard classifiers to imaging-spectrometer data. A second reason is that standard classifiers provide an initial way to organize fraction images for further interpretation.

The first step in applying a standard classifier to fraction images is to define training areas for the desired classes. These may be the same training areas from which image endmembers were selected, or they may be different, depending on the objectives of the analysis. Step two is to create data-attribute (rule) images for each training set (Section 3.4.2). If the same training areas are used for the (non-shade) end-members and for the classes, the data-attribute images for the fraction data will resemble the input fraction images (e.g., compare the synthetic images in Figures 4.6 and 5.14). The third step is to set class thresholds and to calculate the class memberships.

5.2.2 Physical models

In Section 3.4 we discussed standard classifiers as valuable tools for data reconnaissance and organization, but we emphasized that the classes that are produced by these algorithms do not necessarily have any interpretable physical connection to the ground. In contrast, we have approached mixture analysis primarily as a tool for modeling the basic spectral components of scenes and developing thematic maps (Sections 1.4.3 and 7.2). If our goal is to make thematic maps (as opposed to just organizing data), it may not be a good idea to apply standard classifiers to fraction images, because the algorithms may (or may not) arrange the fractions into groups that make sense in terms of scene context.

Fraction images and data-attribute images can reveal different kinds of information, and therefore can be complementary. For example, compare two figures of the Landsat TM image of Manaus, Brazil. In Figure 3.9 we made spectral-angle data-attribute images using six training areas (urban, light-green vegetation, soil, Rio Negro, Rio Solimoes, and primary (terra firme) forest). In Figure 5.11 we defined five image endmembers that are basic spectral components of the same scene, and calculated five fraction images and an RMS image. The image endmembers are: URB, GV, NPV, soil, and shade. In the fraction images, GV represents light-green vegetation. Mixtures of GV and shade model primary forest that has a rough, self-shadowing canopy.

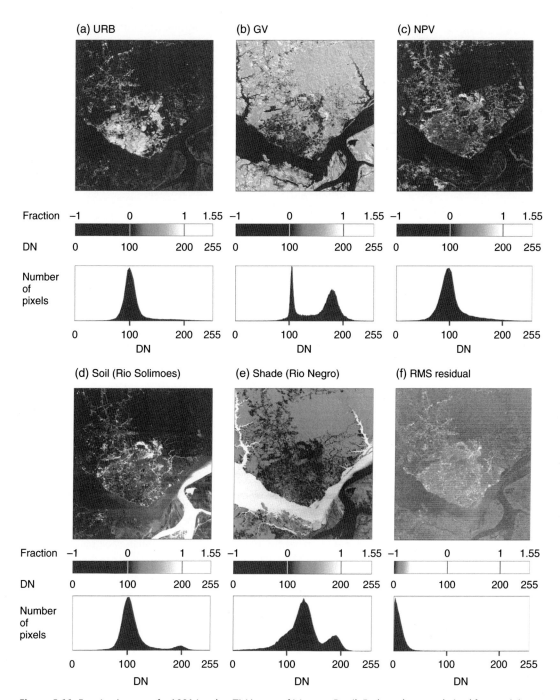

Figure 5.11. Fraction images of a 1991 Landsat TM image of Manaus, Brazil. Endmembers are derived from training areas in the image: (a) URB; (b) GV; (c) NPV; (d) soil; (e) shade; (f) RMS residual. The Rio Negro is a proxy for shade. The Rio Solimoes is a proxy for soil. Lighter tones indicate higher fractions. Primary (terra firme) forest (e.g., upper right quadrant of each image) has high fractions of GV and shade. High fractions of soil (Rio Solimoes, a muddy river) include patches within the urban area where development has exposed bare soil. Compare with Figure 3.9. See also Plate 3.

The fraction images mainly reveal information about canopy texture. The spectral-angle algorithm, however, is insensitive to shade, and it distinguishes between light-green vegetation and primary forest by their slight angular difference in the data space. In this example, the fraction images represent a physical model of illuminated vegetation, whereas the spectral-angle images show a property of the data that still needs to be interpreted. This interesting example is analyzed further in Section 7.2.2.

5.2.3 Setting fraction-class thresholds

Standard classification methods can be applied to fraction images to define the thresholds of class boundaries. The resulting classes may be very similar to those that are derived from the channel-DN data, or there may be significant differences. Examples of comparisons are given in Section 5.2.4. Fraction images provide an effective way to incorporate photo interpretation into decisions about class boundaries. We can manipulate fraction displays interactively as an overlay on a base image, and take advantage of the fact that the human eye–brain system is particularly adept at interpreting changing spatial patterns. We control the values of endmember fractions displayed by setting the histogram adjustments that are used to define contrast stretches.

Proportions of endmembers on an image must change at the boundary between fraction classes (otherwise the classes would have identical fractions and be indistinguishable from one another). By varying the interactive fraction display, we can examine closely the nature of the change. At a sharp boundary, there will be a jump from illuminated pixels to dark pixels as different parts of the fraction histograms are sampled. At a gradual boundary, the fractions change smoothly, giving the visual impression that the interface moves. The image analyst can decide at what fraction threshold to establish each class, using the spatial patterns from the interactive display and knowledge of the image context. A limitation of this method is that it is only practical to vary the spatial pattern of one endmember at a time. If we try to vary two or more endmember patterns simultaneously the visual result is too complicated to interpret easily. This generally is not a serious limitation if the number of endmembers is small. Furthermore, endmember fractions vary sympathetically, and it often is possible to visualize the necessary spatial patterns using only one endmember.

The endmember, shade, requires special consideration when classifying fractions. For some applications we may want to suppress shade by normalizing, because shading obscures the materials that we want to classify. For other applications, though, shade is essential for classifying,

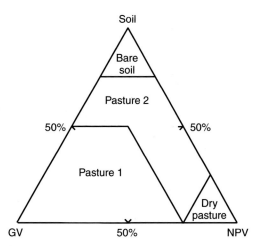

Figure 5.12. Ternary diagram defined by endmembers GV, NPV and soil. The diagram encompasses all possible mixtures of the three endmembers. The four classes of pasture are based on the properties of a hypothetical livestock-range area. Classes and class boundaries can be defined from field experience whenever endmembers are proxies for basic scene components.

because it can reveal differences between surfaces based on their textures (Section 5.1.3). For example, in the Amazon image that has little shading and shadowing from topography, primary forest can be differentiated from other land cover by its higher shade fraction. The boundary between the classes "primary forest" and "other" can be set by varying just the shade fraction, and observing when the boundary coincides with the edge of the forest as judged by photo interpretation.

Arbitrary fraction ranges
Based on field work or other information, we may already have definite opinions about how to characterize classes that are based on endmember fractions. If the spectral endmembers are a good proxy for the equivalent materials on the ground, it becomes evident that we can use the combinations of endmember fractions to define classes as though we were in the field. Whether in the field or using an image, we could decide, for example, that areas having >75% soil (or the soil endmember) will be classified as "bare soil." In fact, we can divide the data space into units any way we want, as long as the result makes sense from the perspective of our mapping objectives. The way we define our units need not be restricted to any particular landscape or image; however the rules do need to be defined for each context (e.g., Adams *et al.*, 1995; Roberts *et al.*, 1998).

As an example, let us consider a hypothetical landscape used for ranging livestock. The objective is to classify and map pasture according to condition (Figure 5.12). Based on field experience, we decide that the best pastures (pasture 1) should have at least 75% green vegetation and no more than 50% exposed soil. All other pastures (pasture 2) are considered to be of lesser quality. Separate categories are defined for

Figure 5.13. Minimum-distance classification of the synthetic image based on endmember fractions. (a)–(c) Data-attribute (rule) images for training areas in A, B, and C. Lighter tones are closer to the means. (d)–(f) Classified images using different standard-deviation thresholds. Black = unclassified; other gray tones designate classes.

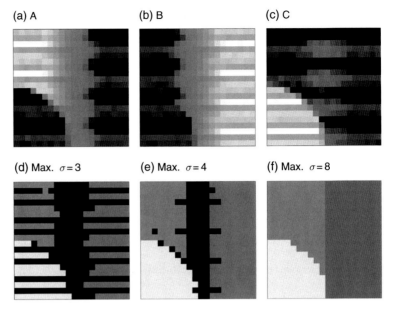

(a) A　　　(b) B　　　(c) C

(d) Max. $\sigma = 3$　　　(e) Max. $\sigma = 4$　　　(f) Max. $\sigma = 8$

Figure 5.14. Spectral-angle classification of the synthetic image based on endmember fractions. (a)–(c) Data-attribute (rule) images for training areas in A, B, and C. Lighter tones are closer to the means. (d)–(f) Classified images using different spectral-angle thresholds. Black = unclassified; other gray tones designate classes.

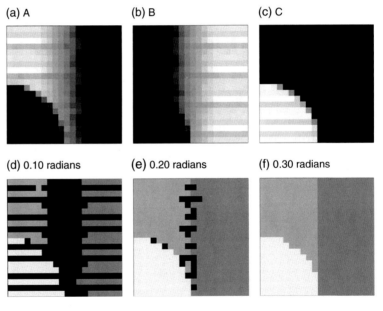

(a) A　　　(b) B　　　(c) C

(d) 0.10 radians　　　(e) 0.20 radians　　　(f) 0.30 radians

areas that are >75% bare soil, and other areas that are >75% dry grass. We model an image of the landscape by three basic spectral components: GV, NPV, and soil. Using SMA, we test a model consisting of [GV, NPV, soil, shade]. Results show minimal RMS residuals and fraction overflow. We remove shade and normalize the fractions of the remaining endmembers that are proxies for the three basic spectral

Table 5.2. *Classification scores for the synthetic image based on endmember fractions.*

Classification algorithm	% Unclassified	% Class A	% Class B	% Class C	% Misclassified
Parallelepiped (5×5)	6.5	21.25	49	23.25	3
Minimum distance (5 × 5)	0	29.5	50	20.5	0
Maximum likelihood (5 × 5)	0	37.75	40.75	21.5	1
Spectral angle (5 × 5)	0	29.5	50	20.5	0
Parallelepiped (25 light)	0	0	24	76	57
Minimum distance (25 light)	0	29.5	50	20.5	0
Maximum likelihood (25 light)	0	26.5	54.25	19.5	0
Spectral angle	0	29.5	50	20.5	0
Synthetic image	**0**	**29.5**	**50**	**20.5**	**0**

Each classifier was tested over a range of threshold settings. The best threshold setting was selected based on the lowest number of misclassified pixels. The percent of misclassified pixels for each classifier is tallied in the right-hand column. Pixels were considered misclassified if: C pixels occurred in the A–B mixed zone; A, B, or C occurred in pure areas of the other classes. None of the algorithms correctly modeled mixtures; therefore, mixed pixels were judged correct if classified as any one of the components. Training areas for A, B, and C are blocks of 5 × 5 pixels or are 25 light (low-shade) pixels. Fewer pixels were misclassified using fractions compared with channel DNs (Table 3.2).

components of interest. The fraction images show that GV, NPV, and soil are mixed throughout the image. Each pixel in the image is then classified according to the fraction limits in Figure 5.12. It is evident that we can partition the fraction space in many ways to suit our mapping objectives, regardless of the distribution of the data points, and regardless of whether the classes have gradational boundaries. For example, we could change any of the class boundaries, add classes or subdivide existing ones. (For instance, we could subdivide pasture 1 into greener and "dryer" sub-classes.) In Chapter 8 we show an example of how of how this approach was used to classify pastures in the Fazendas Landsat image.

5.2.4 Examples of fraction classes

Synthetic image
We tested standard classifiers on endmember fractions of the synthetic image in the same way that we did with channel-DN data (Section 3.4.2). As with the channel-DN data, there is variability from one algorithm to another and from one threshold setting to another (Figures 5.13 and 5.14). Misclassification scores for four classifiers are summarized in Table 5.2.

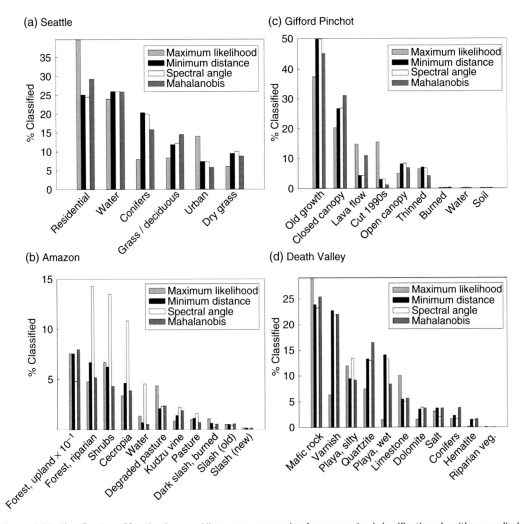

Figure 5.15. Classification of fraction images. Histograms comparing four supervised classification algorithms applied to four Landsat TM sub-images. (a) Seattle; (b) Amazon; (c) Gifford Pinchot; (d) Death Valley. Four histograms are plotted for each training area on the horizontal axis. Percent of pixels classified in the sub-scene is plotted on the vertical axis. Training areas on each image were selected based on knowledge of the landscape from ground studies and photo interpretation. Each classifier was tested with several settings, and the best results for each algorithm are shown in the histograms in this figure. Across all four images the number of pixels assigned to any one class varies from one algorithm to another. Compare with Figure 3.14.

The tests on the synthetic image confirm that standard classifiers give generally equivalent results for channel DNs and for fractions, which is consistent with the fact that the channel data and the fractions are related by Equation (4.2). There are some interesting differences, though, between the misclassification scores for the channel-DN data and the fractions, as is shown by comparing Tables 5.2 and 3.2. Most classifiers performed better on the fraction data in terms of misclassification scores, because simulated variations in illumination are separated from each of the other components by using shade as an endmember. In contrast, the clusters of the channel data for A, B, and C are larger, because of variations in "illumination." However, unsupervised classifiers, and the Mahalanobis classifier, did not work at all on fractions of the synthetic image. This example, of course, does not predict how the same algorithms will perform on other images.

Landsat images

We tested standard classifiers on fraction images of four Landsat TM sub-scenes (Figure 5.15). Mixture models for the four sub-scenes are shown in the Web reference images. The Seattle, Gifford Pinchot, and Amazon images each were modeled by image endmembers: GV, NPV, soil, and shade. The Death Valley image, which is compositionally heterogeneous, was modeled by four endmembers: GV, light rocks (dolomite), red rocks (hematite-colored volcanics), and shade. After calculating the fraction images, training areas were selected from each image for classifying. These were the same training areas that were used for classifying the channel-DN data in Figure 3.14. The fraction-derived classes are superficially similar to the ones calculated from channel DNs, but there are significant differences in the details that are apparent by comparing the histograms in Figure 5.15 with those in Figure 3.14. Based on photo interpretation and our knowledge of the field areas covered by these images, we find that misclassification of the fraction data stems from the same problems that we noted with the tests on the channel-DN data, namely mimicking, and boundaries between classes.

Next

Now that we have evaluated the basic image-processing tools, including those that involve physical models, how do we apply them? In the next chapter, we consider the factors that limit our ability to find what we are looking for in an image when it is not obvious among the spatial and spectral clutter.

Chapter 6
Target detection

Mt. Conness, Sierra Nevada, California.

In this chapter

We explore the factors that influence detection of indistinct targets in spectral images. Detection thresholds depend on the nature of the background, as well as on the properties of the target itself.

6.1 Spectral contrast and target detection

A fundamental reason to use multiple spectral channels is to discover differences among landscape components that would go unnoticed in panchromatic images. Sometimes, though, the information that we are looking for in spectral images is near or below the limits of detection. To find what we are searching for (a target), we need to maximize spectral contrast between target and background, and this requires that we pay attention to target *and* background. Among the tools available for image processing, there is an important distinction between algorithms that gauge spectral similarity (such as the standard classifiers) and those that are designed to facilitate detection. For example, we measure spectral similarity between laboratory reference spectra and image spectra to identify materials, and we compare spectra of training areas

Detection

Discovering something by distinguishing it from its background.

192

with other pixels in the rest of an image to classify land cover. The objective is to find similar spectra – not to enhance subtle differences between a few spectra and their contextual background. When searching for spectral similarity, the background is not defined or taken into account. In this chapter we are interested specifically in targets that are obscure – targets that are not readily found using standard methods for measuring spectral similarity (Section 3.4).

Obscure

Not clear; ambiguous, or uncertain. Hard to perceive. Indistinct, not readily seen. Inconspicuous or unnoticeable. To conceal or conceal by confusing.

6.1.1 Target and background

We define a "target" as a specific thing that we are looking for in a spectral image. The "background" is the whole scene minus whatever we are looking for (Albee's law, again). Background incorporates both spatial and spectral attributes, and it is similar to the term "clutter" that is used more commonly in spatial analysis. A target may be spatially resolved or unresolved; it may be rare or distributed widely throughout a scene. We may know the identity, including the spectral character, of a target and be seeking its location; or we may know neither and merely suspect that an anomaly is present. To be detectable, though, a target must differ from its background in its spatial, tonal SL or spectral characteristics (Section 2.2.1).

Detection depends on the background

To detect indistinct targets, we enhance contrast between target and background. To find spectrally similar pixels, we ignore the background.

As an example of an obscure target and its background, there is a standard test for color vision in which a person is asked to find an object (usually a number) made up of colored dots that are set in a field of other dots of different colors. The dots that make up the number do not differ in size, shape, or texture from those that comprise the rest of the field; therefore, they are spatially obscured in the background. The only contrast between target and background is spectral. Most people see the different colors that define the number, but for those who do not have normal color vision, the familiar spatial pattern of the number remains hidden. By analogy with the color–vision test, targets in spectral images are not necessarily revealed by their spatial properties. They may be lost in spatial clutter, they may not be spatially resolved, or their nature may not be known in the first place. However, even under these circumstances, targets sometimes can be detected by enhancing their spectral contrast relative to the background. Like the numbered dots, once they are detected spectrally they often can be interpreted spatially.

Recognition of a target

To recognize (to re-cognize is to know again) an object or pattern, one needs to know something already about its spatial, spectral, or other properties. A first step is to be able to detect the diagnostic properties

Recognize

To identify something by its appearance or characteristics.

of a target, and a next step is to match the detected properties with the ones that we have in mind as being diagnostic. Recognition is more difficult when a target is difficult to detect (obscure), because the diagnostic properties of the target are similar to those of the background. Not all targets are obscure, though, and a target in plain view can go unrecognized if we do not make the connection between the known properties and the observed ones. Furthermore, we are more likely to fail to recognize something if we are not actually looking for it.

Consider the hypothetical situation where we encounter a friend on a crowded street. The friend calls out "Hi," and our response is "Oh, sorry, I did not recognize you at first." The friend clearly detected and recognized us. Although we detected the other person (she was only a meter away and in plain view), we at first failed to distinguish her diagnostic properties from those of the other people (the background), and, therefore, we failed to recognize (identify) her. In another version of our hypothetical encounter, we might take more notice of our friend and think "That person looks very familiar." In this case, we detect some of her diagnostic features, and we achieve some level of recognition, but we do not have a good-enough match with our stored mental image of the person to be able to identify her by name or context. If we have always known our friend as having dark hair, and today it is blond, an important diagnostic feature – a spectral one – does not fit the expected pattern. Recognition, therefore, is not always complete, and recognizing something does not necessarily mean that we achieve full identification.

In this chapter we are mainly concerned with detection and recognition of things that we are looking for in images (targets). As discussed in Chapter 1, we rely heavily on spatial attributes when doing photo interpretation, and a field investigator who is looking for something often will recognize its diagnostic features when it is spatially resolved *and* it can be detected against the background. Target recognition at sub-pixel scales is more difficult, because we must rely on spectral information alone for detection and recognition (Sabol *et al.*, 1992).

Finding what we are looking for

Our first task is to separate what we are looking for from the background. Paradoxically, it can be difficult to define what the background is unless we already know the diagnostic properties of the target and where it is located. For example, if an obscure target is distributed throughout a large and complex scene, then it could be anywhere, and the whole scene appears to be the background; however, if we know that

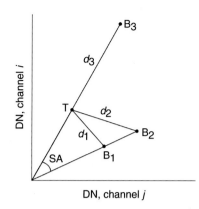

Figure 6.1. Spectral length (SL) and spectral angle (SA) in the data space of channels i and j; SL contrast describes light–dark contrast, and is the normalized difference of the vector lengths. For example, the SL contrast between T and B_3 can be given by $(SL_{B3} - SL_T)/(SL_{B3} + SL_T)$. Spectral angle, SA, is the angle between vectors, and is independent of light–dark contrast. The distance, d, between vectors is a measure of the overall spectral contrast that includes both SL and SA. Target, T, is separated from background, B_1, by the distance d_1; T and B_1 have the same SL: therefore their SL contrast is 0, but they do contrast in SA. Target, T, and B_2 also are separated by SA; however they have greater overall spectral contrast (d_2), because they differ in SL. Target, T, and B_3 contrast only in SL. See also Figure 2.10.

a target is limited to a certain image context such as a forested area, we can restrict the type of pixels that comprise the relevant background. Under some circumstances it may be sufficient to define the background as the average of part or all of a scene. In other cases it may be necessary to characterize the background fully to be able to detect the target. This is especially true when we do not know the location of the target, or, we do not know exactly what the properties of the target are. There even may be occasions when a target is entirely undefined, for example when we want to detect an anomaly in an image, or when we are trying to detect a change that has occurred over time from one image to another.

Contrast between target and background
In Sections 2.1.2 and 2.2.1 we defined two types of spectral contrast: wavelength-to-wavelength (WW) contrast within a single spectrum and pixel-to-pixel (PP) contrast between spectra. Spectroscopic interpretation of an individual pixel usually requires that there be WW contrast (although the absence of contrast in a flat, featureless spectrum still can convey information). Pixel-to-pixel contrast has two distinct aspects: spectral length (SL) contrast that defines the normalized lightness difference between pixels over all wavelengths; and spectral angle (SA) that is the angular separation of vectors in the space defined by the spectral channels (Figure 2.10). To detect a target pixel against its background, the target must differ in lightness and/or in its spectrum (Figure 6.1). By definition, a target pixel cannot be detected when its vector is identical to that of the background; otherwise it is a perfect mimic. Contrast also depends on other factors, including target and background variability and measurement uncertainty, which are discussed in Section 6.2.

Table 6.1. *Data number values on a scale of 0 to 100 in five spectral channels for A, B, C, G, and S.*

Spectra	Channel				
	1	1a	2	3	4
A	25	42	53	59	60
B	68	39	12	55	64
C	70	73	76	82	67
G	36	34	40	58	63
S	2	2	2	2	2

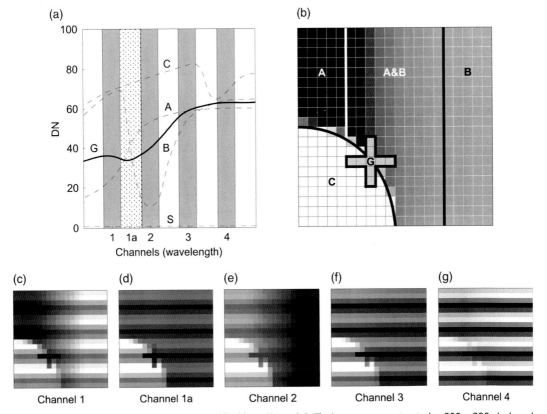

Figure 6.2. Synthetic 20 × 20-pixel image modified from Figure 3.2. The image was constructed as 200 × 200 pixels and re-sampled to 20 × 20 pixels. Spectrum G has been added, and channel 1a (stippled pattern) has been added between channels 1 and 2. (a) Spectra of A, B, C, G, and S. (b) Distribution of A, B, C, and G in the image. Horizontal shading stripes are not shown; G forms a cross that extends into the areas formerly occupied by C and mixtures of A, B, and C. (c)–(g) Images by channel. Shade, S, contributes the horizontal stripes in 10% intervals to a maximum of 50% to simulate shade. See Section 3.4.1 and Table 6.1.

Spectral length and spectral angle in the synthetic image

We have made a case that spectral contrast is essential to be able to detect a target against its background, and that SL and SA are the basic elements of spectral contrast. However, spectral contrast can be difficult to visualize in a multi-channel data space. To gain further insight into target detection, we modified the original synthetic image (Chapter 3) to include a small cross that is difficult to see against parts of the background (Figure 6.2 and Table 6.1).

Spectrum G in Figure 6.2 was invented to nearly (but not exactly) mimic mixtures of A and B, and mixtures of C and S (shade), as is shown in Figure 6.2a. In the scatter plots in Figure 6.3, G lies on or near the mixing line between A and B and the mixing line between C and S in most of the channel pairs. Therefore, in many of the two-channel projections, G lies on the background tetrahedron and mimics the other spectra or their mixtures in both SL and SA. In the plane of channels 1a and 3, however, G is detectable, because it does not mimic the background. The angle between SG and S(AB) in this projection of the data defines the contrast between target and background.

The G cross was placed in the image such that the top and right arms have a background of mixtures of A and B, and the left and bottom arms have C as background. In the individual channel images and in color composites, the pattern of a cross is easily recognized (except in channel 4), because the bottom and left arms contrast in SL with the less-shaded parts of C. However, the top and right arms are much more difficult to detect and recognize, as can be seen by covering up the bottom and left arms. The cross would be difficult or impossible to detect and recognize if we had placed it a little higher in the image against the background of mixed pixels of A and B. We should keep in mind that G could have been assigned any spatial pattern – including ones that are not easily recognized. For example, if G consisted of a horizontal row of pixels, it would be hidden in many parts of the image, including area C, because it mimics the horizontal shade stripes.

6.1.2 Detection by standard methods

Now that we have disclosed that G in the synthetic image is not a perfect mimic of its background, let us examine some of the standard image-processing methods to see how effective they are. Many of the standard methods for analyzing spectral images (Chapter 3) are designed to enhance contrast, and they can be applied directly to the task of detecting targets that are not obvious. If we think we will recognize a target by its spatial pattern, it makes sense to begin our search with simple tools that facilitate photo interpretation (Section 3.3). For example, targets may be recognizable in one or more of the black-and-white images of individual spectral channels that have been contrast-stretched. But for most of the less obvious targets, we need to work with the combined spectral data, and in some cases, to apply models (Sections 3.4 and 3.5).

Searching for G with ratios

For images containing many channels, the number of possible ratio combinations may be too large to permit examination of them all, but the synthetic image has only five channels that allow ten possible combinations. Two of the ratio combinations detect all of the cross (Figure 6.3). Ratio images effectively remove the shading from the image, which resolves the ambiguity in the places where parts of G mimic the striping. Recognition of the cross pattern also is enhanced by removing the visual clutter of the horizontal stripes.

To understand why only certain channel ratios reveal the top and right arms of G against the mixed AB background, it helps to refer to the scatter plots in Figure 6.3. Each of the scatter plots provides a different "view" of a sub-space of the data. For example, the ABC plane is largest when projected onto the channel 1/channel 2 plane, whereas in channel 1a/ channel 3, the ABC plane is seen nearly on edge. A line from the origin through G is the locus of all data pairs having the same ratio values as G. Other lines from the origin that extend through A, B, C (and their mixtures) define the ratio values for all of the other pixels in the image. For example G and C, and mixtures of A and B, all have similar ratio values in channels 1 and 2, and in the image the cross is obscure. However, in channel 1a/channel 3 and channel 1a/channel 4, G has a small but distinctly different ratio from A, B, and all of their mixtures, and there is an even larger difference between C and mixtures of A and C; therefore, the cross is entirely revealed. Indeed, in this view of the data space, G is detected as having a different SA from all other pixels, and the PP contrast between G and the background is maximum for a two-channel projection of the data. Ratios, of course, do not measure SL or SL contrast.

There are potential disadvantages to using ratios for detection. It can be tedious or impossible to analyze images of all ratio combinations when there are many channels. And even when it is feasible to display all ratio combinations, there may not be any that fully capture the SA contrast between target and background.

Searching for G with the PCT

We previously examined the application of a principal-component transformation to the synthetic image in Section 3.4.1 and in Figure 3.5. When a PCT is applied to the G-cross image (Figure 6.4), G is fully

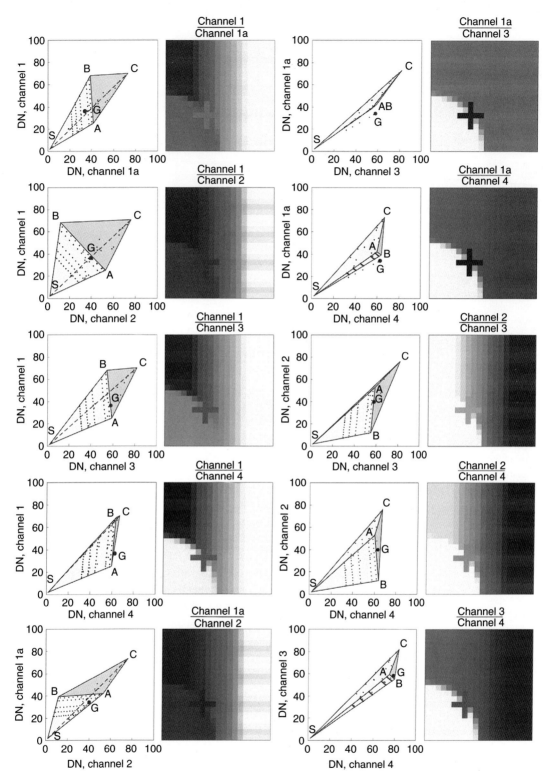

Figure 6.3. Scatter plots and channel-ratio images from the synthetic, G-cross image; G is considered to be the target, and the tetrahedron ABCS is considered to be the overall background. The ABC plane is shown in a gray tone for clarity, as in Figure 4.5. The addition of channel 1a to the G-cross image permits ten channel combinations. In the ratio images there are only two combinations of channels, 1a/3 and 1a/4, in which all of the cross contrasts with the background. These are the same channel combinations for which the scatter plots show a SA >0 radians between G and the ABCS tetrahedron.

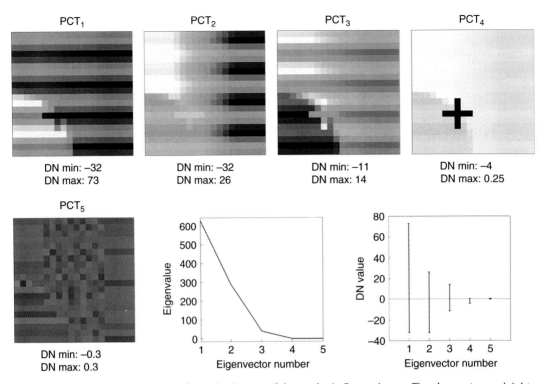

PCT₁ — DN min: −32, DN max: 73
PCT₂ — DN min: −32, DN max: 26
PCT₃ — DN min: −11, DN max: 14
PCT₄ — DN min: −4, DN max: 0.25
PCT₅ — DN min: −0.3, DN max: 0.3

Figure 6.4. Principal-component transformation images of the synthetic G-cross image. The obscure top and right arms of the G cross are revealed in PCT₄, which is close to the noise that appears in PCT₅. See Figure 3.5 for further explanation.

revealed in PCT₄. From our discussion of spectral contrast, it follows that PCT₄ happens to maximize the SA between G and mixtures of A and B. Although this fortuitous result is entirely satisfactory for detecting G, we need to keep in mind that the PCT is governed by the statistical properties of the data that vary from image to image, and that there is no simple way to predict whether a given target will be detected in any given image. Furthermore, small targets in large and spectrally diverse images are less likely to appear among the first few eigenvectors, and may become lost in the image clutter. G is close to being undetectable in the synthetic image, because PCT₄ is near the level of "noise" that characterizes PCT₅. Nevertheless, the PCT can be a powerful tool for enhancing spectral contrast. It is relatively inexpensive computationally, and it employs all channels. It is worth a try, providing that we recognize the target when we see it and we understand the image context.

Finding G by classifying

Tests of standard-classification algorithms on the G-cross image, both unsupervised and supervised, showed that only the spectral-angle classifier

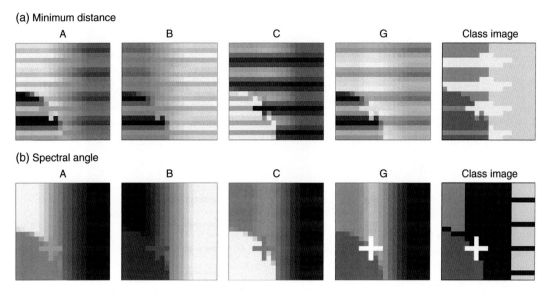

Figure 6.5. Data-attribute (rule) images and classification image for minimum-distance and spectral-angle classifiers applied to channel DNs of the synthetic G-cross image. Spectra A, B, and C are the means of 5 × 5 pixel blocks that include shading. Spectrum G is the mean of all pixels of G. Lighter pixels in the attribute images are closer to the means of the training sets. Gray tones in the class images are separate classes. Black is unclassified. (a) Minimum-distance classifier. All pixels are assigned to the nearest class. The top and right arms of the cross are not revealed in the attribute image of G or in the classified image. (b) Spectral-angle classifier. The SA attribute image is not sensitive to SL, thereby removing the clutter of the horizontal shading stripes. Below a classification threshold of 0.07 radians, G is in a separate class, and mixtures of A and B are unclassified. See also color Figure 6.5 and Figures 6.3 and 3.11.

places the top and right arms of the cross in a separate class from AB mixtures (Figures 6.5, 6.6 and color Figure 6.6; see Web). Other classification algorithms that are based on the *distance* between vectors measure overall spectral contrast that is a function of both SA and SL (Figure 6.1). Because SL is sensitive to shade, distance-based algorithms are not able to detect the G cross against the clutter of shading stripes. The spectral angle classifier, as has been pointed out, is sensitive only to SA and not to SL. The algorithm computes the angle from the mean of each training set (or from any input spectrum) to all other pixels. The data-attribute (rule) image for G defines the optimum SA contrast between G and all other pixels, and similarly, the SA is computed from each of the other training sets to all pixels. The decision as to which pixels belong to which class depends on the threshold angle selected by the analyst. By choosing a very small threshold angle (< 0.07 radians), G can be classified separately from the background of AB mixtures. The unwary image analyst who does not examine the data-attribute images to see that G and AB mixes actually are separated by a small angle, and who just displays the classified image using a larger threshold angle, would not be aware that G could be placed in its

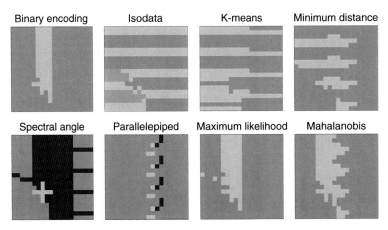

Figure 6.6 and color Figure 6.6 (see Web). Classification of the synthetic G-cross image based on channel DNs. Standard unsupervised and supervised classifiers are labeled (see Section 3.4). The best results of tests with different threshold settings are shown. For the unsupervised classifiers, isodata and K-means, the colors are computer-assigned to classes. For the supervised classifiers, A = red; B = green; C = blue; G = yellow; unclassified = black. Only the spectral-angle classifier correctly detects all of the G cross and places G in a separate class. The maximum-likelihood classifier was applied to an image having 5% random noise added. See also the data-attribute images in Figure 6.5.

own class. However, the small-angle threshold leaves mixtures of A and B unclassified.

6.1.3 Mixture models for detection

Finding G by unmixing

Spectral-mixture analysis (SMA) provides a way to detect target spectra, even at the sub-pixel scale (Chapter 4). Unlike the standard classifiers, SMA optimizes spectral contrast among endmembers by weighting the spectral channels. When an endmember is defined as a target, the other endmembers, along with the residual spectra and noise, make up the background. We applied SMA to the G-cross image (Figure 6.7), using spectra A, B, C, G, and S as endmembers. The cross is fully revealed in the G-fraction image. This is consistent with our understanding of the data and the angular separation of G from the other spectra. By including shade as an endmember, mixture analysis also reduces (but does not entirely eliminate) the influence of the shading stripes on the G-fraction image. Recall from Figure 4.6 that in fraction images the designated endmember has high fractions, and that lower fractions belong to the other (undifferentiated) endmembers. When the endmember fractions are normalized without shade, the striping pattern in the noiseless synthetic image is eliminated, and the G cross contrasts with its background. This, of course, is a best case using a synthetic

> **Endmembers as targets**
>
> If one endmember is the target, the other endmembers, the residual spectra, and the noise become the background.

> **Endmember vectors do not need to be orthogonal**
>
> Endmembers in SMA are defined by their weights in each spectral channel; therefore, endmembers usually are not orthogonal within the conventional framework of the Cartesian data space used to describe the acquired radiance data.

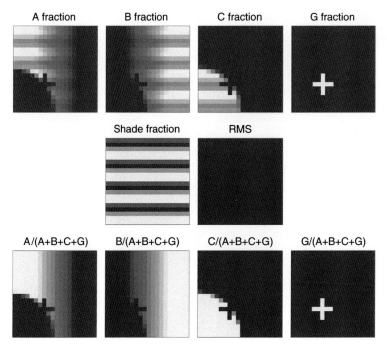

Figure 6.7. Fraction and RMS residual images of the synthetic G-cross image. Lighter tones indicate higher fractions (from 0 to 1) of endmembers A, B, C, and G. Darker tones indicate higher fractions of shade. The mixture model includes all endmembers (A, B, C, G, and S). The cross is fully revealed in the G-fraction image. The RMS image = 0 because the model uses the correct endmembers. Removal of shading stripes by normalizing reduces visual clutter (see also Figures 4.6 and 4.7).

image where all of the endmember spectra are known; however, it illustrates the power of the method, providing that we can construct a suitable model.

If we do not know the spectrum of the target, but we have a good idea of the nature of the spectral components that make up the rest of an image, we can apply SMA, and look for the target in the residual. This approach is shown in Figure 6.8, where the G cross is detected in the RMS residual when we use A, B, C, and S as endmembers. In the common situation where we do not know all of the endmembers, we can build a model by beginning with two endmembers and adding more according to the strategy outlined in Section 4.4.1.

Finding G by classifying endmember fractions

When using fractions (rather than channel DNs) most of the standard classification algorithms detect the G cross in the data-attribute images and designate it as a distinct class (Figures 6.9, 6.10 and color Figure 6.10; see Web). The spectral-angle classifier produces the same class images with channel DNs and with fractions; however, the minimum-distance classifier produces a correct classification only with the fraction input (see Figure 6.5). Interestingly, the unsupervised classifiers, binary encoding and K-means also correctly classify the G cross and the other components of the synthetic image when applied to the endmember

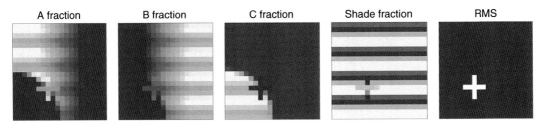

Figure 6.8. Detection of the G cross in the RMS residual of the synthetic G-cross image. Lighter tones indicate higher fractions of A, B, C, and higher RMS residual. Darker tones indicate higher fractions of shade. The mixture model does not include G. The cross is fully revealed in the RMS-residual image.

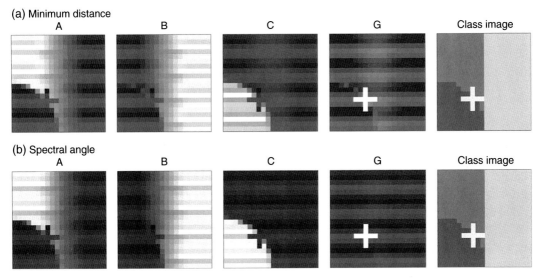

Figure 6.9. Detection of the G cross by classification of endmember fractions. Data-attribute images and classification images for the minimum-distance and spectral-angle classifiers applied to endmember fractions of the synthetic G-cross image. Classes A, B, C, and G are defined from training areas on the respective un-normalized fraction images. Training pixels for A, B, and C are the means of 5×5 pixel blocks. Spectrum G is the mean of all pixels of G. Lighter pixels are closer to the means of the training sets. For the minimum-distance class image, all pixels are assigned to the nearest class. For the spectral-angle class image, all pixels with angles < 1 radian are classified. Classes (class image only) are distinguished by gray tones in the light-to-dark order: G, B, A, C. Both algorithms reveal all parts of the cross in the data-attribute images, and both place G in a class by itself. Notice that the classifiers do not define a class that consists of mixtures of A and B. See also Figures 5.13, 5.14, 6.5 and color Figure 6.6 (see Web).

fractions. The same algorithms when used on the channel-DN data, do not place G in a separate class. These results emphasize the advantage of enhancing spectral contrast by applying a mixing model before classifying.

Limitations of SMA for detection

In Chapter 4 we described SMA as an analytical method that applies simple physical models to complex surfaces. However, simple models

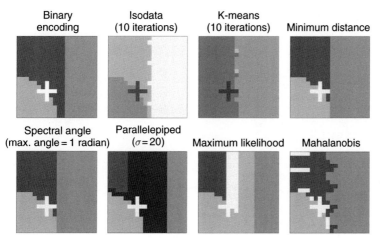

| Binary encoding | Isodata (10 iterations) | K-means (10 iterations) | Minimum distance |

| Spectral angle (max. angle = 1 radian) | Parallelepiped ($\sigma = 20$) | Maximum likelihood | Mahalanobis |

Figure 6.10 and color Figure 6.10 (see Web). Classification of the synthetic G-cross image based on fractions of endmembers A, B, C, G, and S. The best results of tests with different threshold settings are shown. For the standard, unsupervised classifiers, isodata and K-means, the gray tones and the colors are computer-assigned to classes. For supervised classes, A = red; B = green; C = blue; G = yellow; unclassified = black.

that employ only a few (three to five) endmembers cannot account for all of the spectral complexities of most landscapes, and adding more endmembers requires a more complex, iterative analysis, or it results in unstable solutions. If we know the spectrum of a target, the optimum conditions for detection occur when we also know the spectra of all of the endmembers that comprise the background, as is illustrated in Figure 6.7. If we do not know the identity of the target, and we look for it in the RMS residual, the optimum conditions for detection occur when we account for all of the background endmembers, and they are well-modeled (Figure 6.8).

Even when all of the endmembers are known, SMA still does not account for their natural variability, because each endmember is characterized by the mean spectrum of a training area or by a "representative" reference spectrum. Variations within endmembers may mimic the spectra of mixtures between endmembers (Figure 4.4), thereby reducing fraction accuracy and limiting detectability. Smith *et al.* (1994a, 1994b) addressed these limitations of SMA by devising a method known as foreground-background analysis (FBA) that uses matched filters to optimize contrast between a variable target (foreground) and a variable background.

> **Endmember variability**
>
> Spectral-mixture analysis does not account for the natural variability of end-member spectra. Variable endmembers may adversely affect fraction accuracy and detectability.

6.1.4 Matched filtering

In the field of signal processing, matched filtering is used to maximize a signal relative to noise and clutter. The concept has been adapted to remote sensing to detect spectral targets against an image background. A target may be selected either from reference spectra or from a training area in an image. A matched-filter (MF) algorithm can be thought of as

> **Matched filtering**
>
> This analytical method has its origins in the field of signal processing. When applied to spectral images it is one of several ways to find vectors that "match" the mean of a training area.

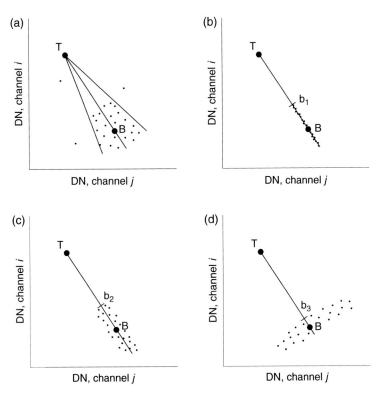

Figure 6.11. Sketches to illustrate matched filtering in the data space of channels i and j; T is a single-spectrum target; B is the mean of a cluster of background spectra. (a) The points that comprise the background are defined as being within some angle originating at T. (b) Background spectra projected onto the TB line; T is given a score of 1 and B a score of 0; b_1 marks the limit of the cluster of background points projected onto the TB line. Points between B and b_1 are part of the background, and have MF scores >0 that are proportional to distance TB. An MF score of b_1 is the detection limit of T, because T is mimicked by background. (c) A different cluster of background spectra is elongated in the TB direction. The orientation of the TB line limits the unambiguous detection of T to b_2. (d) A third cluster places background pixels farther from T at b_3, thereby improving the detectability of T.

creating an imaginary line in data space between target and background that defines the direction of optimum spectral contrast (Figure 6.11a,b). Target and background spectra are projected onto this line. Spectra that fall far off the line, perhaps representing noise or unusual scene components, can be excluded by applying threshold filters. The mean of the target is given a score of 1 and the mean of the background is given a score of 0; thus, pixels having high scores have more contrast with the background and generally are considered to be more like the target; MF scores are rendered as images by assigning higher scores to lighter pixels (e.g., Figure 6.12). Detectability of a target increases with greater spectral contrast between target and background, but detectability decreases as the background becomes more variable. In Figure 6.11b, pixels belonging to a variable background extend to b_1 on the TB line. This is the detection limit of T, because T no longer can be distinguished from background at b_1. However, it is not straightforward to tell at what MF score the variations in the background begin to mimic the target; therefore, just because pixels have MF scores > 0, does not mean that target has been detected.

Spectra that naturally cluster along a target-background line are consistent with (but not proof of) mixtures of two endmembers

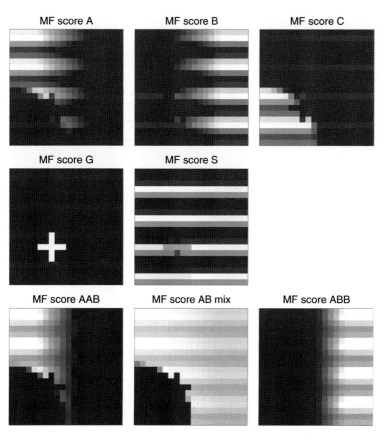

Figure 6.12. Matched-filter images for the synthetic G-cross image. Lighter tones indicate higher scores (on a scale of 0 to 1) for targets relative to the whole-image background. When spectra A, B, C, S, and G are designated as targets, the MF images are able to optimize the contrast between each target and the background, although shade striping remains as distracting clutter. When the target is defined by mixtures of A and B rather than by the individual spectra, the algorithm scores the pure spectra >1 and higher than the mixtures; AAB indicates target fraction A > B; AB indicates fraction A = fraction B; ABB indicates target fraction of B > A.

(Figure 6.11c). In the case of true mixing, MF scores are the same as fractions of a target endmember in a two-endmember mixture, and MF scores >0 imply detection of the target at a sub-pixel scale. However, ambiguity due to mimicking still may exist if the background is variable, as was discussed in Section 4.1.5. One form of matched filtering, FBA, specifically seeks to minimize mimicking that is caused by variability in a restricted background by finding the target-background line that results in the shortest distance through the cluster of background points (compare Figures 6.11c and d).

In other applications of matched filtering, the entire image is defined as the background, and it is unlikely that projections onto the target-background line represent a simple, binary mixing trend. Furthermore, a whole-image background can be highly variable, and may make it even more difficult to interpret MF scores in terms of a detection threshold where target is mimicked by background. Unfortunately, it has become commonplace in matched filtering to refer to any target or background as "endmembers," whether they mix or not. As discussed in Chapters 4 and 5, we reserve the term "endmember" for spectra that mix.

MF and mixing

Matched filtering is a special case of two-endmember mixing in which the target is one endmember and the background is the other.

Foreground-background analysis

Foreground-background analysis is a version of matched filtering designed to minimize fraction uncertainty that is introduced when spectrally variable endmembers mix (Smith *et al.*, 1994a). The method also applies to target detection when target and background mimic mixtures in the data space (Figure 3.12). Spectra, designated "foreground" and "background," are derived from reference spectra or from image training areas. Generally, a single-spectrum foreground is selected (a target), and it is paired with one other spectrum as background or with a composite background. Unlike low-probability detection (discussed below), the whole image is not selected as the background. In FBA, it is expected that foreground and background will be variable if they have been defined from image training areas. The objective is to find channel weightings that define the line(s) of optimum overall spectral contrast between the foreground and one or more backgrounds. The optimum conditions for target detection or for quantifying the fractions of the target, are when there is maximum distance from target to background (Figure 6.1) *and* the amount of mimicking caused by variability is minimum (Figure 4.4). In the case of clusters of foreground and background that are not symmetric, the amount of mimicking is minimized by finding a line in the data space that passes through the least distance inside each cluster (Figures 6.11c,d). By reducing mimicking, FBA improves the detectability of targets, including sub-pixel targets, that are set in highly variable backgrounds. This has proved to be an advantage for detecting and quantifying sparse vegetation on spectrally variable soils in arid regions (Smith *et al.*, 1994a, 1994b; Ustin *et al.*, 1999).

Matched filtering and low-probability detection

An MF algorithm that defines the background as the whole image is included in at least one commercially available image-processing software package. The low-probability detection (LPD) algorithm (Harsanyi and Chang, 1994) is described in the technical literature, but, like FBA, may be less accessible to most image analysts and field investigators. With MF and LPD algorithms, the analyst selects a target, either from reference spectra or from an image training area; however, unlike FBA, the background is defined automatically from the statistical properties of the image as a whole.

Tests of an MF algorithm

We tested an MF algorithm that defines the entire image as background on the synthetic image using A, B, C, and G as targets. Each target correctly received the highest MF scores relative to the rest of the image, and the G cross was fully detected (Figure 6.12). However, when we selected mixtures of A and B as targets, the algorithm produced spurious results. A target defined by mixtures having more A than B scored the pure A pixels >1 and higher than AB mixes, and mixtures having more B than A scored pure B pixels >1 and higher than AB mixes. A target defined by approximately equal proportions of A and B resulted in equally high (>1) scores for A, B, and all AB mixes. These results could be seriously misleading if we make the mistake of inter-preting the highest (>1) MF scores as locating the target. The lowest scores, however, did correspond to those pixels that were most unlike (contrasting with) the target.

Tests on the Amazon Landsat TM image gave results similar to those on the synthetic image. We used the same training areas to define MF targets that we had used previously to define endmembers for a successful mixture model (Figure 5.11). Targets selected from the end-members gave the highest MF scores (Figure 6.13). However, when we defined several different targets in the zone of physical mixing of the dark Rio Negro and the muddy Rio Solimoes, one or both of the rivers received the highest MF scores, rather than the mixed pixels. This result parallels the example from the synthetic image. Another example of a mixed target from the same image is primary forest (terra firme forest) that can be modeled mainly as mixtures of green vegetation and shade (Figure 5.11). The MF-score image for terra firme forest is noisy (scan-line stripes and poor spatial resolution), and it does not consistently discriminate the mature forest from regeneration (light-green vegeta-tion) in forest clearings (Figure 6.13). Inspection of an enlarged and strongly stretched part of the MF image reveals that most patches of regeneration have slightly higher MF scores than the forest itself; however, some patches of green vegetation have similar scores as terra firme forest, and the two are indistinguishable. These results are consistent with the examples from the synthetic image and the mixed rivers, and suggest that matched filtering can be ambiguous if the target consists of mixtures of scene components.

From the above examples we see that matched filtering can opti-mize spectral contrast, and, therefore, it can be effective for detecting targets; however, there are limitations. For best results, a target should not comprise a large part of the background, a caution that is particularly important when the entire image is used as the background. This point is well illustrated in the poor quality of the MF-score image of the terra firme forest that is the main component of the Amazon sub-scene. In addition, a target should not consist of mixtures of the basic spectral components of the background. Unless we have analyzed an image beforehand to determine whether mixing is a factor, it may be difficult

A limitation of matched filtering

Because the nature of the background is not always evident beforehand, it may not be possible to predict whether matched filtering will give the desired results.

(a)

Amargosa Range

Death Valley

Panamint Range

Figure 6.19

(b)

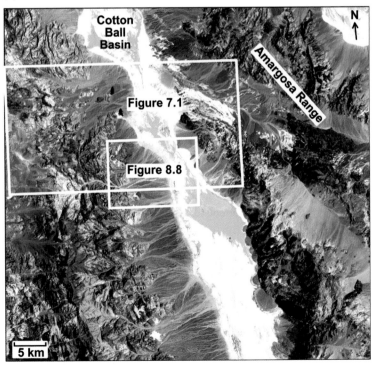

Cotton
Ball
Basin

Amargosa Range

Figure 7.1

Figure 8.8

N

5 km

Plate 1. June 7, 1984 Landsat 5
TM image of the Death Valley
area, California and Nevada.
(a) Location image showing b
and Figure 6.19. (b) Sub-image
of Death Valley showing the
locations of Figures 7.1 and
8.8. Red = Band 5;
green = Band 4; blue = Band 3.
Path/row 040/035.

Plate 2. August 10, 1992 Landsat 5 TM image of Seattle, Washington State. (a) Location image showing Puget Sound and the Seattle area. The larger box is the location of b. The smaller box locates Figure 7.20. (b) Sub-image of the Seattle urban area that is used in several examples in the text. Red = Band 5; green = Band 4; blue = Band 3. Path/row 046/027.

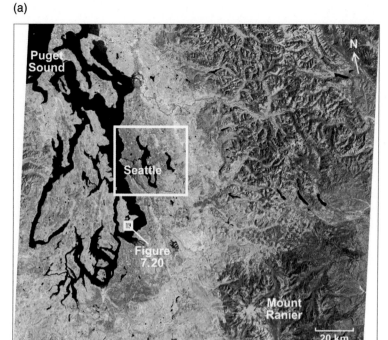

(a)

(b)

(a)

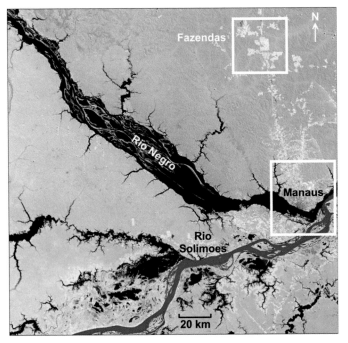

Fazendas

N

Rio Negro

Manaus

Rio
Solimoes

20 km

(b)

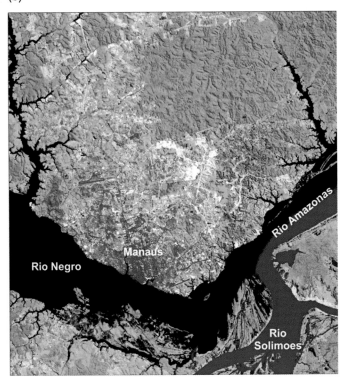

Rio Amazonas

Manaus

Rio Negro

Rio
Solimoes

Plate 3. August 8, 1991 Landsat 5 TM image of Manaus, Brazil. (a) Location image showing the city of Manaus at the confluence of the Rio Negro and the Rio Solimoes that join to make the Rio Amazonas. The location of the Fazendas sub-image in Plate 4 is shown in the top box. The location of b is shown in the lower box. All images are used as examples in the text. (b) Manaus city sub-image. Red = Band 5; green = Band 4; blue = Band 3. Path/row 231/062.

Plate 4. Fazendas sub-image. See Plate 3. The larger box locates Fazenda Dimona; the smaller box locates Figure 7.26.

(a)

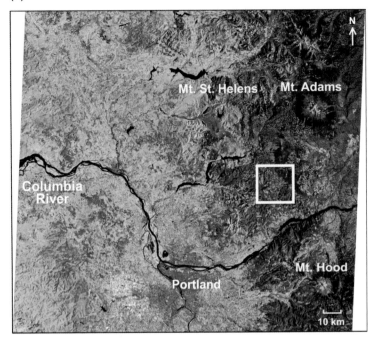

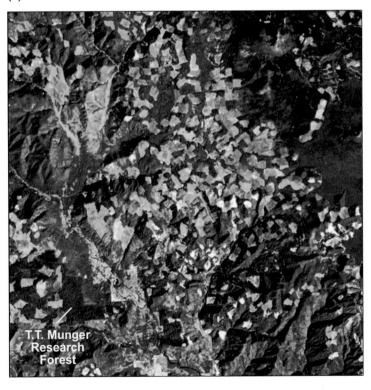

Plate 5. July 31, 1994 Landsat 5 TM image of Gifford Pinchot National Forest, Washington State. (a) Location image showing the Columbia River and Portland, Oregon. The white square is the location of image b. (b) The sub-image is used in several examples in the text. Red = Band 5; green = Band 4; blue = Band 3. Path/row 046/028.

(b)

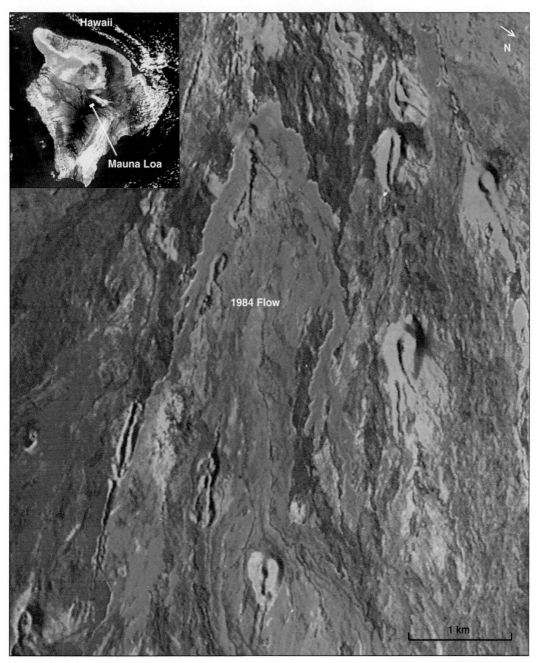

Plate 6. NASA aircraft image of basalt flows on Mauna Loa, Hawaii from the Thermal-Infrared Multispectral Scanner (TIMS). Youngest flows are blue, but silica coatings have accumulated on the older rocks causing their spectral response to change from blue to red and then to green. See Section 8.12, Figure 8.7, and color Figure 8.7 (see Web).

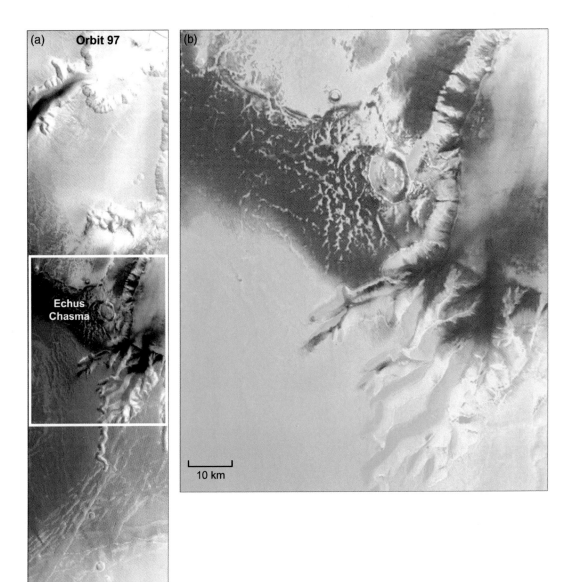

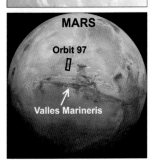

Plate 7. February 9, 2004 Mars Express image of the Echus Chasma region of Mars (north of Valles Marineris). (a) Color composite image of part of orbit 097. Red = channel 3 (0.75 μm); green = channel 2 (0.53 μm); blue = channel 1 (0.45 μm). The area outlined in white is enlarged in b. (b) Color fraction image based on spectral-mixture analysis. Magenta and yellow areas are interpreted as smooth at the sub-pixel scale (<50 m). See Section 7.2.3. Images derived from data supplied by the European Space Agency and G. Neukum.

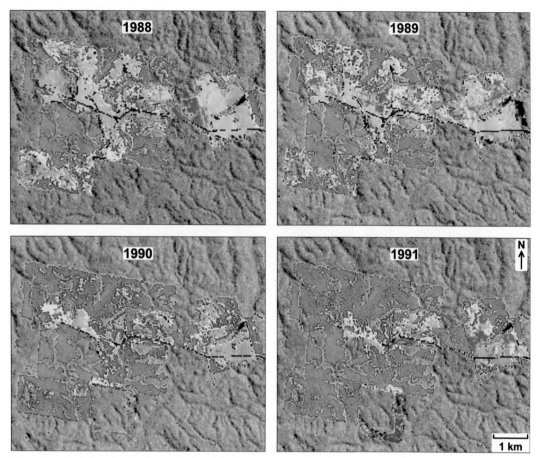

Plate 8 Land-cover change at Fazenda Dimona, central Amazon Basin, Brazil (Plates 3 and 4). Classes of fractions of endmembers are derived from August 1988, 1989, 1990, and 1991 Landsat TM calibrated to reflectance. Gray = primary forest; green = closed-canopy regeneration; blue = open-canopy regeneration; yellow = pasture; light orange = sparse pasture or partially cleared slash; dark orange = dry pasture or slash; black = soil. See Section 8.3.3. From Adams *et al.* (1995) with permission from Elsevier.

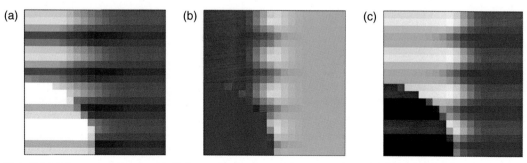

Plate 9 (a) Color composite of synthetic image. Red = channel 4; green = channel 2; blue = channel 1. (b) Color fraction composite. Red = A; green = B; blue = C. (c) Color fraction composite. Red = (1 − shade); green = A; blue = B. See Section 3.4.1 and Figures 3.2, 5.7, and 5.9.

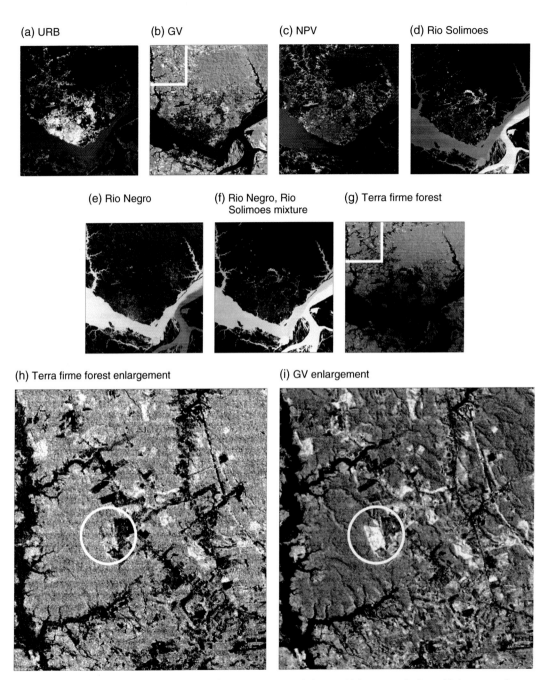

(a) URB   (b) GV   (c) NPV   (d) Rio Solimoes

(e) Rio Negro   (f) Rio Negro, Rio   (g) Terra firme forest
                          Solimoes mixture

(h) Terra firme forest enlargement   (i) GV enlargement

Figure 6.13. Matched-filter images of the Landsat TM Amazon sub-image. Lighter tones indicate higher scores for targets relative to the whole-image background. (a)–(e) The targets are the image endmembers: URB, GV, NPV, Rio Solimoes (a muddy river used as a soil proxy), and Rio Negro (a shade proxy) that were used in the mixture model for Figure 5.11. Image endmembers produce the highest MF scores. (f) The target is the zone of mixed waters at the confluence of the Rio Solimoes and Rio Negro. The target of mixed pixels does not get the highest MF scores; instead, one or both of the endmembers score highest, depending on the mixing ratio. (g) and (h) The target is terra firme forest (h is an enlargement of part of image g) – mainly a mixture of GV and shade. The target of mixed pixels does not have the highest scores. Instead, some patches of light vegetation (GV) score highest; other patches of GV score the same or lower than the target (circle in h). (i) The same patch as circled in h has a high MF score in an enlargement of the GV image.

to tell whether a proposed target represents discrete scene components or spectral mixtures of some components. Therefore, it may not be possible to predict whether matched filtering will give the desired results.

6.1.5 Prospecting for targets

Resolved vs. sub-pixel targets

When a target spectrum mixes with other spectra, the mixed spectra will differ from the target in SL and/or SA. With progressively larger pixel footprints, a target vector will mix with the background spectra, and, referring back to Figure 6.1, the target spectrum (T) will move along one of the mixing lines toward B_1, B_2, or B_3. We can detect the target, even though it is not spatially resolved as a "pure" pixel, providing that we can interpret the mixtures of T and its background. Thus, un-mixing pixel spectra is central to target detection. Unlike SMA, though, detection of a target against its background does not require that we quantify the fractions of the endmembers. First, we need to detect what we are looking for – then, if it is important, we may try to find out how much of it is there (Clark's law). In fact, in many instances we may be satisfied just to be able to narrow the selection of those pixels that are most likely to contain the target. This is the prospector's approach: look for gold in the most likely places.

Spectral-length (albedo) contrast

Recall from Section 2.2.1 that SL is the length of a vector, whereas SL *contrast* is defined as the normalized difference in length between two vectors. In most spectral images, PP contrast consists of both SA and SL contrast (Figure 6.1). Although two or more spectral channels are needed to define SA contrast, we can determine SL contrast between pixels in each of the black-and-white images measured by individual channels. Some channel images will have more SL contrast between a selected target and its background than others. The maximum SL contrast occurs in one of these channels.

Individual channel images are SL images, but they typically include the confusing effects of shading and shadowing. We want to measure the contrast between tangible scene components, and not the effects of topography or surface texture. Therefore, to measure SL contrast we want to find the maximum SL of each known component in each channel. As an example, consider the SL contrast between A and C in the synthetic image (Figure 6.2). Table 6.2 lists the maximum SL values (DNs of the least-shaded pixels) for each component. The maximum SL contrast between A and C, based on the least-shaded pixels, occurs in channel 1. Unlike SA, there is no advantage in using multiple channels

> **Maximum SL contrast**
> The maximum SL contrast between two components occurs in a single channel. Algorithms that use multiple channels are *less* sensitive to an albedo contrast between target and background.

Table 6.2. *Spectral length and spectral-length contrast from the synthetic image.*

Spectra	Channel*						
	1	1a	2	3	4	1 + 2	All channels
SL$_A$	24.3	40.7	51.3	57.2	58.1	56.8	159.8
SL$_B$	65.9	37.8	11.7	53.3	62.0		
SL$_C$	67.8	70.7	73.6	79.4	64.9	100.1	107.4
SL$_G$	34.9	33.0	38.8	56.2	61.0		
SL$_{AC}$ contrast	0.47	0.27	0.18	0.16	0.06	0.28	0.20

* Spectral length (SL) in each channel is the DN of the least-shaded pixels. Spectral length in multiple channels is the square root of the sum of the squares of the channel DNs. $SL = \sqrt{\sum_i DN_i^2}$; i = channel number. SL$_{AC}$ contrast = (SL$_C$ − SL$_A$)/(SL$_C$ + SL$_A$).

to maximize SL contrast, because the contrast between two components over multiple channels cannot be greater than that in the single channel having the maximum SL contrast. In Table 6.2, there is less SL contrast between A and C in the combination of channels 1 and 2 than in just channel 1, and, when all five channels are used, SL contrast decreases by more than half that in channel 1 alone. Virtually all of the standard methods for analyzing spectral images make use of multiple channels, thus, they are designed primarily to exploit SA contrast. However, if target and background differ mainly in albedo, the best approach for detection may be to find the single channel in which SL contrast is greatest. Algorithms that use multiple channels actually may make it more difficult to detect a target.

Detection using SL (albedo)

Spectral length can be used to detect the sub-pixel presence of one or more targets, providing certain conditions are met. The first requirement is to identify the components that rank highest and second highest in SL within the target prospect area. The second requirement is that we know the shading pattern. In the simplest case, imagine a flat, evenly illuminated surface that contains only two surface materials, one lighter and the other darker. The only way the darker material can appear lighter than its maximum SL is when it is contaminated by the lighter material. The lighter material can have less than maximum SL when it is mixed with the dark material *or* when it is mixed with sub-pixel shade (Table 1.1). We can use this simple, but often overlooked, approach to identify all of the pixels of C in the synthetic image that are mixed with

A along the AC contact. However, first let us be clear about what we mean by the target prospect area, because this determines which channels to analyze.

If the target prospect area is restricted to that part of the synthetic image where we know that the only components are A and C, then we can select channel 1 that has the maximum SL contrast. By restricting the search to areas containing only A and C we eliminate the ambiguity that is introduced by other components and their mixtures. If we do not restrict the target search in this way, and we use the whole synthetic image, mixtures of A and C may mimic other components and their mixtures, as is shown in the histogram of channel 1 (Figure 6.14a).

It may not be possible to identify a search area that we are sure contains only two components: a target and a single-component background. In this case, we need to examine the rankings of the maximum SL values of each component in each channel, and select a channel that avoids ambiguity. This is easily done using Table 6.2, where C ranks first and A ranks second in maximum SL in channels 1a, 2, and 3. For example, in the histogram of channel 2 in Figure 6.15a, all pixels between C and A contain only these two components plus shade. These pixels are displayed in Figure 6.15b by setting pixels for which DN is $\leq DN_A$ to 0 DN, and pixels for which DN is $>DN_A$ to 255 DN. Pixels having DNs between A and C contain at least some fraction of the highest-ranking component, C; however, the image does not show shaded pixels of C that have DNs $< DN_A$.

The two largest spikes on the channel 2 histogram between A and C are rows of shaded C (rows 11, 13, 16, and 18). Other pixels are mixtures of A and C, and these are labeled as %A in Figure 6.15c. The gray tones of the AC mixtures are governed by the fractions of the endmembers, as can be seen by the graph in Figure 6.15d. The linear nature of the mixing is consistent with the way the synthetic image was constructed. In one respect this is a trivial example, because all we have done is select a single channel and apply a contrast stretch. However, by paying attention to the shading pattern and the ranking of the components, we are able to detect trace fractions (1%) of A and of C unambiguously.

In Figure 6.16 we show an example from a Landsat TM scene in which SL contrast is useful for detection. The area of interest is a relatively flat, alluvium-covered valley floor in arid, southwestern Nevada, USA. From field work nearby, we know that older, undisturbed alluvial surfaces are relatively dark, owing, among other factors, to the accumulation of dark desert varnish on rock surfaces and the removal of exposed, fine materials by wind. The rocks that are exposed in stream channels have been abraded, and are not so dark as those between

Figure 6.14. Spectral-length images and histograms of the G-cross image; DNs are normalized to the least-shaded pixels of C. In the histograms the labels A, B, C, and G indicate the least-shaded pixels of these components. (a) Spectral-length image and SL histogram for channel 1; SL contrast between A and C is 0.47. (b) Spectral-length image and normalized histogram for channels 1 and 2; SL is the square root of the sum of the squares of the channel DNs; SL contrast between A and C is 0.28. (c) Spectral-length image and normalized histogram for all five channels; SL is the square root of the sum of the squares of the channel DNs; SL contrast between A and C is 0.20. See Table 6.2.

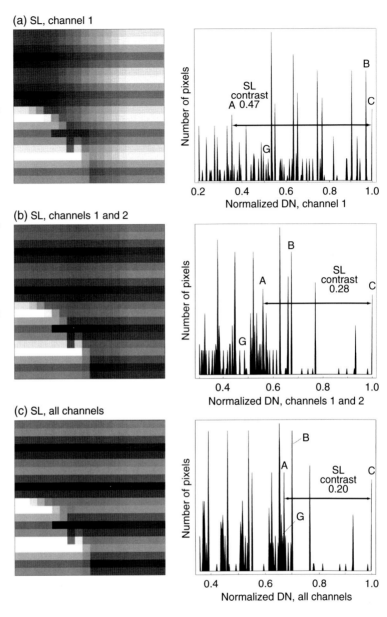

(a) SL, channel 1

(b) SL, channels 1 and 2

(c) SL, all channels

channels. Less active or abandoned channels darken with time to blend with the inter-stream areas. Stream channels, new and old, are distinguished by their sinuous, braided patterns. Disturbance of all but the youngest alluvial surfaces exposes lighter sub-surface materials.

Our immediate objective in analyzing this Landsat TM sub-image is to detect patterns of human activity. There are prominent light/dark variations in the image that include regular linear and geometric

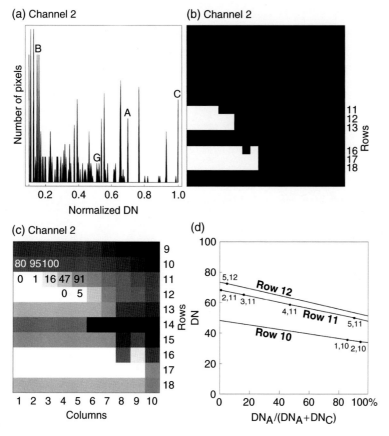

Figure 6.15. Detection of AC mixtures in channel 2 of the G-cross synthetic image. (a) Histogram of channel 2, normalized to the DN of the least-shaded pixels of C. (b) Image of pixels having DNs $>DN_A$ and $\leq DN_C$. (c) A 10×10 pixel subset of channel 2. Numbers on pixels are $DN_A/(DN_A + DN_C)$ in %. (d) Graph of mixtures of A and C for rows 10 (0.40 shade), 11 (0.20 shade) and 12 (0.10 shade). Pixels having DNs $< DN_A$ are ambiguous, because they could contain shaded C or C mixed with B.

patterns created by roads and other structures. These obvious patterns are evident in all channels. But there also are a few subtle patterns of disturbance that require maximum spectral contrast to be seen. In this example, we focus on a faint elliptical shape (the "ellipse") marked by an arrow in the right center of the sub-image in Figure 6.16b. We need to maximize SL contrast between disturbed and undisturbed desert surface, because the valley floor shows little spectral (channel-to-channel) variability. It is possible to measure SL contrast on the valley floor, because the terrain is relatively flat, thereby relaxing the ambiguity caused by topographic shading. From the previous discussion, we know that maximum SL contrast will occur in a single channel, rather than using multiple channels, and we know that we must find a channel where the target is lighter than its immediate background. We examined images of each of the six VIS-NIR Landsat TM Bands, and found that the ellipse is not evident in TM Bands 1 and 2; it is faint in Band 3; and it is obvious in Band 4 and, to a lesser extent, in Bands 5 and 7. By contrast-stretching the high-DN range of TM Band 4, the ellipse is clearly

(a)   (b)

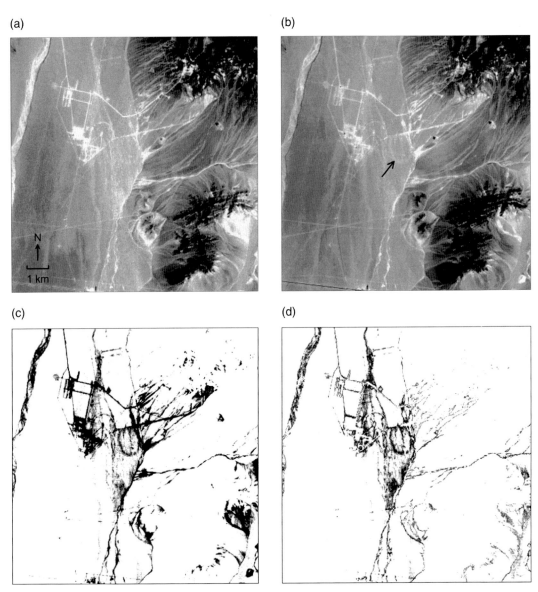

Figure 6.16. Landsat TM sub-image of Amargosa Valley, Nevada. Subtle patterns of human disturbance are detected by enhancing SL contrast between light sub-surface soil that is exposed underneath the darker, undisturbed surface. The target is the light elliptical pattern (arrow in b). (a) Thematic Mapper Band 1 image; the target is not evident. (b) Thematic Mapper Band 4 image; the target is evident. (c) Complemented TM Band 4 image in which the ellipse is dark. Pixels <115 DN are set to 255 DN; pixels >121 DN are set to 0 DN. (d) Minimum-distance data-attribute image. The reference vector is a single pixel of the highest DN in TM Band 4 on the ellipse. Darker pixels are closer in Euclidean distance to the reference vector. Notice that the minimum-distance image does not have as many pixels that mimic the ellipse as does the complemented TM Band 4 image in c. See also Figures 6.19 and 6.20.

distinguished from its immediate, lower-SL background. As in the example with the synthetic image, this is a simple detection strategy; however, it is the one that takes advantage of the maximum PP contrast, in this case, SL contrast, and it is effective for resolved and sub-pixel target material.

By contrast-stretching the TM Band 4 image, the ellipse is detected against its immediate background; however, other parts of the image also have similar lightness in Figure 6.16c, and we would like to find another way to measure SL contrast that is more specific to the ellipse itself. A solution is to select a training set from the ellipse and to use it to make a minimum-distance data-attribute image. Although there is a potential ambiguity, because the minimum-distance algorithm is sensitive to shading and shadows (see Figure 6.5), the problem is minimal in this image of the evenly illuminated valley floor. In Figure 6.16d the ellipse is clearly detected, and there is less mimicking in other parts of the image. The reference vector that we selected as input to the minimum-distance algorithm is a single pixel of the highest DN in TM Band 4 from the ellipse. This choice is important, because high-DN pixels in this context imply more pure samples of the disturbed surface. A more representative sample of many pixels from the ellipse includes more mixtures with the background, with the result that the mean spectrum has a smaller Euclidean distance from the background and less SL contrast.

In Figure 6.17 and color Figure 6.17 (see Web) we show additional examples of SL contrast from the Fazendas Landsat TM image (Plate 4). The subset image covers an area recently cleared from tropical forest for a palm-oil plantation. The clearings are mainly confined to the flatter areas between drainages, therefore shading does not affect SL values significantly. The main scene components in the recently cleared areas are soil, second-growth vegetation and woody debris (slash) from clearing of the forest. In this example, the main scene components are spectrally distinct from one another, and the background is spectrally variable. Because soil, green vegetation and woody debris differ spectrally, we can distinguish them by SA; however, we also can select individual TM Bands in which high fractions of the individual components can be separated by SL. We selected the Bands from this image based on the spectral properties of the materials on the ground. In other words, we applied a simple physical model. Soil that is freshly exposed along roads has maximum SL contrast with its immediate surroundings in Band 3. The lightest green vegetation, kudzu vine (*Pueraria lobata*) that is used as a ground cover, has maximum SL contrast with its background (including other green vegetation) in Band 4, and woody debris has maximum SL contrast with its background in Band 5. By contrast-stretching each of these Bands to enhance the high-DN pixels, and interactively observing the image patterns, we select the DN ranges

(a) TM Band 3 > 50 DNs   (b) TM Band 4 > 105 DNs   (c) TM Band 5 > 102 DNs

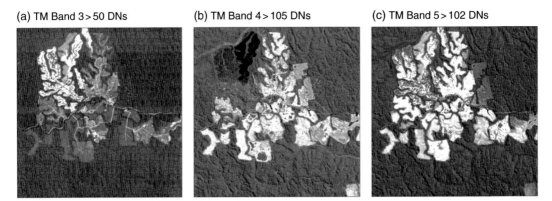

Figure 6.17 and color Figure 6.17 (see Web). Spectral-length detection of targets in a 400 × 400-pixel sub-image from the Landsat TM Fazendas image. See also Plate 4. The image depicts irregular cleared areas surrounded by primary tropical forest. (a) Band 3; red, >50 DN. Red pixels represent soil exposed along access roads and in recent clearings. (b) Band 4; red, >105 DN. Red pixels represent kudzu-vine ground cover. (c) Band 5; red, >102 DN. Red pixels represent woody debris (slash) from recent clearing of forest.

that show the highest fractions of each component. In color Figure 6.17 (see Web), the pixels in the selected DN ranges (density slices) are colored red. The low-DN threshold is determined by photo interpretation.

Spectral-angle contrast

The angle between vectors (SA) generally is the most important spectral metric for distinguishing target from background. Indeed, this is why multiple channels are used to make spectral images; the more channels available, the more likely that target and background will differ in their DN values in two or more channels. Ratio-based measurements such as the NDVI, describe variations in spectral angle in the domain of two spectral channels. With three or more channels, the most powerful SA tools for target detection are SA data-attribute images, PCT, SMA, and match filtering. All of these image-processing tools were able to detect G against the background of AB mixtures (Sections 6.1.2 and 6.1.3). As long as there is SA contrast between a target and its background, and the angle can be measured above the system noise, the target can be detected. However, which tools should we apply, and under what circumstances?

If we know the spectrum of the target, or we can train on the target in one part of an image and we want to detect it in another area, a simple approach is to make a data-attribute image using the SA classifier. We select the reference spectrum or one spectrum from a training area, and apply the algorithm. We are not interested in the class image, but we do want to work with the data-attribute image and its histogram. We prefer to display the smaller angles as lighter pixels, because it is easier to see subtle differences between target and background. The target spectrum that we select is the only input to the classification algorithm, and it will have SA = 0; the rest of the pixels that are not perfect SA mimics will have SA > 0. Using the contrast-stretch sliders on the histogram, we can set the higher-angle pixels to 0 DN. As the zero slider is moved to lower angles, it reaches the threshold angle at which all background pixels are extinguished on the image, and only the target pixels are still illuminated. In Figure 6.18 we illustrate this technique to separate the G cross in the synthetic image from the AB background.

When we use only one spectrum (target) as input for the spectral-angle classifier, it becomes the reference vector, and the maximum spectral angle is measured to each of the pixels in the rest of the image (background). The background pixels that are closest in SA to the reference vector are the ones that are most difficult to distinguish from the target – mixtures of 25% A and 75% B in the case of the synthetic image. However, the conventional way to use the SA classifier is to select *several* spectra from pixels at training sites. Each input spectrum is used to make its own data-attribute image, and each of these images displays the angles (as DNs) from the input spectrum to all other pixels. But, for >2 channels, the angular relationships between vectors may vary, depending on the choice of input spectrum. For example, the SA between A and B is 0.21 radians if G is the input spectrum, and it is 0.53 radians if A is the input spectrum, because the direction of the angular measurement changes. This is just another way of saying that the direction in the data space that has the maximum SA contrast between any two vectors is not necessarily the maximum contrast between any other vectors. Unfortunately, it is difficult to visualize the spatial arrangement of vectors in a multi-channel data space; therefore, it usually is not intuitively obvious which pixels are closest to the target and, therefore, the most likely to obscure the target. Because the data space is defined by the channels, it also follows that the spectral angles between vectors depend on which channels are selected and how channels are weighted. In the simple case of two channels, each with a weighting of 1, we can measure the SA on the scatter plots in Figure 6.3. The plot for channels 1a and 3 has the largest SA (0.10 radians) between G and the rest of the pixels, and this is shown in the ratio image of channels 1a and 3 where G has maximum SA contrast with its background, and A and B have the same angular and ratio values.

An advantage in using the SA to measure contrast is that, as long as we know the spectrum of the target, we do not need to know the spectra of each part of the background. The situation is different for matched filtering where, besides defining the target, we either have to specify

SA contrast

Spectral angle between target and background depends on the choice of channels.

Figure 6.18. Spectral angle data-attribute images and histogram for the synthetic G-cross image. See also Figure 6.12. Lighter pixels in the images have smaller angles. The lightest pixel of G (top of cross) is the reference vector.
(a) Histogram of spectral angles. The gray background on the histogram indicates pixels ≥ 0.20 radians that are set to DN $= 0$ in the images. The top pixel of G has SA $= 0$. Shaded pixels of G vary from the reference vector by < 0.01 radians which is the limit of angular resolution. Pixels of G have the most SA contrast with a background of B (0.38 radians). Pixels of G have the smallest SA contrast with a background of 25%A and 75%B (0.06 radians); AB mixes are shown under the horizontal arrow. (b) Image in which pixels > 0.07 radians have DN $= 0 =$ black; G is detected against all backgrounds. (c) Image in which pixels > 0.17 radians have DN $= 0 =$ black; AB mixes mimic G. (d) Image in which pixels > 0.20 radians have DN $= 0 =$ black. Pixels of A and mixtures of AC, AB, and ABC mimic G.

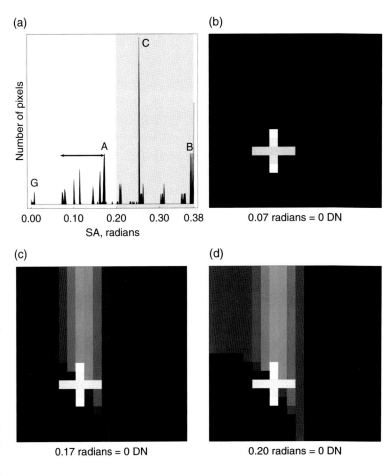

(a)

(b)   0.07 radians = 0 DN

(c)   0.17 radians = 0 DN

(d)   0.20 radians = 0 DN

the local or the overall background, or the background is defined as the mean spectrum of the image. An alternative is to characterize the background using SMA. If the target is not a basic spectral component of the local scene, it may appear among un-modeled pixels in the RMS residual (Figure 6.8). Spectral-mixture analysis, therefore, is a way to characterize the background *and* a way to find otherwise obscure targets. It has the further advantage of being able to suppress the clutter of shading by normalizing endmember fractions. A more hit-or-miss approach to defining the background and finding obscure targets is to look for anomalous pixels in ratio images, PCT images and SA data-attribute images of (non-target) scene components. For example, the relatively obscure top pixel of G stands out as an anomaly in the channel 1a/channel 3 ratio image and scatter plot (Figure 6.3), in

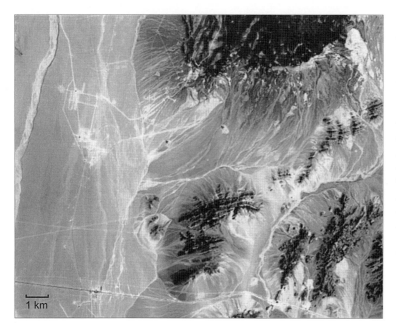

1 km

Figure 6.19 and color Figure 6.19 (see Web). Landsat TM sub-image of Amargosa Valley, Nevada. See also Plate 1. The color image is a composite of red = Band 5, green = Band 4, blue = Band 3. Rock layers in the right part of the sub-image differ in albedo and in their spectra. Albedo differences are ambiguous, because there is topographic shading and shadowing. Subtle compositional differences among layers are detected by measuring SA contrast. See Figure 6.20.

the PCT$_4$ image (Figure 6.4), and in the SA data-attribute image of C (Figure 6.5). All three of these analytical tools have the advantage of suppressing the confusing effects of shading. Once a candidate target has been discovered against one part of the background, its spectrum can be used as input for the SA classifier or other algorithm to achieve maximum SA contrast with other parts of the background, as described above.

Let us return to the Amargosa Landsat TM sub-scene for examples of how SA contrast can be used to detect targets. In the first example, we focus on layers of rock on the right-hand side of the image (Figure 6.19 and color Figure 6.19; see Web). In all six reflective channels, the rock layers are darker than the alluvium that dominates the valley floor, and the contrast among layers is partially obscured by topographic shading and shadowing. The first task is to minimize the confusing effects of shading by examining channel ratios. The Band 3/Band 5 ratio, for example, reveals numerous thin (at the scale of ~30 m/pixel) layers that differ in their radiances in these two channels. A PCT$_2$ image happens to resemble the Band 3/Band 5 ratio, confirming that the rock layers are spectrally diverse when all channels are used, and that it makes sense to explore SA contrast among the layers with SA data-attribute images. Individual layers are selected as targets, and the other layers and the surrounding alluvium become the background. Each target layer is the reference vector (SA = 0), and SA contrast between each target and the other layers can be measured on the SA image and its histogram.

Figure 6.20 was made by training on single pixels that represent three types of rock layers. The objective is to train on one or a few pixels that are most likely to represent "pure" samples that are relatively uncontaminated by adjacent layers or by alluvium. In Figure 6.20a,b,c the three SA images distinguish three types of rock layers that differ in spectral properties (not just in albedo) at the scale of individual pixels. Further detail could be extracted from these outcrops by using other layers as reference vectors and making additional SA images.

In the second example of SA contrast (Figure 6.20d), we return to the ellipse shown in Figure 6.16. In the previous discussion we stated that there is little SA contrast between the ellipse (target) and its background, and that it can be detected best by its SL contrast. If we examine the scatter plots of TM Bands, for example Bands 4 and 5, and highlight a collection of pixels from the ellipse and another set from the immediate background, we find that they lie along a single vector, with the ellipse having a slightly greater SL. When the same training set for the ellipse is used as input to the SA algorithm, the ellipse is not detected. The reason is that the input vector is the mean of a representative collection of target pixels, and the angular distance between this mean vector and the ellipse background is too small relative to the combined resolution of the software and the Landsat TM system ($\sim$0.01 radians). However, we can approach this detection limit by selecting a different reference vector for the ellipse. Our rationale is that each pixel in the target area is a spectral mixture of disturbed and less-disturbed alluvium, and this means that we can use the SL contrast in TM Band 4 to guide us to the lightest pixel, which is likely to be a relatively pure sample of the most disturbed surface. When a single such pixel is used as the reference vector for an SA image, the ellipse is detected in a contrast stretch from 0.00 to 0.02 radians. Only a few pixels in the ellipse have SA values close to this reference vector, but they differ enough from the background for the visual pattern of the ellipse to be recognized. An enlarged SA image of the ellipse in Figure 6.20d is "noisy"; nevertheless, the relative spatial concentration of lowest SA values (rendered as dark tones) is easily seen as a pattern by the human eye. In this example the SA contrast between target and background is just above the limit of spectral resolution, and it is just above the "noise" of the image as expressed by variability from one scan line to another. Later in this chapter (Section 6.3) we explore the effect of "noise" thresholds on spectral contrast and detection.

Summary of detection guidelines
Let us summarize the important factors that affect our ability to detect targets. In all of the examples that we presented, the targets are obscure

(a)   (b)

(c)   (d)

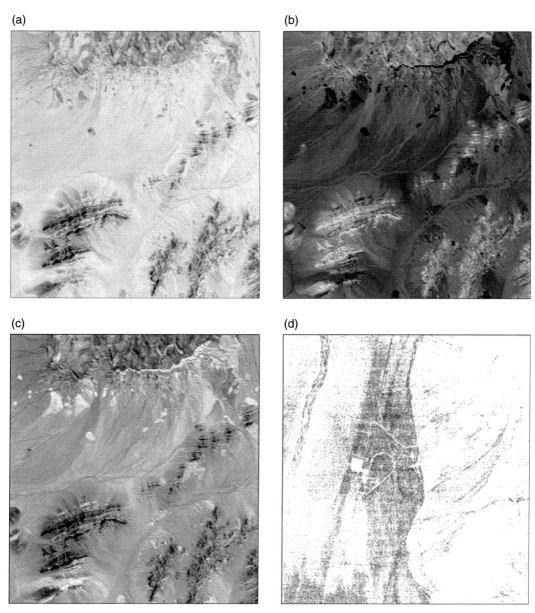

Figure 6.20. Spectral-angle contrast in layered rocks in the Landsat TM sub-image of Amargosa Valley, Nevada. See also Plate 1. Lighter pixels are closer in SA to single-pixel reference vectors in a, b, and c. (a) Complemented SA image; the reference vector is the highest-DN pixel in TM Band 5, which is taken from a rock layer. Contrast stretch: SA > 0.27 radians = 0 DN. (b) Complemented SA image; the reference vector is the lowest-DN pixel in TM Band 5 taken from a different rock layer than a. Contrast stretch: SA > 0.33 radians = 0 DN. (c) Complemented SA image; the reference vector is a pixel sampled from a "red" layer in the Band 5, Band 4, Band 3 color composite (color Figure 6.19: see Web). Contrast stretch: SA > 0.31 radians = 0 DN. (d) spectral-angle image of the larger subset (Fig. 6.16). Darker pixels are closer to the reference-vector pixel, which is the highest-DN pixel in TM Band 4 on the "ellipse." Contrast stretch: 0.00 to 0.02 radians; SA is just able to detect the ellipse, considering that the resolution of the measurement is ~0.01 radians.

because their spatial patterns are not apparent against the clutter of the local background. The G cross, although easily seen against most parts of the synthetic image, can be difficult to distinguish against the spectrally similar background of AB mixtures. Furthermore, the shading stripes in the synthetic image cause ambiguity in detecting the cross, and it is necessary to use methods that suppress shading and exploit SA contrast. The ellipse in the Amargosa image is not obscured by shading, and it is detected against its local background by its SL contrast alone; however, this target is barely detected on the basis of its SA contrast. The rock layers in the Amargosa image contrast with one another and the alluvium in SA, but variations in shading made it difficult to distinguish them by albedo.

A variety of techniques can enhance PP contrast, but a hit-or-miss approach can easily result in a missed target. Problems to watch for are:

1. Shading and shadowing can confound target detection that relies on SL contrast (albedo differences). In the absence of shading and shadowing, the minimum distance between vectors measures the maximum PP contrast, because it combines SL and SA contrast. However, minimum distance does not distinguish between the contributions of SL and SA, and when contrast depends heavily on SL, it may be mimicked by variations in shading and shadow. Example: minimum-distance data-attribute images do not detect the G cross in the synthetic image or the dark layers of rock in the Amargosa sub-image.
2. Unguided analyses such as PCT that rely on the statistical properties of the data may overlook a target if it comprises only a small part of an image. Example: The ellipse is detected by PCT in a 400×400 pixel subset of the Death Valley Landsat TM image, but it is not as easily detected (or detected at all) in larger, more complex sub-images.
3. Targets that themselves are spectral mixtures may be difficult to detect against mixed backgrounds, especially when a target comprises a significant proportion of the background. Example: matched filtering does not detect the ellipse on the Amargosa Valley floor. It was necessary to select relatively pure pixels of disturbed alluvium as input for the minimum-distance and spectral-angle algorithms in order to detect the rest of the ellipse against its mixed background.

Clutter

'Clutter' is that part of PP variability – both spatial and spectral – that expresses real variability in a scene. 'Clutter' is distinct from 'noise' which refers to variability in an image that is introduced by sensors or the data system.

6.2 Detection limits

6.2.1 Pixel-to-pixel clutter and noise

In the previous section we discussed how spectral contrast is a limiting factor in detecting a target against its background, and we explored the utility of a variety of image-processing tools for detection. In the

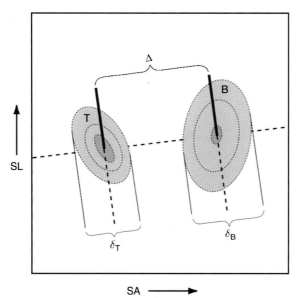

Figure 6.21. Sketch of PP contrast between two clusters of spectral vectors, where contrast is defined by SL and SA. The centroids (or means) of clusters T and B are separated by distance Δ, and δ_T and δ_B are the respective widths of the clusters projected onto the line between the centroids. Detectability is proportional to

$$\Delta/\sqrt{(\delta_T)^2 + (\delta_B{}^2)}$$

simplest case, PP contrast is described by the distance between two vectors in the data space defined by the spectral channels. We also know that simple cases are unusual in images of landscapes, and that, in fact, we often have to deal with targets and backgrounds that themselves are variable. The general case of a variable target and background is sketched in Figure 6.21. If one cluster is the target and the other is the background, the detectability of the target is a function of (1) the distance (Δ) between the cluster means, and (2) the sizes of the clusters (δ_T and δ_B). We can easily visualize that when the clusters become larger relative to the distance between them, they eventually begin to overlap. Vectors in a zone of overlap share the same or similar SL and SA values. Identical vectors (mimics), of course, cannot be distinguished from one another. Similar vectors within an overlap zone, although not spectral mimics, are spectrally ambiguous, because they could belong to either cluster. Two principal factors affect the size of data clusters: natural variability (sometimes referred to as "scene clutter") and measurement imprecision ("noise"). The clusters in the cartoon of Figure 6.21 do not tell us whether their variability describes the properties of the scene or the measurement imprecision. In either case, the detection limit is where target and background vectors overlap.

However, spectral images are more than just spectral vectors. Detection also can be influenced by the spatial arrangement of pixels. Detection limits based on patterns are much more difficult to specify than for spectral contrast. Standard spectral classifiers (Chapter 3) do not take into account spatial information. Instead they resolve the ambiguity of pixels in cluster-overlap zones by using statistical parameters such as distance to means or their likelihood of belonging to one class or another. However, assigning an ambiguous pixel to a vector cluster is not the same thing as detecting it unambiguously as the target. In the following sections we explore examples of how image patterns can assist in detecting spectrally obscure targets.

Noise, clutter, and PP contrast

Pixel-to-pixel clutter can be difficult or impossible to distinguish from noise. Both clutter and noise can be spatially random, in which case there is no simple way to separate the two without tracking down and eliminating the source of the noise. Commonly, though, we can recognize spatial patterns of either clutter or noise that help us to distinguish the two. In fact, the human eye–brain system is remarkably adept at distinguishing superposed patterns. (Examples are given in Figures 6.22, 6.23 and 6.25.) Recognizing the patterns of clutter or noise in some cases can lead to detection of an otherwise obscure target, but this depends on the image context and on the photo-interpretive skills of the analyst. Just because we recognize patterns, still does not mean that individual pixels (or small clusters that lack patterns) cannot hide undetected in the background. Detection limits are not easy to specify.

A loss of fidelity in encoded radiance is expressed as unwanted variability in the recorded DN values, and it is referred to as "noise." Noise may be random or it may appear as striping, banding, or other patterns. Fortunately, in modern electro-optical systems, sensor noise and other imperfections that are introduced by various steps of calibration and data processing, usually do not pose serious problems for extracting information from images. The exception, of course, is when we want to push target detection to the noise threshold. Sometimes the uncertainties that are associated with noise can mask the signal of a target, and then we can no longer rely on our interpretations. Noise is a topic dear to the hearts of remote-sensing scientists and engineers, and the interested reader will find a rich literature on the subject (e.g., Schowengerdt, 1997, p. 288.).

One of the most familiar examples of noise is the striping and banding which occurs in images that are made by systems that use linear arrays of detectors, such as Landsat Thematic Mapper. Even slight differences between the gains of different detectors are expressed as small differences in image DNs. Examples from the Seattle and Amazon Fazendas Landsat TM images are shown in Figures 6.22 and 6.23. In many cases, striping and banding can be removed, or at least suppressed, by spatial filtering techniques (e.g., Schowengerdt, 1997, p. 299). Although this approach involves several steps, it can be

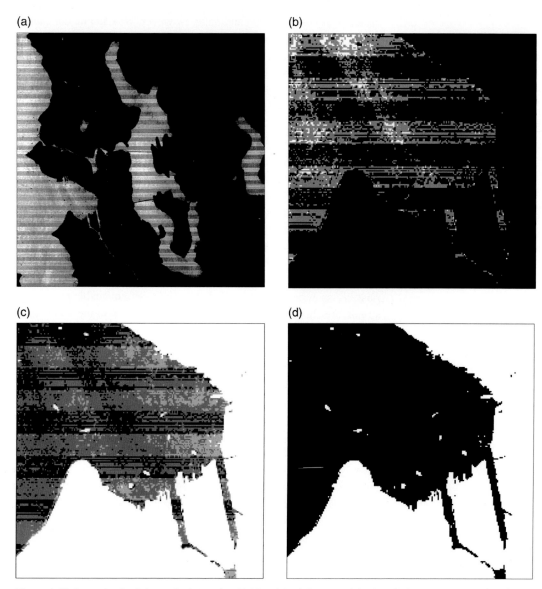

Figure 6.22. Example of striping noise in Landsat TM Band 4 sub-images of the Seattle 1992 image. See also Plate 2. (a) Complemented Band 4 image, making land areas dark, and water areas light. Contrast stretch: pixels ≥ 32 DN are set to 0 DN, and those ≤ 26 DN are set to 255 DN. Striping is prominent in the areas of water (e.g., Puget Sound and Lake Washington). (b) A 200 x 200 pixel enlargement of Elliot Bay and the port of Seattle (lower left part of image a). Band 4 is complemented. Stretch: pixels ≥ 8 DN are set to 0 DN; pixels ≤ 6 DN are set to 255 DN. In addition to the striping pattern, additional variability is caused by unspecified sensor noise and diffuse, vertical streaking. The streaking records real contrast in water reflectance which is low in Band 4. Several ships (dark pixels) are lost in the noisy background. (c) Band 4 as in a, but not complemented. Stretch: pixels ≥ 10 DN are saturated at 255 DN; pixels ≤ 7 DN are set to 0 DN. Diffuse streaks are dark in this image. In this stretch, the horizontal striping and banding are still prominent; however, ships and ship wakes are detected as light pixels. (d) Same as c, but striping and banding are suppressed by setting all pixels ≤ 9 DN to 0 DN. Ships and wakes are more evident.

effective at "cleaning up" an image. However, we reiterate our caution from Chapter 1 that cosmetic procedures such as spatial filtering, which modify the basic data, usually are best applied after information (such as detecting a target) has been extracted. Alternatively, there are relatively simple ways to suppress noise by masking and contrast-stretching that do not risk the loss of information.

In the example of the Seattle image (Figure 6.22), horizontal striping and banding in TM Band 4 is apparent for dark parts of the scene, especially water, where the noise variability of a few DNs from line to line is a relatively large proportion of the low overall radiance. However, the water areas also contain two targets of interest, and their spatial patterns are more easily recognized when we can suppress the noise pattern. One target consists of diffuse northwest-trending streaks that we know are properties of the water surface, because they are present in other Bands. The origin of the streaks is not confirmed, but they probably are areas where the water surface is relatively smooth, possibly as a result of faint oil slicks. The most noticeable parts of the streaks are about 2 DN darker in Band 4 than the rest of the water, and it is this DN difference that can be exploited by contrast-stretching to reveal a diffuse, vertical pattern that is not consistent with the horizontal noise stripes. The streaks cannot be entirely separated from the background by contrast-stretching, and parts of the streaks may be hidden in the noise.

In the same Landsat TM scene, several ships and ship wakes in the port of Seattle can be isolated from the noisy background in Band 4. This is done by entirely suppressing the striping with a contrast stretch (or density slice) that makes pixels ≤ 9 DN black, and pixels ≥ 10 DN white. However, without careful stretching to the 1-DN level, the ships can be overlooked, because they comprise only one or a few pixels and, therefore, do not have evident spatial patterns (Figures 6.22b, c, d).

In the Amazon Fazendas image (Figure 6.23) there is unusually prominent striping, banding, and other noise in Band 3. The primary forest is relatively dark in Band 3, owing to shading and shadowing in the rough canopy and absorption by chlorophyll. In addition, there is little dynamic range in Band 3, apparently because of a low gain setting for the detectors. The DN range of PP variability in the forest coincides with that of the noise, and it is not possible to suppress the noise by contrast stretches. In this example, there is little visible pattern to the variability of the forest to distinguish it from random noise.

The Seattle and Amazon images illustrate how clutter and noise can influence detection that is based on recognizing spatial patterns. Another approach is to examine the data structure of an image to find out to what extent data clusters of target and background overlap. Noise

(a) 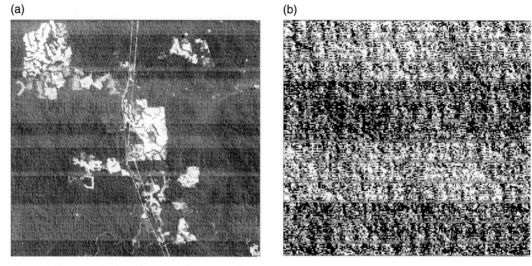 (b)

Figure 6.23. Example of striping, banding, and within-line noise in TM Band 3 of the Amazon Fazendas image. (a) Stretch: 26 to 32 DN. (b) A 200 x 200 pixel enlargement of the forest background from the left side of image a. Stretch: 27 to 30 DN. There is fine-scale striping within the broad bands. The forest itself is heterogeneous on the scale of individual pixels; however it is difficult to distinguish this scene variability from random and patterned noise in the image.

adds its own variability to the natural variability of each data cluster, and this tends to increase the amount of any cluster overlap. When clusters overlap, only those target pixels outside the overlap zone can be unambiguously detected.

In the Amargosa sub-image (previously shown in Figure 6.16), the ellipse is revealed in some Bands by its slightly higher-DN pixels that are arranged in a recognizably different pattern from the otherwise largely random variability of the background. Data from target and background pixels are shown in the scatter plot and histograms of Figure 6.24. The means of the target and background clusters are separated by 5 DN in both Band 3 and Band 4. The standard deviations of the Band 4 clusters are relatively low (0.72 for the target, and 1.17 for the background), and the overlap between clusters is only 1 DN (8% of the total DN range). However, in Band 3 the standard deviations are higher (2.86 for the target, and 2.46 for the background), and the overlap between clusters is 8 DN (42% of the total DN range). As can be seen graphically in Figure 6.24, in Band 4, nearly all of the target pixels fall in a 5-DN range that is outside the overlap zone with the background (gray pattern); however, in Band 3, less than half of the target pixels are outside the zone of overlap with the background (gray pattern). Unambiguous target pixels that are outside the overlap zone in Band 3 are white in the image in Figure 6.25b. The different amounts of overlap

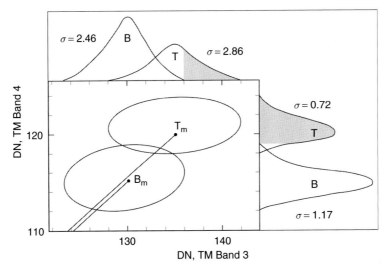

Figure 6.24. Simplified sketch derived from the scatter plot and histograms of TM Bands 3 and 4 for the "ellipse" in the Amargosa Landsat TM sub-image (see also Plate 1). T and B represent the target (the ellipse) and background, respectively; T_m is target mean DN, and B_m is background mean DN. The gray tone on the target histograms defines those target pixels that do not overlap in DN range with the background, and, therefore, are unambiguously detected. Greater DN overlap in Band 3 than in Band 4 is consistent with the ellipse being visually less distinct in Band 3 (Figure 6.25) than in Band 4 (Figure 6.16). Lines from T_m and B_m point toward the origin, and illustrate graphically the very small amount of SA contrast. The ellipse contrasts with its immediate background primarily in SL.

between target and background explain why the ellipse is more easily seen in Band 4 (Figure 6.16) than in Band 3 (Figure 6.25).

In Figure 6.25 we also see that noise (striping, banding, and within-line noise) contributes to the variability of target and background in Band 3, and that this noise cannot be entirely suppressed by contrast stretching to the level of 1 DN. Although the overlap between target and background in Band 3 appears to be the result of a combination of natural variability on the ground and noise, Band 1 has only very faint striping and banding noise. The ellipse nevertheless is obscure in Band 1, because target and background clusters overlap almost completely. The PP variability in Band 1 may be entirely due to the nature of the ground surface, but we cannot rule out some contribution by noise, especially random noise that would be difficult to isolate in this sub-image.

6.2.2 Sub-pixel detection limits

Most of our discussion so far has been about spatially resolved targets. However, targets that are smaller than pixel footprints are of considerable interest in remote sensing. What comes to mind first for many people is the example of a single pixel in an image in which some small target, such as a vehicle or a special type of rock is hidden. A more

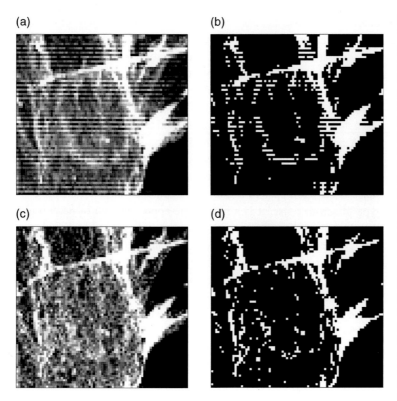

(a) (b)

(c) (d)

Figure 6.25. Landsat TM Band 1 and 3 images of the 'ellipse' in an Amargosa sub-image (80 × 80 pixels). (a) Band 3 image. Contrast stretch: DNs ≤ 125 are set to 0; DNs ≥ 138 are set to 255. Contrast stretching enhances both the target and the striping and banding noise. (b) Band 3 image. Contrast stretch: DNs ≤ 135 are set to 0; DNs ≥ 136 are set to 255. Background clutter and noise are suppressed by making black those target pixels that overlap with background pixels, and making the remainder white (see gray pattern in Fig. 6.24). Noise remains in the pattern of the target. (c) Band 1 image. Contrast stretch: DNs ≤ 184 are set to 0; DNs ≥ 194 are set to 255. The ellipse is obscure against the background clutter, because target and background have nearly the same DN ranges. Patterned noise is not evident. (d) Band 1 image. Contrast stretch: DNs ≤ 192 are set to 0; DNs ≥ 193 are set to 255. Although patterned noise is less in Band 1, and clutter has been suppressed, fewer unambiguous pixels of the ellipse are detected than in Band 3.

common need in remote sensing, though, is to detect the presence of a given target at the sub-pixel scale in many pixels. Perhaps the best example is mapping sparse vegetation cover on arid landscapes. In this case, we expect any, and perhaps all, pixels to contain some target (vegetation). If we want to know how much vegetation is there, we also need to know what the detection limit is. Sub-pixel targets cannot be detected directly, because their spectra are blended with the spectra of the other components within pixel footprints. We can, however, apply spectral-mixing models that evaluate the spectrum of each pixel as to whether it is consistent with the presence of some fraction of a target spectrum. If we do not know the target spectrum, but we can estimate the spectra of the other components, we can look for evidence of the target in the RMS residual or in channel residuals (Chapters 4 and 5; Figure 6.8).

The limit of sub-pixel detection is the smallest fraction of a target endmember that can be measured to a specified degree of confidence in a given spectral mixture. (It does not depend on the PP contrast between a pixel that contains the target, and the other background pixels. Any pixel in which the target has been detected is differentiated from all other pixels.) The smallest fraction of a spectral endmember that can be

---

Sub-pixel detection limit

The limit of sub-pixel detection is the smallest fraction of a target that can be detected in a spectral mixture.

detected depends, then, on the contrast between endmembers at each wavelength, *and* on the uncertainty of the measurement. In Chapter 4 we discussed several sources of uncertainty that affect fraction accuracy. One is the initial choice of endmembers, and we already have shown how endmembers can be improved by evaluating RMS residuals, channel residuals, and image context. Other sources of uncertainty are the natural variability of endmembers, and noise.

A convenient way to estimate detection limits is by modeling the sensitivity of the fractions of one endmember (the target) to degrees of fraction uncertainty that are represented by random noise. Random noise is easy to apply to a mixing model, and, besides modeling one type of sensor noise, it is a reasonable approximation of natural endmember variability. This approach has been developed by Sabol *et al.* (1992) to estimate "best-case" detection limits (Figure 6.26).

In Figure 6.26 the sub-pixel detection limit is represented as the fraction of the target spectrum that must be present to be detected at a specified confidence level. In this case, there must be a fraction of 0.07 of the target for us to be confident (at a statistical confidence level of 68.3%) that the target actually is present. Of course, to have a higher confidence level, there must be even more target present. For example, for 100% confidence, the fraction must be greater than the high-end tail of the probability distribution.

Notice that in Figure 6.26 the spectra that represent materials in a landscape are measured in reflectance and then transformed to fractions by a mixing model in order to allow discussion of detection limits. Fractions of two spectra that are sampled by one channel, along with a noise function, can be represented graphically on a line; however to show three spectra that are sampled by two channels, a ternary diagram is needed, and this distorts the shape of the noise distribution. Figure 6.27 illustrates the fractions of three components, and shows that a circular zone of uncertainty in channel-DN space becomes an ellipse in fraction space. The long axis of the ellipse is the direction of the most uncertainty. The example shows that mixtures including spectrum C fall within the zone of uncertainty that surrounds a mixture of 0.5 A and 0.5 B. This means that if we are trying to detect sub-pixel amounts of C against a background of half A and half B, there is a detection limit, below which we cannot be sure that C is really there at the specified confidence level.

Sub-pixel targets that cannot be distinguished by their continuum spectra sometimes can be distinguished by residuals, especially those caused by narrow resonance bands, that are expressed in only a few channels in high-resolution spectra (Chapter 4). To find a potential target using residuals, we need to compare a modeled spectrum with a measured one. If the two are identical, the residual is zero at all wavelengths. If the

(a)

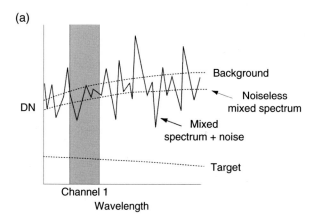

Background

Noiseless
mixed spectrum

Mixed
spectrum + noise

DN

Target

Channel 1
Wavelength

(b) Noisless mixed spectrum

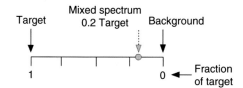

Target

Mixed spectrum
0.2 Target

Background

1   0 ← Fraction
of target

(c) Mixed spectrum + noise

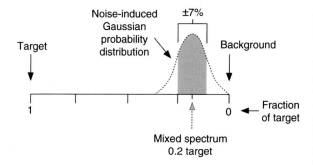

Target

Noise-induced
Gaussian
probability
distribution

±7%

Background

1   0 ← Fraction
of target

Mixed spectrum
0.2 target

Figure 6.26. Uncertainty in sub-pixel fractions of a target, with and without noise. (a) Spectra of target and a single background, sampled by one channel (channel 1). The target spectrum is shown as noiseless and with simulated random noise. (b) Target fraction without noise. The fractions of both target and background are proportional to the DN contrast in the sampled channel under conditions of linear mixing. In the example, the target has a fraction of 0.20. (c) Target with an amount of random noise (normally distributed about the real fraction) sufficient to produce the probability distribution shown in the gray overlay. A standard deviation of 1 in the probability distribution corresponds to an uncertainty in the target fraction of ± 0.07. There is a 68.3% probability that the true target fraction is ≥ 0.13 and ≤ 0.27. If the fraction of the target were 0, the same measurement uncertainty would create false fractions up to 0.07. Therefore, we would not be confident (at the 68.3% level) that fractions of 0.07 are real. This is the detection limit. Figure modified from Sabol *et al.* (1992).

measured spectrum has structure that we have failed to model, it will appear as a residual, preferably in several adjacent channels, thereby minimizing the possibility of an instrumental anomaly in one channel. An example of how residual analysis can enhance a faint absorption band was given in Chapter 2 (Figure 2.7). The uncertainty in calculating a residual is defined by the uncertainty of the measurements in each channel (Figure 6.28). To estimate detection limits, we can model the sensitivity of residuals to specified levels of noise and to desired levels of confidence in the same way that was described for continuum analysis. However, a linear mixing model, although sufficient for continuum

Figure 6.27. Transformation of noise uncertainty from channel-DN space to fraction space. (a) Reflectance spectra of endmembers A, B, and C (from Figure 3.2) and a mixed spectrum of 0.5 A and 0.5 B. Addition of random noise to the AB mixture (not shown) produces uncertainty in the fractions. (b) Scatter plot showing the endmembers and the mixing area in the space defined by channels 2 and 3. The AB mixture lies at the mid point of the AB mixing line. In the example, the added noise is assumed to affect both channels equally, and forms a cluster, shown here as a circle. (c) Ternary diagram of the fractions of the endmember spectra. Transformation from channel-DN space to fraction space distorts the circular noise cluster into an ellipse. The long axis of the ellipse is the direction of lowest WW spectral contrast. Figure modified from Sabol *et al.* (1992).

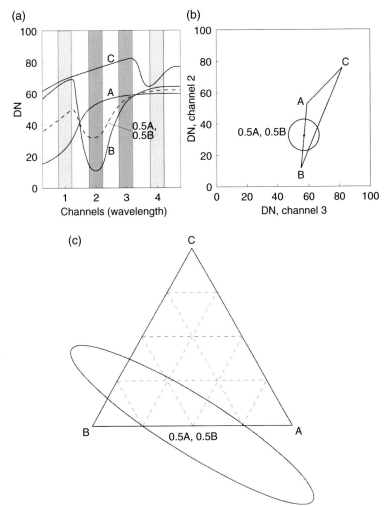

analysis, is not as accurate when applied to spectral mixtures within resonance bands (Chapter 2). More accurate estimates of detection limits require using a non-linear model, once residuals caused by resonance features have been identified by linear modeling.

6.2.3 Predictive modeling and sensitivity analysis

We can predict detection limits by the method of Sabol *et al.* (1992) using laboratory or image spectra as endmembers, providing that we can model the main sources of uncertainty – noise and natural scene variability. The ratio of signal-to-noise (SNR) varies from channel to channel for each sensor system, and SNR values that are measured prior

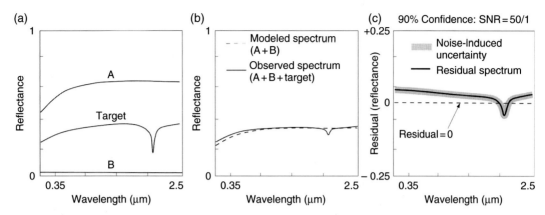

Figure 6.28. Detection limit of a sub-pixel target by residuals. (a) Noiseless spectra of A, B, and target. (b) Noiseless mixed spectra. Solid line: observed spectrum. Dashed line: spectrum modeled by endmembers A and B. (c) Residual spectrum. Small differences between the observed and modeled spectra are amplified. The presence of the target in the observed spectrum of b is revealed by the absorption band and by the departure of the continuum from the zero-residual line. The gray pattern represents the envelope of the modeled level of noise and the confidence level (for example: signal-to-noise 50/1, and 90% confidence). The detection limit for the target (at each wavelength) is the non-zero residual that is outside the gray noise and confidence envelope. Adapted from Sabol *et al.*, 1992.

to flight are published for systems that are in the public domain. (For some systems, updated in-flight data are available.) Models of detection limits that are based on the assumption of random sensor noise in each channel define a limited, best-case scenario. Other sources of variability such as patterned noise (e.g., striping/banding) and end-member variability will make the detection limits worse. Endmember variability in a scene can be the main source of uncertainty in measuring fractions, as we have discussed (Section 6.1.3). To simulate the natural variability of a chosen endmember, we can introduce additional noise, beyond that used to model sensor noise. This allows us to test how different levels of uncertainty affect detection limits for different sensor systems and for different endmembers.

The basic idea of predictive modeling is illustrated in Figure 6.29, using laboratory endmember spectra of green vegetation, soil, and shade. The endmembers represent the basic spectral components of common types of vegetated landscapes (Chapter 2). The example models the sub-pixel detection limits of soil against backgrounds of GV and shade and their mixtures. A SNR of 10/1 is assumed in this case, and the noisy spectra are repeatedly unmixed (linear model) to produce the scatter of points for a mixture of [0.5 GV and 0.5 shade] (Figure 6.29a). The data are contoured into confidence levels in Figure 6.29b. At the 90% confidence level, the detection limit of soil (the target) in a mixed pixel of [0.5 GV, 0.5 shade] is 0.08. Other detection limits are: soil in a GV background = 0.12; soil in a shade background = 0.06. In this

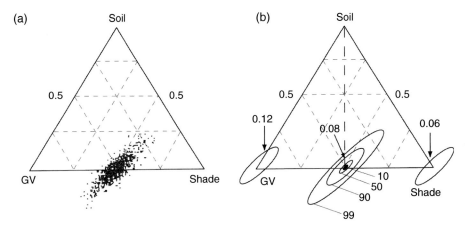

Figure 6.29. Uncertainty in endmember fractions simulated by addition of random noise (10/1 SNR) to laboratory reflectance spectra. (a) Noise-induced uncertainty in the fraction 0.5 GV and 0.5 shade. Fractions of soil that fall above the GV/shade line imply that soil is present when it actually is an artifact of the noise. Fractions that fall below the GV/soil line indicate negative fractions of soil. (b) Contoured confidence intervals. At 90% confidence, the detection limit of the soil fraction is 0.08 in a mixed pixel that contains 0.5 GV and 0.5 shade. The detection limit of soil against a background of GV is 0.12; and the limit against a shade background is 0.06. Adapted from Sabol *et al.* (1992).

model it is apparent that twice as much soil is required for detection against a background of GV than against a shade background. We also can see how the detection limits will become smaller with less noise or if we relax the confidence level. Conversely, if we have underestimated the noise, or if we wish to simulate endmember variability by increasing the modeled noise, a larger fraction of soil is needed for detection.

Sometimes we want to know which imaging system is the best one for detecting a specific type of target against a particular background. How do channel wavelength and width affect detection limits? Should we use all of the channels? Which channels are essential, and which ones only add to the uncertainty? To address these questions we need to construct models that can be varied to include the properties of different imaging systems, including the noise characteristics of each channel. We also want to be able to add noise to simulate endmember variability. The examples presented by Sabol *et al.* (1992) suggest that detection limits, although influenced by system noise and endmember variability, also can be highly sensitive to channel placement relative to endmember spectra. Furthermore, the effects of the combined variables are not necessarily intuitive. For example, Sabol *et al.* (1992) found that using continuum analysis, the detection limit of soil in the [soil–GV–shade] model actually is worse using AVIRIS than for Landsat TM (or even the old MSS); and using residual analysis, Landsat TM Band 7 has a lower detection limit for soil than does AVIRIS at 2.2 μm, at least with the SNR of AVIRIS when the data were acquired. Results such as these

emphasize the desirability of developing software tools that allow analysts to conduct "what if" tests on images routinely to determine the
sensitivity of target-detection limits to various sources of uncertainty.

6.3 Spectral contrast and spatial scale

Based on our visual experience, we already have an intuitive grasp of
how spatial scale affects our ability to discern objects. In addition to the
obvious fact that objects become more difficult to resolve spatially at
greater distances, we also know that colors are harder to pick out when
they comprise only a small part of a scene, whether it is the red, green,
and blue dots that populate a TV screen or the colors of people's clothes
in a crowd seen from a distance. Patches of color that are small relative
to the scale of a scene tend to blend into the background. As we zoom
in on a scene, though, the same patches begin to contrast with their
surroundings. It is natural, therefore, for field investigators using remote
sensing to prefer images that have small pixel footprints and high
spectral resolution. Realistically, though, we also know that if we select
an image that has small pixel footprints and many channels, we face
a tradeoff with areal coverage. What is the nature of the tradeoff, and
what happens to spectral contrast and detection limits when we enlarge
pixels? As we will see, these questions are surprisingly difficult to
answer.

Pixel footprints

The size of a pixel footprint projected onto the ground generally is fixed for specific spectral channels of
remote-sensing systems. For nadir-viewing remote-sensing systems that have narrow angular fields of view
(e.g., Landsat TM) the pixel footprint for the acquired data is nearly constant across an image. Pixel
footprints vary in size across images that are made by camera systems having wide fields of view (such as
AVHRR) and off-nadir viewing systems (such as MISR). To keep things simple, we will use examples
where pixel-footprint size is the same throughout an image.

Let us start with two simple thought experiments. Consider the case
of a single, small target, for example, a small oasis or a vehicle in a
large, homogeneous desert. With meter-size (or smaller) pixel footprints, the target is at first spatially resolved, and PP contrast between
target and background is at a maximum. By progressively enlarging the
pixel footprints, parts of the target, starting with its edges, will be
incorporated into the larger pixels. At first, some of the target is still
resolved (occupies a whole pixel) and some of it is mixed with the
background in other pixels. The small target eventually will be swal-
lowed entirely by the larger pixels and become incorporated in a spectral

mixture. Pixel-to-pixel spectral contrast decreases until eventually the target is below the detection limit that is imposed by measurement uncertainty. The example becomes more complicated if the larger pixel footprints begin to incorporate additional targets (more vegetation or more vehicles) or if the background is variable. Detection of the target would improve by increasing its fraction, but by incorporating new background that mimics the target, detection could get worse. Contrast and detection limits, therefore, will depend on the particular scene.

In the second thought experiment, consider that the target is distributed evenly throughout the background. An example would be sparse vegetation cover such as meter-diameter shrubs evenly spaced in a desert. One-meter or smaller pixel footprints would give the maximum PP contrast, because some would contain just vegetation (neglecting gaps in the vegetation itself) and some would contain just the background desert surface. As pixel footprints are enlarged, mixing will cause the spectral contrast to decrease until the pixels have approximately equal proportions of vegetation and background, at which point the spectral contrast will remain constant. Eventually, perhaps on the scale of kilometers or tens of kilometers, PP contrast may increase, because pixel footprints begin to overlap with new landscape elements. Again, the effect of the size of the pixel footprint on spectral contrast depends on the nature of the scene.

The thought experiments suggest that the size of a pixel footprint (often referred to as the instantaneous field of view, or IFOV) and the size of the field of view (often referred to as FOV) of an image both influence spectral contrast, but in opposite ways. This idea is sketched in Figure 6.30 that shows PP contrast generally getting worse with larger footprints, and PP contrast increasing with larger fields of view. (Contrast becomes minimal in the extreme case where the field of view is very small and the pixel footprints are very large.) We have drawn bumpy, irregular curves in Figure 6.30 to convey the idea that contrast usually is not a smooth function of either pixel-footprint size or field of view, but rather depends on the details of each scene in ways that are not easily predicted. We can illustrate this point by drawing on specific examples based on the synthetic image and on images of real scenes.

Wavelength-to-wavelength and PP contrast and pixel size in the synthetic image

We used the 200×200-pixel version of the synthetic G-cross image to test what happens to spectral contrast and detectability with increasing pixel size (Figure 6.31). The first test was to assess changes in WW spectral contrast. We averaged the DN values for blocks of pixels, 1×1, 3×3, 5×5, and so on, in effect making the samples larger to simulate a

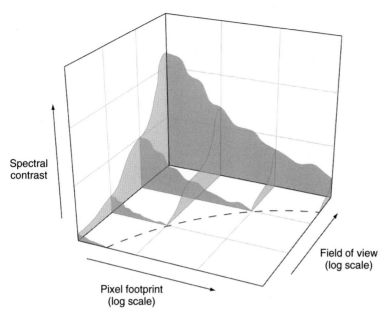

Figure 6.30. Cartoon illustrating the effect of field of view and pixel-footprint size on spectral contrast. As the size of a pixel footprint becomes larger, spectral contrast decreases, owing to sub-pixel mixing. As the field of view becomes larger, spectral contrast increases, because new landscape components are encountered. The bumpy curves illustrate the idea that landscapes are spectrally complex, and that the interactions of the scale variables are not easily predicted.

range of pixel footprints. Starting with one un-shaded pixel of A, B, C, or G, we calculated the changes in each spectral channel. As surrounding pixels were added to enlarge the simulated footprint, other scene components became incorporated, and they diluted the properties of the starting pixel. The spectra of simulated pixels centered on G and B are plotted in Figure 6.31c,d. The spectrum of the pixel centered on G changed shape the most as pixel size increased. This is consistent with the fact that G, that occupies a relatively small area of the image, mixed initially with C, making the mixture lighter, and then incorporated shade and B, making the mixture darker. There was less change in the spectrum of B over the same range of pixel sizes, because mixing only occurred with shade. Both of these examples illustrate that WW contrast diminishes as the simulated pixel footprint becomes larger.

A second test assessed how PP spectral contrast changed with increasing pixel size. In this case, we simulated larger pixels by adding successively larger blocks of pixels over the whole image. We then measured the SL and SA of the new pixels that were centered on the locations of the starting pixels of A, B, C, and G, respectively. An example of how SL changed with pixel size is shown in Figure 6.31e, using channel 1. The pattern of SL change with pixel size is sensitive to the channels that are selected and to the structure of the image. For this reason the SL change of any component is not easily predicted; however, the experiment illustrates that SL contrast decreases as pixels become larger.

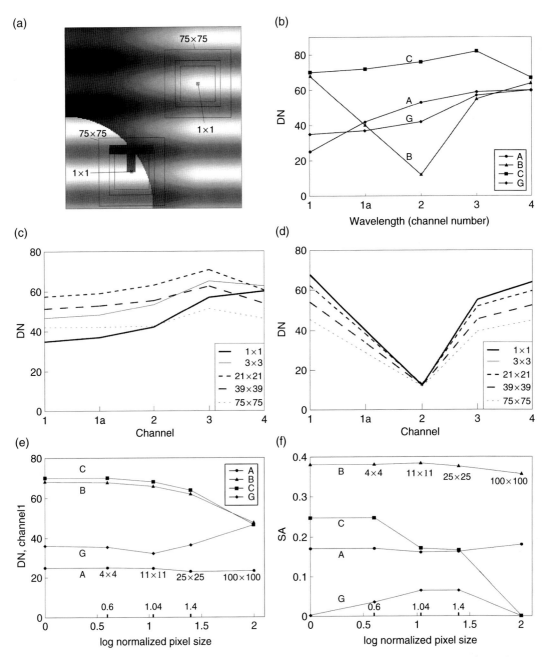

Figure 6.31. Change in WW and PP spectral contrast in the synthetic G-cross image with increasing pixel size. This is a 200 × 200-pixel version of the same image that is defined in Figure 6.2. (a) Image of channel 1 showing the locations of starting pixels of G (bottom of cross) and B. Boxes outline enlarged pixels that are centered on the starting pixels. (b) Spectra of A, B, C, and G. (c) Spectra centered on G with increasing pixel size; DNs in each channel are the average for the pixels in each size range. (d) Spectra centered on B with increasing pixel size. (The 3 × 3 spectrum is coincident with the 1 × 1 spectrum.) (e) Spectral length of A, B, C, and G in channel 1, measured in DN, with increasing pixel size (log scale); SL was measured at the starting (1 × 1) locations of samples of A, B, C, and G. (f) Spectral angle with increasing pixel size; SA was measured relative to G. Pixel size was increased as described for SL.

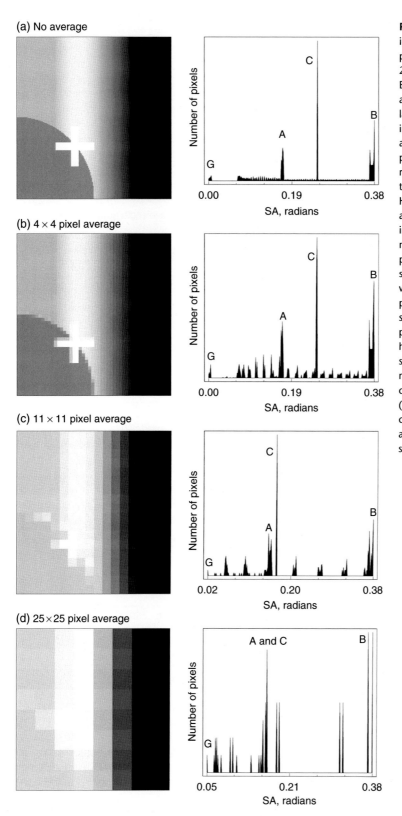

(a) No average

(b) 4 × 4 pixel average

(c) 11 × 11 pixel average

(d) 25 × 25 pixel average

Figure 6.32. (a)–(d) Change in spectral angle with increasing pixel size for the synthetic 200 × 200 pixel G-cross image. Blocks of original pixels were averaged into new pixels as labeled. Images are SA attribute images, using the original G vector as the reference (SA = 0). Lighter pixels indicate angles closer to the reference vector. Histograms show the number of pixels vs. SA. Histograms correspond to the adjacent image. (a) Original image. G is separated from AB mixtures by 0.03 radians. (b) 4 × 4 pixel average. Some pixels of G are still pure, but others are mixed with the background. (c) 11 × 11 pixel average. The G cross is spatially indistinct, and all pixels of pure G (SA = 0) disappear from the histogram. Pixels having the smallest SA (relative to pure G) are mixtures of G and other components and have SA = 0.02. (d) 25 × 25 pixel average. The G cross no longer is recognizable; A and C now have identical angular separations from pure G.

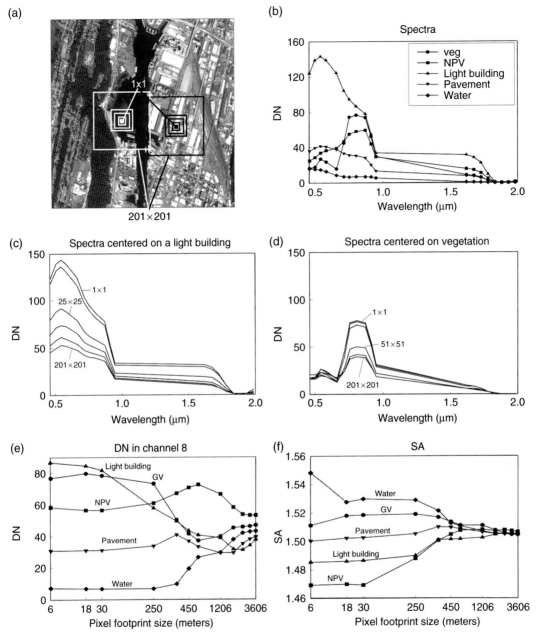

Figure 6.33. Spectral contrast variations with pixel footprint in MODIS-ASTER simulator (MASTER) sub-image of an industrial area in Seattle, Washington. Footprint size is simulated by adding blocks of pixels. Measurements are made on pixels centered on the address of the original, starting pixel. (a) Channel 8 (0.8 μm) sub-image. Scale: sub-image is approximately 3.7 km across. North is at the top. Individual buildings and roads are resolved with ~6 m pixels. (b) Reflectance spectra of spatially resolved objects in the sub-image. (c) Variability of WW contrast in spectra with pixel-footprint size. Pixel spectra are centered on a light-toned building shown in the right boxes in a. (d) Variability of WW contrast in spectra with pixel-footprint size. Pixel spectra are centered on an area of green vegetation shown in the left boxes in a. (e) Variability of SL contrast with pixel-footprint size, as represented by reflectance in channel 8. (f) Variability of SA contrast with pixel-footprint size; SA is measured from an arbitrary reference vector. For this sub-scene, PP spectral contrast becomes severely degraded when pixel footprints are a few hundred meters in size.

Shown in Figure 6.31f is the variation in SA with pixel averaging. The SA for each averaged block of pixels was measured relative to a pixel centered on the original address of pure G. Without pixel averaging, the SA of G is zero, but with increasing pixel size, G becomes mixed with surrounding components; A, B, and C also mix with one another as pixels enlarge, depending on the image structure, and eventually as SA contrast disappears, fewer of the original components can be detected.

Another way to show the loss of spectral contrast as pixel footprints become larger is presented in Figure 6.32. The reference vector (SA = 0) is the spectrum of pure G (Table 6.1). For the pixel-averaged images, this simulates a situation where we know the spectrum of a target, although there are no pure pixels of it in an image. Without pixel averaging, G appears in the histogram at SA = 0; however, when pixels are averaged to 11×11, there are no pixels at SA = 0, because there are no longer any pure pixels of G. Instead, the low angles (>0.03 radians relative to pure G) correspond to mixtures of G and AB. Thus, G no longer is detectable against an AB background, even though we know the spectrum of pure G. With further averaging of pixels, the angular positions of the new, larger pixels at the original addresses of A, B, and C change in response to mixing, until, eventually, the G cross is no longer recognizable.

Another example of how spectral contrast changes as the pixel footprint is enlarged is shown in Figure 6.33 using a 201×201 pixel sub-image of the Seattle docks area acquired by the aircraft-borne MODIS and ASTER simulator (MASTER). Changes in WW contrast of an area of a light-toned building in an industrial area are plotted in Figure 6.33c, and changes in pixels centered on green vegetation are shown in Figure 6.33d. In both cases the WW contrast diminishes as pixels enlarge and incorporate different surroundings. The loss of SL and SA contrast is illustrated in Figure 6.33e, f, respectively. In this sub-scene, spectral contrast is greatest at the maximum spatial resolution of ~6 m, but contrast between scene components is essentially lost in footprints >200 m. As in the synthetic image, the general trend is toward reduced spectral contrast as pixel footprints are made larger. However, the details of how WW and PP contrast degrade depend on the choice of a target and on its background context.

Next

Once we have established that the things we are looking for in a spectral image are detectable, how do we go about defining their spatial extent and how they interface with the surroundings? How do we integrate spectral and spatial information into image maps that illustrate selected ideas? The objective is a thematic map, and this is the topic that we address in the next chapter.

Chapter 7
Thematic mapping of landscapes

Gobi Desert, Mongolia. Photograph by A. Mushkin

In this chapter

Thematic maps that are made by investigators in the field are based on an understanding of the physical properties of landscape materials. When we make thematic maps from spectral images, we need to demonstrate how the image data are connected to the properties of the ground.

7.1 Field maps and image-derived maps

Thematic maps of landscapes contain information about the nature of the land surface, in addition to showing where things are located in two- or three-dimensional space. Thematic maps are interpretive. They convey selected information, and, perhaps most importantly, they deliberately omit other information. Field investigators are fully aware of the selective nature of thematic maps. They do not try to map everything that they see; instead they focus on one or a few subjects.

Image analysts also must focus on one or a few subjects to make thematic maps. Their task is to extract only the relevant information (a map of what?) by interpreting encoded radiance and derived data attributes in a spatial context. For data-attribute images to qualify as thematic maps they must be based on a physical understanding of the connection between the image measurements and the ground surface. In some cases, attribute images can be organized into units (Section 1.4.3). Image-derived thematic maps usually are judged by whether they contain information that is useful to a field investigator.

Unfortunately, the term "thematic map" is widely used in remote sensing without being clearly defined. For example, the Landsat Thematic Mapper does not do what its name implies. It does not map themes, because it cannot decide what is important and what to leave out. Alas, Landsat TM just measures radiance from each pixel footprint. Under certain conditions, though, radiance measurements can be interpreted to make thematic maps of materials and conditions on the ground. This requires an analyst's understanding of physics and spectroscopy, and an interpretation of constraints imposed by spatial patterns and scene context. The idea that Landsat TM can map themes has its roots in an early assumption that machine processing would make it possible automatically to produce the same kinds of maps that investigators make in the field. However, most geologists, soil scientists, foresters, botanists, ecologists, urban planners, and others who make thematic maps in the field would never consider trying to automate this interpretive process. If spectral images are proxies for being in the field (Chapter 1), we are still a long way from making field-like thematic maps from images without applying human guidance and experience.

In this chapter we propose that image-derived thematic maps must conform to the same general requirements which apply to thematic maps that are derived from field observations and measurements. In addition to being necessarily selective (theme-based), image-based thematic maps should present a clear interpretation that is accompanied by an explanation of the evidence. Thus, images that leave the main burden of interpretation to the user do not qualify as thematic maps, any more than a color photograph taken on the ground qualifies as a thematic map for a field study.

Thematic maps from images

We define two kinds of image-based thematic maps (Section 1.4.3):

1. *Attribute maps* that show continuous properties of the image data such as reflectance or DN distance from a mean value of a training area.
2. *Unit maps* that have labels and boundaries.

Interpretation

Thematic maps are interpretive. They require the analyst to define what is important in a scene or landscape and to leave out the rest.

Definitions of terms	
Theme	A unifying, dominant, or important idea or motif.
Thematic map	A representation of an area of the land surface showing one or more themes. Thematic maps are based on physical properties of the ground surface and the interpretation of the map maker; they do not require the user to supply the interpretation (see rules in Section 7.2.1).
Attribute	A property, characteristic, or quality.
Data-attribute image	An image in which the pixel values represent properties of the data; for example, the radiance in Band 4, or the SA from the mean vector of a training set. If data-attribute images can be linked to physical properties of a scene, they may qualify as thematic maps, providing that they are interpretive and satisfy the rules in Section 7.2.1.
Physical-attribute image	An image in which the pixel values represent physical attributes of the ground surface; for example, radiant temperature, relative amount of absorption by chlorophyll, or fraction of exposed soil. Physical-attribute images are interpretive, and qualify as thematic maps if they satisfy the rules in Section 7.2.1.
Unit	A basic organizational entity used in thematic mapping on the ground and in image interpretation. Units have boundaries, labels, and defined physical attributes. In general use, any group of things interpreted as an entity.
Class	A grouping of per-pixel data attributes according to specified thresholds. Classes of pixels may exist in the data space (e.g., defined by the spectral channels or fractions of endmembers), or in image form. Image classes do not qualify as themes unless they have a physical basis for interpretation and satisfy the rules for thematic maps. Also used casually to describe a number of things in a group having common properties.
Classify	To make per-pixel classes from digital-image data. In general, to arrange or organize by classes; to organize numerical or textual data into groups.
Category	Any general or comprehensive division.

7.1.1 Thematic mapping in the field

In certain disciplines in the natural sciences, thematic maps are essential tools. On the ground, an observer typically is faced with an enormous amount of information, too much to be assimilated readily and much of it irrelevant to the task at hand. In order to produce a thematic map one must select only a small portion of the potential observations and measurements – that portion of the data that expresses a unifying idea. A successful field investigator decides what is important and what is not, and assembles information into a coherent interpretation that is supported by physical evidence. Experience, skill, and judgment are essential. Thematic maps use appropriate symbols or notations to reveal any places where there is insufficient physical evidence to map the themes confidently. Thus, a thematic map is a best interpretation by the author; interpretation is not left up to the user.

Most thematic maps that are based on field observations and measurements are unit maps. Units are portrayed by labels, colors, or patterns, and they are separated by lines or patterns that convey information about the nature of mutual unit boundaries. Unit maps are accompanied by a key that explains what the labels, colors, or patterns symbolize in terms of attributes. The number of units and the map symbols are most effective when they are easy to understand so that we quickly grasp the ideas that the author wants to convey. Field units vary as a function of spatial scale as is discussed in Section 7.1.3. Unit boundaries also vary as a function of spatial scale. The classic dilemma of where to define the edge of the sea (Mandelbrot, 1967) illustrates the general problem of scale in thematic mapping. At all scales, from continental shorelines down to sand grains on a beach, any definition of the edge of the sea is an approximation of a fractal-like structure, in the sense that the boundary always has more detail at each finer scale. In the same way, boundary lines between units in the field always are approximations. Nevertheless, field investigators draw lines or patterns separating units that, although approximate, are good enough for the mapping purposes at some defined scale. Units and unit boundaries are judged by whether they pass the practical test of conveying useful information.

Thematic mapping in the field entails editing the available information and then *aggregating* the information in order to tell a story. In remote sensing, all information within the field of view is *integrated* without editing, which has the effect of diluting the information that we are seeking with a clutter of information that we do not care about. It is for this reason that remote-sensing measurements at regional and global scales cannot be interpreted without establishing a link to measurements at finer scales, and eventually to the scale of field observations. Landsat TM owes much of its popularity and value to the fact that the pixel footprint is within the range of critical scales that links field observations and the measurements by AVHRR and MODIS.

> **Map units**
> A map unit has consistent attributes within its spatial limits and is distinct from its neighbors. Above all, map units are meaningful to the investigator.

> **Fractal**
> A geometrical or physical structure having an irregular or fragmented shape at all scales of measurement between the greatest and smallest scale.

7.1.2 Different mapping cultures

Image analysts sometimes are puzzled and disappointed when image units do not coincide with units mapped on the ground, and field investigators sometimes wonder why image analysts cannot make thematic maps that are correct. Much of the problem lies in unrealistic expectations that image data can be processed somehow to produce the same units that one would define on the ground. Field investigators who have heard about the wonders of remote sensing often expect that spectral images will discriminate and identify land cover *and* land use as they see them on the ground. These expectations overlook the fact that land use per se has no spectral attributes, and must be inferred by the observer. On the other hand, image analysts often expect classification algorithms to make field-like maps, even though there is no a-priori reason why statistically distinct spectral classes should be the equivalent of field units (Chapter 3). Unfortunately, image and field data are not easily inverted. Analysts cannot make field maps from images, and investigators in the field cannot predict channel-to-channel radiances for pixels measured by a satellite imaging system.

Land-cover and land-use hierarchies

There are various schemes for organizing names and attributes of the land surface. Grouping names and attributes into categories is a form of classifying textual data, but should not be confused with classifying numerical image data or mapping units on the ground. Just because there is a textual class called "horse pasture" does not mean that there is any corresponding grouping of digital image data. And mapping "horse pasture" on the ground requires interpretation and additional information.

In today's world of specialization it is unusual for a skilled field investigator to know the intricacies of processing spectral images, although many field investigators are experienced photo interpreters who use aerial photographs and other images to facilitate thematic mapping on the ground. Specialists in photo interpretation and in processing spectral images, however, are not necessarily skilled in making thematic maps on the ground. Thus, there tend to be distinct cultures with different ways of understanding landscapes. From the perspective of a field investigator, it is essential to be able to translate among field observations, photo interpretation, and spectral data. All of the techniques can lead to thematic maps. The success of each technique ultimately is judged by what observers find on the ground, as is implied by the term "ground truth" (Section 1.5).

What gets in the way of this translation? How do thematic maps made from spectral images compare with thematic maps made by photo interpretation and from ground observations? Part of the answer is that remote-sensing analysts make images of radiance values, or those derived from radiance measurements (e.g., albedo, temperature, NDVI, greenness), or they try to organize pixels into patterns based on the statistical properties of the data. In the field, however, we rarely make maps of data attributes that are derived from radiance measurements, and we seldom use statistical properties to define units beyond local scales. In the field, for example, we might map the amount or type of green vegetation, but we would never try to map NDVI, greenness, albedo, or temperature. An additional difference between images and the field is that image data are limited by the scale of a pixel footprint. We can *model* what is at the sub-pixel scale, but because we cannot measure directly, it is more difficult to sort out important (thematic) information from the unimportant.

A further obstacle in the way of translating between spectral images and field data is that spectral properties of the land surface are *not* the primary basis for thematic mapping in the field. Even though colors sometimes are useful to fill out descriptions (Figure 1.3), most field mapping can be done in black and white. It would be unusual, indeed, to employ a spectrometer in the field as a primary mapping tool. Instead, field spectrometers generally are used to help understand how a spectral remote-sensing system might respond to radiance from the ground surface. (Interestingly, portable spectrometers also can be used to identify some minerals in the field that otherwise would escape visual detection; but this is another story.) On the other hand, some of the most common ways of extracting information from spectral images are based on organization of the *spectral* data. The problem is compounded by the differences in spatial resolution and perspective between remote-sensing images and the field.

How, then, can we process images so that they reveal field units? In the strict sense we cannot, because the two ways of organizing and interpreting information are fundamentally different. By recognizing the differences, though, we can approach thematic mapping using images without unrealistic expectations. Instead of asking how we can make image maps look like field maps, we can try to answer the more relevant question: how can we make thematic maps from spectral images that are accurate and meaningful to an observer on the ground?

> **Pixels integrate all information**
> In field mapping we can edit and aggregate information, but in remote sensing, each pixel footprint integrates everything – both the information that we want and the unwanted "clutter."

7.1.3 Spatial scale and thematic maps

Descriptions of the land surface traditionally are organized around human visual perception, which, as discussed in Chapter 1, is overwhelmingly dominated by spatial, not spectral, information. Furthermore, the way we describe landscapes is strongly influenced by our most familiar perspective and scale – standing on the ground itself. There are various comprehensive systems for classifying terrestrial land surfaces that are widely used by field investigators and image analysts alike. The organizational framework of these classification systems usually is based on different types of land *use* that would be apparent to an observer on the ground, or, perhaps in an aerial photograph. In the hierarchical system of Anderson *et al.* (1976) (see also Sabins, 1996), for example, the

> **Names change with scale**
> The names that we attach to map units must change as the spatial scale changes.

coarsest scale is Level I, which includes names such as Urban/Built-Up Land, Agricultural Land, Range Land, and Forest Land. At finer scales, each level is further subdivided. Agricultural Land (Level I) is subdivided into Level II categories of Cropland and Pasture, Orchards, Groves, Vineyards, Nurseries, and Ornamental Horticultural Areas. At Level III, Orchards is subdivided into Citrus Orchards, Non-citrus Orchards, etc. It is implicit in this kind of classification that the thematic-map units must change in name and in attributes as the scale changes. Clearly, we cannot apply names and attributes that are appropriate at fine scales to coarser scales unless we also have access to the information at the finer scales. However, one of the main purposes of remote sensing is to organize images without having to do the tedious work of examining the land surface at high resolution. For most remote-sensing applications, therefore, we need to be exceedingly careful to distinguish the attributes that actually are measured from those that we wish to find, and when we assign names, they must be consistent with the measured attributes, reserving the use of field names for situations where there is adequate information.

> **Wishful thinking**
>
> Names have to be consistent with measured attributes – not expressions of what we wish to find.

Names of things

Consider the Anderson *et al.* (1976) category of Agriculture (Level I) which includes the Level II category of Orchards, Groves, Vineyards, etc. On the ground we might know we were in the sub-category Orchard by the presence of cultivated, fruit-bearing trees that are arranged in a regular manner. At ground level we have sufficient visual resolution to identify the types of trees and to measure the spacing and arrangement of the trees. Furthermore, we have rich contextual information such as tractors, ladders, fruit boxes, etc. that make it perfectly obvious how the land is being used. When we move back from an orchard to the perspective of aerial photography we lose some of the detail that we had on the ground. Even though our photograph might show individual trees, we no longer have enough resolution to tell apples from pears, and many of the contextual clues are no longer evident. The critical attributes that remain are the size, shape, uniformity, spacing, and arrangement of the trees. Even these attributes are lost as we change scales to the perspective of most satellite images. If no trees are resolved, the orchard might be characterized by a few pixels that are lighter or darker than the surroundings and that have spectra reminiscent of green vegetation. It certainly is not justified to name these pixels Orchard, unless we also have access to ground observations or high-resolution images that supply additional information.

7.2 Thematic mapping with spectral images

7.2.1 Rules for making thematic maps

Thematic maps that are derived from field observations and measurements are judged by how well they convey selected information about the land surface. We expect field observations and measurements to be accurate within defined limits, and we expect the interpretations to be intelligible and useful. Anyone who takes a thematic map into the field should be able to find the mapped materials and understand how the map maker interpreted the physical evidence. We suggest that it is reasonable to have these same expectations for thematic maps that are derived from images. The rules in thematic mapping are derived from our understanding of how various types of field-based thematic maps have been made, and from our own experience in making geologic, soil, and vegetation maps. We propose that the same rules provide useful criteria for making image-based thematic maps.

Rules in thematic mapping

1. A thematic map, by definition, focuses on one or a few themes. A theme is an interpretation, and a thematic map is a visual expression of one or a few important ideas that the author of a map wants to convey to others. A thematic map is not a map of everything in a scene.

2. Interpretation is done by the author of a thematic map. The responsibility for interpretation is *not* left to the user of the map.

3. A thematic map requires physically based evidence, and a description and evaluation of that evidence. Data attributes, including statistical properties, do not qualify as themes until they have been evaluated and interpreted by the map maker.

4. A thematic map is predictive. Each part of a map predicts the physical properties of the land surface that the author of the map considers to be important.

5. A thematic map correctly portrays the specified objective properties of a landscape at the time the map was made. Uncertainty in mapping the properties are specified on the map by appropriate symbols or notations.

6. A thematic map expresses the interpretations of the author. Consistent with the scientific method, however, all interpretations are subject to later revision in light of new information or ideas.

These rules are deceptively simple, but they can help us to understand how to go about making thematic maps from spectral images. Rule 1 reminds us that we must decide up-front what information we want to communicate. Then we extract that information from the less-interesting background clutter in an image, and present it in a way that is easy to understand. If we just make a color composite or other derivative images that we use for photo interpretation, we have not yet selected any specific information, and we have not called out something that we think is important; therefore, we have not produced a thematic map. Similarly, images that are based on the statistical properties of the data space are not thematic maps, because they are not interpretive, and they do not convey any idea of the relative importance of the information content. Principal-component transformation images and classifications

Statistical properties
By themselves, statistical attributes of image data are not interpretive. The user must supply the interpretation and decide what is important.

are examples. In both cases, the reader is left with the task of interpretation, which is not consistent with rule 2. Rules 1 and 2 remind us that if we do not understand an image, then we are not ready to make a thematic map.

Rules 3 and 4 specify that we cannot make a predictive map until we know how the properties of an image, spatial and spectral, relate to the properties of the materials on the ground. To make predictions that are based on spectral data, we need to construct physical models that relate radiance to the materials on the ground. The most robust physical models are based on resolved absorption or emission bands that are diagnostic of specific materials. However, other physical models can be powerful predictors when applied in conjunction with spatial and contextual information that narrows the possible interpretations.

Rule 5 says that we expect field maps accurately to portray the materials and conditions on the ground that a map maker claims were there when the map was made. If a map maker is uncertain about what or how much of something is in a particular place, we expect that we will be told that. We do not expect to visit a place on the ground and to find that a field map is incorrect, especially when there is no explanation for the error. When there is uncertainty about the observations or measurements, field investigators reveal this in thematic maps by symbols or notations. For example, an area might be described as consisting of "heterogeneous, undifferentiated soils." In this case, the map maker makes clear that it was not important to try to map all of the separate soil types in the area, and that anyone using the map should not expect this part of the map to be accurate in detail. However, it is still expected that a visitor to the site would indeed find an area of "heterogeneous, undifferentiated soils," rather than something else. Unfortunately, similar qualification of mapping accuracy usually is absent from thematic maps that are derived from images, and this can be a source of frustration for investigators who try to apply remote-sensing data to field problems.

Accuracy of interpretations is a separate but related issue (Section 1.5.2). Rule 6 reminds us that the interpretations that are imbedded in thematic maps can change as ideas evolve and as new information becomes available, but that the objective observations and measurements should reflect what was there on the ground at the time the map was made. Clearly, it is essential that we be able to distinguish between interpretations and objective observations and measurements.

Begin with a reconnaissance
If we have only a limited understanding of an image and what information it contains about a landscape, we need to begin with a reconnaissance. Presumably we already know what information we want as our

theme(s); therefore, our task in the reconnaissance stage is to become familiar enough with an image to determine how to extract that information. As discussed in Section 3.4, standard image-processing programs include several tools such as scatter plots, the PCT and unsupervised classifiers that can be useful for visualizing the data structure and for relating it to the spatial context of an image. However, these tools operate unselectively on the data, and do not produce themes, because they do not know how to extract and evaluate information. Unfortunately, inexperienced analysts sometimes limit their work on a spectral image to an unsupervised classification, and mistakenly consider that they have made a thematic map if they can relate some of the classes to familiar materials or activities on the ground. Such classifications, though, are not consistent with rules 1 and 2.

Supervised-classification images also commonly are confused with thematic maps. As discussed in Section 3.4.2, the training sites for supervised classifiers are based on places in an image that the analyst considers to be important; therefore, training areas already are themes. However, there are two critical issues here. First, training areas for classifying are not necessarily selected to convey a few important ideas, as is the case for thematic maps (rule 1). There is no fixed limit to the number of training areas, and, typically, many classes are created by analysts when classifying images. Second, as emphasized in Section 3.4.2, training is driven by desired map themes, but these may or may not match the data structure. It is circular reasoning to claim that we can use a desired (training) theme to find comparable pixels (data structure), and then turn around and use the defined class (data structure) to make a map of the same theme. In our view, per-pixel classifications are not thematic maps, unless they are constrained by physical models. Accordingly, we treat unsupervised and supervised classifiers as tools that are most useful for data reconnaissance and organization (Section 3.4).

Physical models

Rule 2 imposes the constraint that a thematic map must be predictive, and that it must be useful to a field investigator. Physical models provide the best connection between image and ground (Section 1.4.2), because they require insight into the criteria used to map each theme, and can be revised and improved when more information is available, such as from field testing of thematic maps. However, physical models do not have to be elaborate or complicated to be effective. As an example, the NDVI is a simple model that represents green vegetation. An NDVI image of a scene that is known to contain green vegetation is a thematic map that predicts the relative absorbing power of chlorophyll from pixel to pixel. By applying contextual constraints, such as knowing that a vegetation canopy is open, we can make a thematic map that further predicts the relative cover of green foliage. On the other hand, if we select a training area that we think represents green vegetation, and we use it to classify an image, we only map other pixels that are similar to the training site within selected thresholds. If we make a poor choice of training pixels, the class map will be incorrect, and it will not conform to rule 2 as being predictive and useful. However, we could constrain the class map by a physical model, simply by verifying that the mean spectrum of the training area matches that of chlorophyll, or by training on pixels having high values of NDVI, thereby qualifying the class map as a physically based thematic map.

7.2.2 Thematic maps from physical-attribute images

A first step in making a thematic map from an image is to prepare one or more physical-attribute images that we can link to physical properties of the land surface. If we can confidently predict the nature of the land surface in ways that express a theme of interest, then we can make a thematic map. Below we explore different examples, including one in which it was *not* possible to make a satisfactory connection between data attributes and physical attributes of the scene.

Example 1: Death Valley

Day and night TIMS images, acquired by a NASA aircraft (Figure 7.1 and color Figure 7.1 on the Web), were used to make a thematic map that shows the main rock types in the Furnace Creek area of Death Valley, California (e.g., Gillespie *et al.*, 1984). Thermal-IR laboratory emissivity spectra of field samples are linked to the daytime TIMS image by lines. Color Figure 7.1 (see Web) is a decorrelation-stretched color-composite image. As discussed in Section 3.3.1, a decorrelation stretch visually enhances spectral (WW) contrast, while suppressing the otherwise dominant light–dark variations that arise from temperature variations in the scene (Gillespie *et al.*, 1986).

> **Decorrelation stretch**
>
> As a reminder, in images dominated by light–dark variability, such as daytime TIR images, a decorrelation stretch can be used to display color information effectively (Section 3.3.1).

Red areas (relatively high emissivity in Band 5) define quartz-rich rocks, green areas (high emissivity in Band 3) are salts, blue-green areas (high emissivity in Bands 1 and 3) are limestones and dolomites, and magenta areas (high emissivity in Bands 1 and 5) are shales, basalts, and andesites. Desert varnish locally coats rock surfaces, and is especially noticeable on some of the visually light rocks (quartzites) of the alluvial fans. Varnish incorporates wind-deposited dust derived from other lithologies, and increases the emissivity of the coated quartzite in Band 1, thereby shifting the red of the quartzite toward magenta.

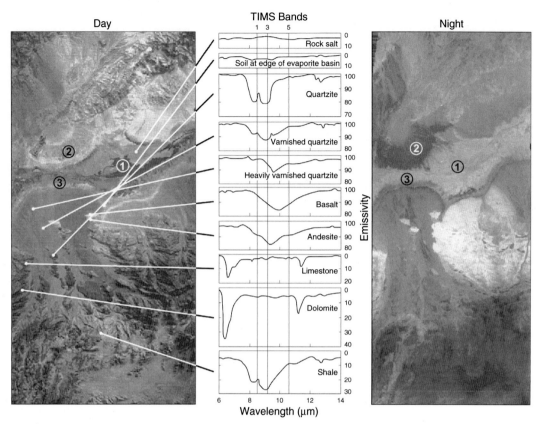

Figure 7.1 and color Figure 7.1 (see Web). Daytime and nighttime TIMS images of the Furnace Creek area of Death Valley, California, and TIR spectra of rock types from the scene. See also Plate 1 to locate the Furnace Creek sub-image within Death Valley. The TIMS images are decorrelation-stretched color composites of TIMS Bands 1 (blue), 3 (green), and 5 (red). Lighter tones indicate warmer temperatures. Colors in the images are interpreted in terms of rock types by comparing the keyed TIR spectra. The day image includes shading and shadowing, whereas the nighttime image does not include this information. Circled numbers 1, 2, and 3 identify areas in the scene where day/night temperature differences are controlled by soil moisture (see text). NASA/JPL.

In the day image, illuminated surfaces are warmer than shaded or shadowed ones, giving the visual impression of topography. The night image does not show these effects, and the emitted radiance more faithfully expresses the composition of the surface materials. The day and night image pair also shows some interesting differences in the thermal properties of the surface materials. Lake Manly (circle 1) is cooler than its surroundings in the daytime, whereas the springs at the base of the Tucki Fan (circle 2) and the Furnace Creek Fan (circle 3) are cooler than their surroundings at night. The cool (dark) region at the bottom of the night image is an adiabatic effect of the relatively cool atmosphere over the Panamint Range.

The images in Figure 7.1 and color Figure 7.1 (see Web) meet the criteria for a thematic map that are defined in Section 7.2.1. They focus on the theme of rock composition. Emitted radiance images are linked to the physical properties of the surface (rock composition) by a spectroscopic model that uses laboratory measurements of samples collected from the scene. Potentially confusing temperature effects are suppressed by color-enhancing small variations in emittance, and by measuring daytime and nighttime emittance. A second, but subordinate, theme is the variation in day–night temperature contrasts.

The images are predictive, in the sense that we could reasonably expect to take copies into the same field area and be guided to the named rock types. Furthermore, because the results were field-tested, we would expect that similar images would be useful to a field geologist in the imaged scene and perhaps in other areas in the Death Valley region (Plate 1).

Example 2: Seattle

A 1992 Landsat TM image (Plate 2) was used to make a thematic map of the urban core of the city of Seattle, Washington. Knowing the extent and location of an urban core is relevant for a variety of city-planning and hydrology studies (Section 8.2.2 and Greenberg, 2000). At first glance this seems like a simple objective; after all, using photo interpretation, the city of Seattle is obvious in the Landsat TM image (e.g., Plate 1). But images of Seattle are not the same as thematic maps, because they are not selective, and spectral interpretation does not have a physical basis. The term "urban core" actually describes land use, and urban land use is defined by familiar features such as tall buildings, industrial complexes, roads, and parking areas (Section 7.1.2); however, it is not clear whether urban core can be characterized uniquely by a single, representative spectral sample.

Spectral identification

Even without collecting samples from the Seattle area for laboratory spectral measurements (as was done in the Death Valley example), we can link some of the six-channel Landsat TM spectra to laboratory reference spectra. Water is easily identified from its relatively flat, dark spectrum and from its obvious spatial context. We can predict the main spectroscopic properties of green vegetation, and there is no need to collect GV samples from the scene to assist with a generalized identification. An NDVI image of the Seattle area (e.g., reference images on the Web and Figure 8.3) is a predictive model of those pixels likely to have high proportions of GV. It is a simple matter to examine the image spectra of candidate-GV pixels to test whether the responses

> **Interpretation of image spectra**
>
> The key is to relate image spectra to known materials on the ground using spatial context.

in all six channels are consistent with reference spectra, taking into account calibration factors (Section 3.2.1). Similarly, we see that a TCT (Section 3.5.2) of the Seattle image (reference images on the Web and Figure 8.4) shows an image pattern for GV that is consistent with the NDVI image. The spectroscopic identification of GV is further supported by the image context that clearly shows GV in city parks and in surrounding residential and agricultural areas.

Dry grass (a high-albedo variety of non-photosynthetic vegetation) dominates some parks, fields, and lawns in Seattle in August, and image spectra of these areas are consistent with reference spectra of NPV (Figures 2.3, 5.1, 8.8). But NPV spectra are notoriously variable, and from experience we know that NPV spectra can mimic the spectra of some soils and rocks. Thus, we have higher confidence in interpreting the image spectra of water and GV, and, although the image spectra of known areas of dry grass are diagnostic, we are cautious about possible confusion between dry grass and other materials such as leaf litter and soil.

Image spectra of the city core, though, are not easily identified among reference spectra. Some image spectra are consistent with "urban" materials such as pavements and roofs, but many also are similar to spectra of soils and rocks. However, in the context of the Seattle image with its background of water, GV and NPV, the image spectra of the urban core and of the main industrial area stand out as being distinctive. Nevertheless, we need to be cautious here, because it is the image *context* that supports our interpretation of the "urban" image spectra, not the unique character of the spectra themselves. For example, the scatter plot in Figure 3.4 would look much the same if it had been derived from a scene in which there was exposed soil and rock (Figure 8.1), but no urban area at all.

> **Spectra of the urban core**
> These are not diagnostic by themselves. Identification depends on the image context.

Spectral mimicking of the urban core

Mimics of the urban-core spectrum can be seen by inspection of Figure 7.2. The figure includes a spectral-angle attribute image that was trained on the downtown Seattle core. Dark pixels are closer (in the data space) to the mean spectrum of the training set. In addition to highlighting the urban core, the other spectrally similar areas include the main industrial area, docks and marinas, commercial centers, roads, bridges, parking areas, beaches (especially along Puget Sound), shallow water, and fallow fields in agricultural areas. Mimicking of the mean urban-core spectrum presents a serious problem if we are trying to make a thematic map that is predictive.

> **Spectral mimics**
> Spectra of different types of physical objects may resemble each other, especially at low spectral resolution.

Spectral variability of the urban core

The potentially wide range of albedo of an urban core presents an additional problem for making a thematic map. We expect a downtown

Figure 7.2. Spectral-angle attribute image and histogram of a sub-image of the 1992 Landsat TM image of Seattle. (a) Spectral-angle attribute image. Darker pixels are closer in SA to the mean of a training set centered on the Seattle urban core. (b) Histogram showing the distribution of SA values. Most pixels in the training area plot within the gray-toned part of the histogram.

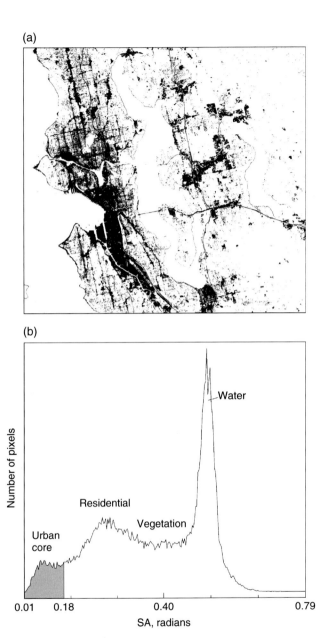

(a)

(b)

area to have both light and dark structures, and light and dark sidewalks and road surfaces. Shadows cast by tall buildings will create additional contrast. In the Seattle example, these factors are expressed in the linear distribution of urban pixels in scatter plots (Figure 3.4), in the individual channel images (reference images on the Web), and in the patchy nature of the minimum-distance attribute image shown in Figure 7.3. (Recall

(a)

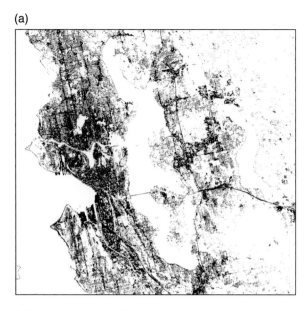

Figure 7.3. Minimum-distance attribute image and histogram of a sub-image of the 1992 Landsat TM image of Seattle. (a) Minimum-distance attribute image. Darker pixels are closer in distance to the mean of a training set centered on the Seattle urban core. (b) Histogram showing the distribution of minimum-distance values. Most pixels in the training area plot within the gray-toned part of the histogram.

(b)

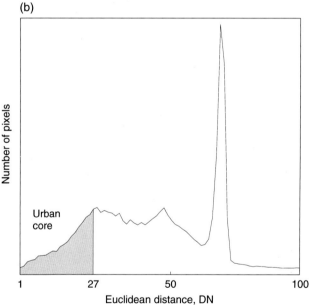

from Chapter 3 that the minimum-distance method is sensitive to SL contrast, whereas the spectral-angle method is not.)

In addition to the inherent spectral variety of an urban core area, there is sub-pixel mixing with those other scene components that we associate with non-urban areas. Small amounts of green vegetation are present along streets and in small parks in the downtown Seattle area, and, in the August image of Seattle, dry grass and weeds are ubiquitous in the industrial areas and along roads. Mixing of urban-like spectra with the spectra of GV and NPV further increases the apparent spectral variability, and, of course, raises the question of how and where to define the boundary between urban and surrounding urban-residential, suburban and rural areas.

Limitations of an urban thematic map

From the above discussion, we conclude that the spectral and albedo attributes of the Seattle urban core are not sufficient in themselves to make a thematic map. There is no well-defined set of reference spectra that defines "urban," and the image spectra of the downtown core are much too heterogeneous to be characterized by the mean value of a single training set, or even by several subsets. Nevertheless, various attribute images of the urban core, while failing to meet the require-ments of the rules for thematic maps, may potentially be useful for a field investigator. Visually, the "urban" areas seem to stand out clearly, because, in the Seattle case, the urban materials and their mimics are easily distinguished from the other basic spectral components of the scene – water, GV and NPV. Therefore, attribute images that are based on training on "urban" areas may be useful for certain purposes, providing that the potential ambiguities are clearly understood. One solution is to label the attribute images (e.g., Figures 7.2 and 7.3) as "urban and urban mimics." The objective would be to make clear to any users that *they* must interpret the image context, and *they* must decide whether a given pixel or area qualifies as "urban." This is an important point, because it means that the user decides what is important and makes the judgments, rather than the person who prepared the image. In the case of a thematic map, the one who creates the map provides the interpretation.

Example 3. Manaus, Brazil

The objective in this example is to make thematic maps of the main vegetation types in a tropical forested landscape from the data in a Landsat TM image of the area near Manaus (Plate 3). The vegetation types of interest, and their properties, have been defined on the ground by localized field studies, primarily in a few areas near roads and along rivers. The large area and the limited and difficult access make it especially valuable to be able to extend local observations to larger regions using remote sensing.

Our analysis of the Manaus image involves most of the concepts that we have presented in this book, and the detailed discussion that follows is a practical example of an analysis that succeeded in its thematic-mapping objectives. The Manaus image itself was acquired by Landsat 5 in 1991. The image quality is only fair, as Bands 1, 2, and 3 have noticeable horizontal striping and the dynamic range of Band 3 is seriously limited. To our knowledge there still are no high-resolution digital topographic data for the area. The example shown is based entirely on one image that has not been corrected for atmospheric effects; however, in other work (Adams *et al.*, 1995) we corrected this image and others of the Fazendas area to reflectance to study changes in vegetation cover with time (Plate 8). In presenting this example, we make the general point that it is possible to extract useful information from older spectral images of modest quality, and in some cases, without applying atmospheric and topographic corrections.

We begin with the Fazendas sub-image (Plate 4) that includes primary forest and vegetation that has grown back into deforested patches. We and our colleagues have studied the area for several years on the ground, using low-flying aircraft and with remote sensing (e.g., Adams *et al.*, 1995; Lucas *et al.*, 1993, 2002). The site was selected because of previous intensive ecological studies (e.g., Bierregaard *et al.*, 2001 and references therein; Kapos *et al.*, 1997; Tanner *et al.*, 1998; Logsdon *et al.*, 2001) in an area that is undergoing rapid deforestation and development. Work by Bohlman (2004) in Panama and Bohlman *et al.* (1998) in the Brazilian Amazon has further developed the physical basis for interpreting VIS-NIR images of tropical forests, and in particular has shown the importance of measuring canopy NPV remotely to understand deciduousness.

The color composites in Plates 3 and 4 are convenient reference images that also show the context of the Fazendas sub-image within the larger TM scene. From previous experience with similar color composites, we expect that green vegetation will appear green, and that non-vegetated areas will be red or blue. From field experience in the imaged area, we know that the image distinguishes patches of bare ground and re-grown vegetation (commonly referred to as "regeneration") from the spatially extensive background of primary forest. In keeping with the rules for making thematic maps, we need to find specific evidence that links the spectral attributes of the image to the physical properties of the land cover. In this example we focus on those spectral attributes in the VIS-NIR that allow us to distinguish different types of green vegetation (Chapter 2), specifically, absorption by chlorophyll at red wavelengths (TM Band 3), scattering, transmission and reflection by leaf structures in the NIR (TM Band 4), and absorption by water in the NIR (TM Band 5).

Table 7.1 shows mean DN values (uncorrected for atmosphere) for eleven vegetation types, soil, slash (cut trunks, branches, and dry leaves) and water in each of the six reflective TM channels. The table lists four types of primary forest and seven types of regeneration. All are

Fazendas

The name "fazenda" means "ranch." The study area includes several ranches and other clearings in the primary forest about 70 km north of Manaus.

"Regeneration"

Regeneration is a process, but foresters also use the word to describe the results of re-growth, and we use the term in this sense.

Landsat TM Bands

Band 1	0.45–0.52 μm
Band 2	0.52–0.60 μm
Band 3	0.63–0.69 μm
Band 4	0.76–0.90 μm
Band 5	1.55–1.75 μm
Band 7	2.08–2.35 μm

Table 7.1. *Mean DN values in the VIS-NIR Landsat TM Bands for vegetation and substrate types in the Fazendas sub-image (Plate 5).*

Vegetation type		TM Band						NDVI	$\frac{(4-5)}{(4+5)}$
		1	2	3	4	5	7		
Closed-canopy primary forest	Riparian	72.3	31.6	28.9	74.3	47.3	8.2	0.44	0.22
	Caatinga	77.6	32.2	30.4	61.1	46.3	9.0	0.34	0.14
	Terra firme	72.2	30.4	28.6	64.1	48.2	9.0	0.38	0.14
	Varsea	76.4	31.9	30.0	63.4	51.6	9.9	0.36	0.10
Closed-canopy regeneration	Cecropia green	72.5	30.6	28.1	85.6	59.4	10.4	0.51	0.18
	Vismia	73.0	31.5	28.9	86.9	62.8	11.6	0.50	0.16
	Cecropia average	72.7	31.1	28.7	88.1	64.7	11.7	0.51	0.15
	Mature regeneration	72.8	31.8	29.5	85.9	65.8	12.1	0.49	0.13
	Cecropia yellow	72.9	31.2	28.8	88.6	68.9	13.0	0.51	0.13
	Kudzu	75.4	34.5	31.4	108.4	96.6	21.8	0.55	0.06
	Pasture	81.1	37.9	42.6	70.5	103.3	29.8	0.25	− 0.19
	Slash	87.1	41.8	49.4	59.1	99.0	30.4	0.09	− 0.25
	Soil	100.8	50.7	65.5	69.7	135.1	42.5	0.03	− 0.32
	Water	72.4	30.3	29.0	43.1	28.7	5.0	0.19	0.20

closed-canopy (forest, shrubs, or ground cover), with the exception of pasture which has some exposed soil, as well as a significant fraction of dry grass (NPV). Training sites were keyed to field observations, mainly in the Fazendas area; however, caatinga and varsea forests are only present outside the sub-image. The DNs for each vegetation type in Bands 3, 4, 5, and 7 are plotted in Figure 7.4, and spectra for selected vegetation types are given in Figure 7.5. Images for each channel are shown in Figure 7.6 and in the reference images on the Web. Locations of each of the vegetation types are shown on the NDVI image in Figure 7.7.

Initial inspection of the data in Table 7.1 shows that most of the regeneration cannot be distinguished from primary forest in Bands 1, 2, and 3, as can be verified in the channel images in Figure 7.6. Pasture, soil, slash, and water differ from all other closed-canopy vegetation in these channels. In each of Bands 4, 5, and 7, regeneration has higher DNs than does the background primary forest, as is evident in Figures 7.4 and 7.6. For example, the large patch of cecropia labeled on the NDVI image in Figure 7.7 is not visible against the background of primary forest in Bands 1, 2, and 3, but it is clearly seen in Bands 4, 5, and 7.

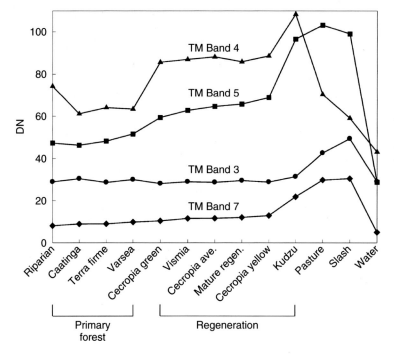

Figure 7.4. Mean DN values (uncorrected for atmosphere) in Landsat TM Bands 3, 4, 5, and 7 for vegetation types in the 1991 image of Manaus, Brazil. Except for pasture, all vegetation types are sampled from closed canopies (no substrate exposed). Slash and water are added for reference. Vegetation types and canopy closure were verified by observations in the field.

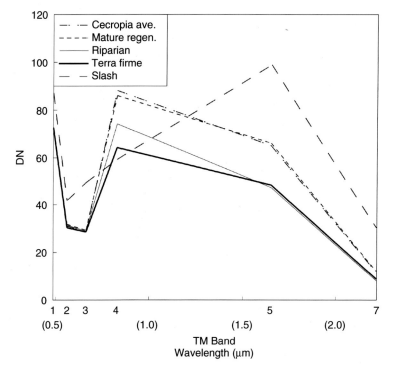

Figure 7.5. Six-Band spectra of selected vegetation types from the Fazendas sub-image, keyed to Table 7.1 and Figure 7.4.

TM Band 1

TM Band 2

TM Band 3

TM Band 4

TM Band 5

TM Band 7

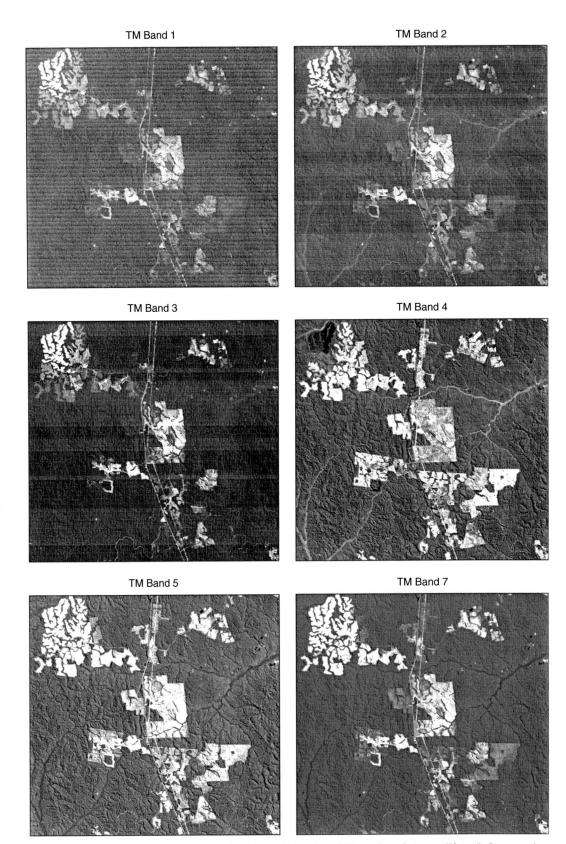

Figure 7.6. Images of the six VIS-NIR channels of the 1991 Landsat TM Fazendas sub-image (Plate 4). Regeneration vegetation has low contrast with the background of primary forest in Bands 1, 2, and 3, and, therefore, is not easily distinguished in the human visible range.

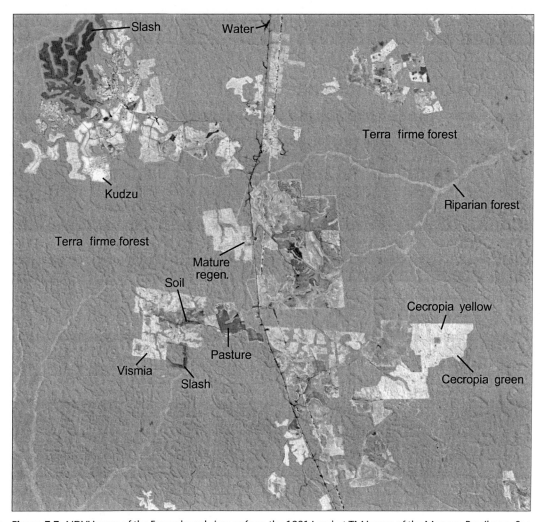

Figure 7.7. NDVI image of the Fazendas sub-image from the 1991 Landsat TM image of the Manaus, Brazil area. See also Plates 3 and 4. Lighter tones have higher NDVI values (Section 3.5.1). Uncorrected for atmosphere. Darkest pixels indicate exposed soil and/or NPV. Intermediate and high NDVI values indicate the presence of green vegetation; however, interpretation of vegetation types (gray tone) is ambiguous without also knowing whether the canopy is open or closed. Labels and arrows indicate representative sites of the vegetation types that are listed in Table 7.1.

Data numbers in each channel are not reliable indicators of the vegetation types unless the illumination is uniform. Although the overall topography of the sub-scene is gentle, SL contrast varies with canopy structure and where topography is locally steep along stream channels. For areas where we can rule out topographic effects, regeneration and primary forest are reliably distinguished simply by the fact that the primary forest is darker in Bands 4, 5, and 7. Spectral-length contrast alone, however, is not adequate to distinguish among the individual vegetation types, with the one exception that kudzu vine is unusually reflective in Bands 4, 5, and 7 relative to other types. In Chapter 6 and Figure 6.17 we used the example of kudzu to show how SL contrast could be used to detect otherwise obscure materials under certain circumstances.

Channel ratios

With six channels and multiple types of land cover, it is not immediately obvious how to prepare a thematic map of vegetation types. To sort out the variables, and to establish a physical basis for interpretation, we will follow the rationale presented in Section 3.5, and begin with an examination of ratio-based images. In the NDVI image (Figure 7.7), as expected, patches of bare ground, recently cut patches of forest, roads, and water have low NDVI values and appear dark. The highest NDVI values occur for patches of regeneration, and intermediate values include several of the cleared areas as well as the primary forest. To interpret the intermediate and high NDVI values we need to know whether the vegetation canopies are open or closed (Sections 3.5.1 and 8.1.2). In open-canopy areas, we expect the NDVI to increase mainly due to increasing cover of GV relative to exposed leaf litter and/or soil on the ground. In closed-canopy areas (substrate not exposed), the NDVI will vary with the relative amounts of absorption by chlorophyll in the red (Band 3) vs. the scattering/reflecting power in the NIR (Band 4). But in closed canopies, the NDVI also can be lowered by exposed branches and/or senesced leaves that produce higher DNs in Band 3.

Because all of the closed-canopy vegetation types have essentially the same DNs in TM Band 3 (Table 7.1 and Figure 7.4), the chlorophyll absorption band in the red is saturated, and NDVI variations are controlled entirely by the NIR (Band 4). (This is consistent with the image of Band 4 in Figure 7.6.) Saturation in Band 3 signifies that the NDVI in this case is not related to the amount of chlorophyll absorption, and that the index has no significance regarding primary productivity.

Now we can understand why in Figure 7.7 the terra firme forest has a *lower* NDVI relative to all of the closed-canopy regeneration: there is less light scattered and reflected in Band 4 by the rough, self-shadowing canopy of the terra firme forest, relative to the reflectance of the smoother canopies of the regeneration (see ahead to Figure 7.14). This also explains why the NDVI of the terra firme forest mimics that of many areas that have only partial vegetation cover.

What do we learn about mapping vegetation types from the NDVI in this example? As usual, topographic effects are strongly suppressed in the NDVI image. The lowest values of NDVI reliably indicate areas

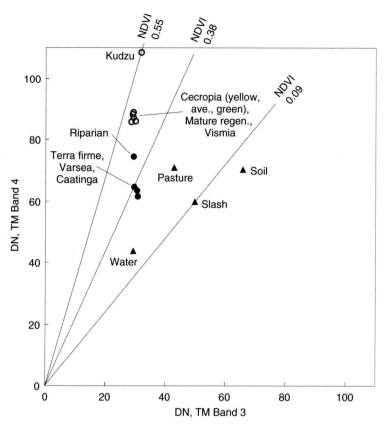

Figure 7.8. Scatter plot of TM Bands 3 and 4 showing vectors of the mean values of vegetation types, slash, soil, and water. All closed-canopy vegetation types (open and solid circles) (Table 7.1 and Figure 7.4) have the nearly the same DNs in Band 3; however, primary-forest types (solid circles) have lower DNs in Band 4 than regeneration (open circles). The NDVI values of primary-forest types mimic NDVI values of mixtures of regeneration and substrate.

where GV is sparse or absent. Primary forest can be distinguished from regeneration, *providing* that we can be sure that both canopies are closed. The reason for the ambiguities in interpreting the NDVI image are illustrated in the scatter plot in Figure 7.8. The range of NDVI values of the primary forest (terra firme and riparian) overlaps that of mixtures of regeneration with substrate (soil, slash). Therefore, the NDVI image by itself is limited in its predictive value for mapping vegetation types, unless we already know whether or not substrate is exposed at the sub-pixel scale. Because of these ambiguities, the NDVI image does not qualify as a thematic map.

As discussed in Section 3.5.1 and in Figure 2.11, higher DNs in Band 5 in the context of closed-canopy vegetation can signify less absorption in the adjacent, broad water bands at 1.4 and 1.9 μm. Less absorption by water implies the presence of "dryer" components of the canopy, namely NPV as woody tissue and/or senesced or dehydrated foliage. In Table 7.1 and Figure 7.4 regeneration has higher DNs in Band 5 than does the primary forest (as is the case for Band 4). But in this context, DN values in

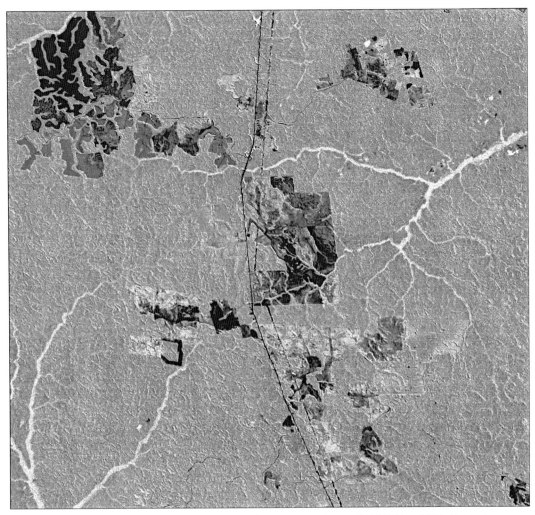

Figure 7.9. Normalized-difference image of the Fazendas sub-image: $(DN_4 - DN_5)/(DN_4 + DN_5)$. Lighter tones are pixels having relatively lower DNs in Band 5 that are interpreted as due to more absorption by water in green foliage. Darker tones indicate areas of exposed substrate (soil and slash), or, in closed-canopy vegetation, more exposed branches and/or senesced leaves.

Normalized difference

The normalized difference for Landsat TM Bands 4 and 5 is $\frac{DN_4 - DN_5}{DN_4 + DN_5}$.

Band 5 express mainly the amount of shading and shadow in the canopies, and high DNs for regeneration relative to primary forest do *not* mean that regeneration has more exposed NPV. To evaluate the relative effects of water absorption (or lack of it), we need to normalize the data, and a convenient way to do that is by a ratio of DN_5/DN_4, or, as a parallel to the NDVI, by the normalized difference of Bands 4 and 5 (Figure 7.9).

Although the DN values in Band 5 for primary-forest types are all about the same, DNs for the regeneration types have a wider range of

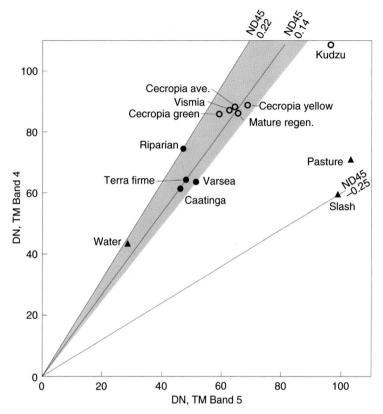

Figure 7.10. Scatter plot of TM Bands 4 and 5 for the Fazendas sub-image. Lines are values of the normalized difference for Bands 4 and 5 (Figure 7.9). The gray pattern defines the range of values of the normalized difference in Bands 4 and 5 for closed-canopy primary-forest types (solid circles). The range of values for primary forests overlaps the range for closed-canopy regeneration (open circles), which is consistent with the observation that regeneration is difficult to distinguish from terra firme forest in Figure 7.9. However, the normalized difference varies within the forest and regeneration categories.

values, as is shown in Table 7.1 and in the scatter plot of Band 4 vs. Band 5 (Figure 7.10). It is difficult to interpret these channel DNs alone, because of the ambiguities due to canopy roughness and topography; however, we can reduce the ambiguity by working with the normalized difference of Bands 4 and 5 that provides a physical model of variability in water-absorption (or dryness) properties.

In Figure 7.10 higher values of the normalized difference of Bands 4 and 5 imply more absorption by water in the canopy, or, conversely, less woody material and/or senesced or dehydrated leaves. The data are consistent with field observations of the relative amounts of NPV in closed-canopy forests. For example, low-altitude observations and aerial photographs confirmed that riparian forests (typically dominated by palm species) exposed less woody tissue than the surrounding terra firme forest. The varsea forest, which was flooded at the time of the field observations and the acquisition of the Landsat image, had more leafless trees than did the terra firme forest. Among the regeneration types, ground observations confirmed that there was relatively more woody material exposed in areas dominated by vismia than in areas dominated by cecropia, which are characterized by large, broad leaves that make a relatively flat canopy. Low-altitude aerial observations confirmed that the canopy of cecropia yellow had been damaged by wind storms, and exposed more woody material than the areas of cecropia green.

For thematic mapping, therefore, the normalized-difference values of Bands 4 and 5 give us a physical basis for separating closed-canopy primary-forest types from one another, and closed-canopy regeneration types from one another, but, as is evident from the image in Figure 7.9, primary forest and regeneration are mimics or near mimics. However, we already have established that all of the primary-forest types are all darker than all of the regeneration types in Bands 4, 5, and 7. As long as there is minimal topographic shading, we can map the primary forest simply by its lower albedo that is the result of shading and shadowing in the rough canopy at the sub-pixel scale.

So far we have focused on Bands 3, 4, and 5. Next, we explore two other approaches to physical modeling of vegetation, the tasseled cap and spectral-mixture analysis, that incorporate all six of the VIS-NIR channels.

The tassled cap

The tasseled-cap transformation (TCT) is a fixed model that was designed for agricultural landscapes. In Section 3.5.2 we discussed the physical basis for interpreting TCT images, and in Figure 3.19 we showed a plot of the TCT coefficients for Landsat 5. The Fazendas TCT images (Figure 7.11) are generally consistent with the results discussed above; however, we need to be cautious about how we interpret each image in terms of closed-canopy forest types. The image of TCT_1, soil brightness, (Figure 7.11a) depicts in light tones the areas of exposed substrate: soil and mixtures of soil, NPV, and GV. The same image portrays patches of regeneration in intermediate tones, and primary forest in dark tones. Clearly, though, the soil-brightness image should not be used to interpret the closed-canopy forest where there is no substrate exposed. The image of TCT_2, greenness, (Figure 7.11b), in which Bands 3 and 4 receive the heaver weightings (Figure 3.19), is generally similar to the NDVI image (Figure 7.7). We already know that Band 3 is saturated ($DN_3 \sim 30$) for the closed-canopy vegetation (Figure 7.4), therefore "greenness" closely resembles the image of Band 4 alone (Figure 7.6), and the lighter tones indicate more NIR scattering and reflectance.

The $(1 - TCT_3)$, dryness, image (Figure 7.11d) is weighted heavily by Band 5 (Figure 3.19). In fact, the $(1 - TCT_3)$ image closely resembles the Band-5 image (Figure 7.6). The lightest tones are known areas of pasture where senesced grass is exposed, and areas of slash and

(a)

(b)

(c)

(d)

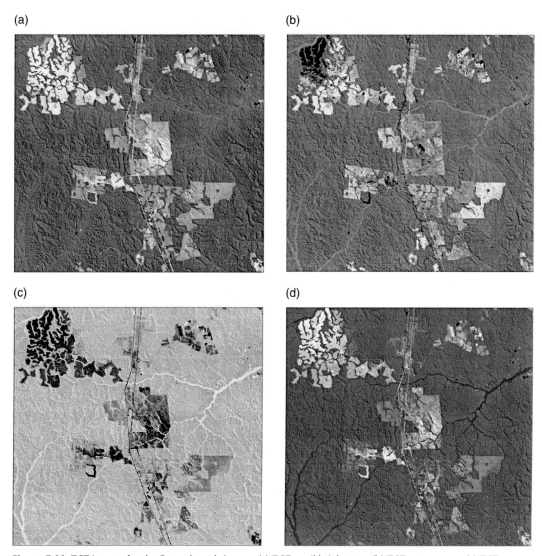

Figure 7.11. TCT images for the Fazendas sub-image. (a) TCT_1, soil brightness; (b) TCT_2, greenness; (c) TCT_3, wetness; (d) $(1 - TCT_3)$, dryness. See also reference images on the Web. Lighter tones indicate more of the designated attribute.

mixtures of slash with soil and GV. Intermediate and darker tones are areas with green vegetation, where absorption by water is greater. However, as previously discussed, and as illustrated in the Band 4 vs. Band 5 scatter plot in Figure 7.10, unless shade has been normalized, higher DNs in Band 5 do not necessarily mean less absorption by water. Therefore, to use the $(1 - TCT_3)$ image to map dryness relative to greenness, we must confine our interpretation *within* the two closed-canopy forest categories, but we cannot compare those attributes *between* regeneration and primary forests. For example, in Figure 7.11d, both cecropia green and cecropia yellow appear lighter than the adjacent terra firme forest, implying that both regeneration types have more NPV than does the primary forest. This is not consistent with field observations or with the scatterplot in Figure 7.10. However, Figure 7.11d correctly shows that cecropia yellow is lighter than the adjacent cecropia green. This presents a problem for thematic mapping according to our rules, because a TCT-based map would require the user already to be able to distinguish reliably open and closed canopies, and primary-forest types from regeneration types.

Lessons from the TCT
Let us pause and review why the TCT does *not* allow us to make a satisfactory thematic map of closed-canopy vegetation, even though it incorporates information from all of the TM channels and it is based on a physical model. This review will set the stage for a discussion where we apply a spectral-mixture model that does lead to a thematic map.

In keeping with the original purpose of the TCT, the three TCT images correctly discriminate areas of exposed substrate from those of closed-canopy vegetation. That is not our main objective, but it is helpful for identifying those areas that are entirely covered by green vegetation. Soil brightness is not relevant to our objective, thus our analysis is limited to the greenness and dryness images. Greenness, as we see in Figures 7.8 and 7.10, is driven almost entirely by Band 4 (because Band 3 is saturated), and dryness is mainly influenced by Band 5. Greenness and dryness are spectral attributes. We would not (and could not) map these attributes on the ground; however, we are exploring whether we can use greenness as a proxy for canopy and leaf architecture, and dryness as a proxy for the relative amount of NPV within the canopies. These are properties that are meaningful to an observer on the ground, and that are variable from one vegetation type to another.

Ranking attributes
Greenness and dryness are defined only in terms of relative channel weightings. This means that DN values for these attributes cannot be inverted to yield an amount of NPV or a unique description of the canopy architecture. The best we can do is to make a consistent relative ranking of the different vegetation types according to the attributes. Ranking, however, requires that we make two critical decisions. First, we need to decide whether the two attributes are independent or

dependent variables. Second, we need to decide on a vegetation type that we use as a reference (e.g., greener, or less green, than what?).

Are the greenness and dryness attributes independent for the closed-canopy vegetation? Apparently. The canopy architecture does not predict the amount of NPV exposed, neither does the amount of NPV predict the canopy architecture. This is expressed in the scatter plot in Figure 7.10, because DNs in *both* TM Bands 4 and 5 tend to increase with smoother architecture, but that does not mean that the amount of NPV increases, too. If we want to isolate NPV, we need to normalize for canopy architecture (shade) that affects all channels. We already have displayed part of this information using the normalized difference in Table 7.1 and Figure 7.10. The TCT, however, does not normalize for shade. As a result, in Figure 7.11d cecropia yellow has a higher $(1 - TCT_3)$ value than does the surrounding terra firme forest, implying more NPV, whereas Table 7.1 and Figure 7.10 indicate that both vegetation types have essentially the same normalized difference values. This in turn implies the same amount of exposed NPV. In effect, the TCT does not adequately separate the effects of exposed NPV and canopy architecture. However, let us be fair, we have asked the TCT to do something that it was never designed to do.

Spectral-mixture analysis
As we discussed in Chapters 3, 4, and 5, spectral-mixture analysis is a more flexible modeling approach than the TCT, and it retains the advantage of using all spectral channels. Importantly for the task of mapping the vegetation types in the Fazendas image, SMA allows us to model shade, and to remove the effects of sub-pixel shade by normalizing fractions (Sections 5.1.2 and 5.1.3). One of the basic guidelines for SMA is to use the fewest endmembers possible. Accordingly, we model the closed-canopy forests in the Fazendas image in terms of three basic spectral components: shade, GV, and NPV. The image is not calibrated to reflectance; therefore, we select image endmembers, taking advantage of prior knowledge from field observations. The TM spectrum of water serves as a proxy for shade; slash is a proxy for NPV; and cecropia green represents GV. These spectra are the same as the ones in Table 7.1. The locations of the image endmembers are shown in Figure 7.7. Fraction images using these endmembers are displayed in Figure 7.12, along with the RMS-residual image. Notice that there is no endmember for soil; therefore pixels that include soil (roads, pasture, and recently cleared areas) do not fit the model well and are partitioned into the RMS-residual image. As discussed in Chapter 4, it is advantageous analytically not to have any endmembers for the closed-canopy forests that we know are not actually there.

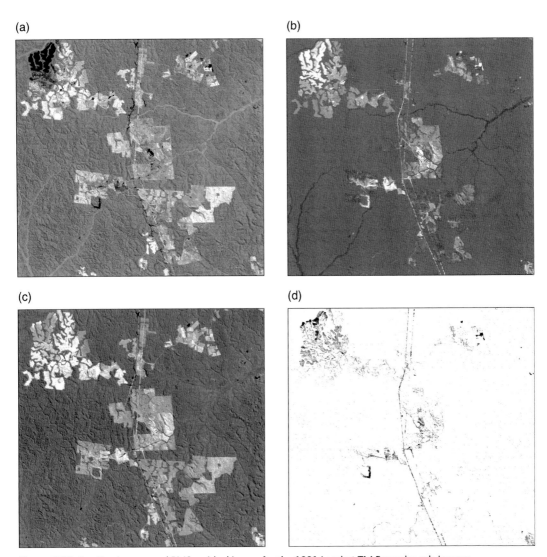

Figure 7.12. Fraction images and RMS-residual image for the 1991 Landsat TM Fazendas sub-image:
(a) GV = cecropia green; lighter pixels = higher fractions; (b) NPV = slash; lighter pixels = higher fractions;
(c) shade = water; darker pixels = higher fractions; (d) RMS-residual; darker pixels = higher residuals. Soil
was not included in the model as an endmember; therefore, soil is in the RMS-residual image. The spectra are
derived from Table 7.1, and their locations are shown in Figure 7.7.

Spectral relativity and overflow fractions

Three-endmember mixture models define triangles within the six-dimensional space of the Landsat TM channels. Projections of two different triangles onto the planes of Bands 3 and 4, and Bands 4 and 5, are shown in Figure 7.13. A GV endmember by definition has zero fractions of the other endmembers, in this case NPV and shade. It makes

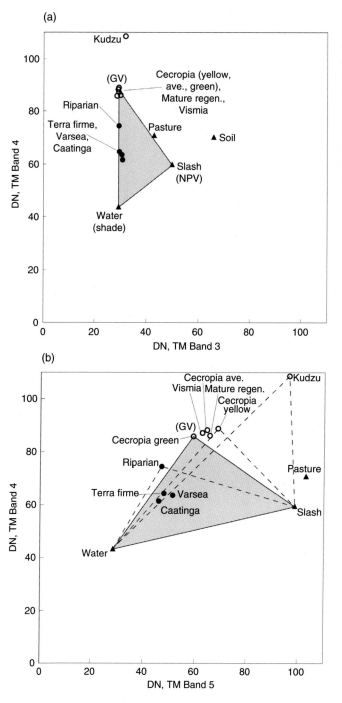

Figure 7.13. Scatter plots for the Fazendas sub-image showing the mixing triangle (gray fill) defined by the image endmembers GV = cecropia green, NPV = slash, and shade = water. (a) TM Bands 4 and 3. (b) TM Bands 4 and 5. Other triangles (no fill) define different mixing models using cecropia yellow, kudzu, and riparian.

Figure 7.14. Oblique aerial photograph of closed-canopy forest in Fazenda Dimona, north of Manaus, Brazil. The dark, rough canopy is terra firme forest (upper left part of the image) and the smoother, more reflective canopy is regeneration forest consisting of cecropia (bottom and right side of image). The cecropia-green image endmember is derived from an area similar to the cecropia in this photograph. Image spans ~100 m across.

sense, therefore, to select a spectrum for GV that represents a type of vegetation that has little or no exposed NPV in the canopy, and that has little shading and shadowing in the canopy. Cecropia green meets these criteria, as is illustrated in the aerial photograph in Figure 7.14. Other choices of spectra for GV are less satisfactory. For example, cecropia yellow is not as useful, because other regeneration types then have negative fractions of NPV. (Recall that field observations and a high Band-5 value indicate that cecropia yellow has *more* NPV than the other types of regeneration, kudzu excepted.) The purpose of the mixing model is to define an interpretive framework that is intuitive and easily understood in terms of field observations. Negative fractions of shade are not a problem. They simply mean that the vector is "lighter" than the non-shade endmember; furthermore, shade will be normalized to arrive at the true fractions of GV and NPV. However, negative fractions of GV or NPV indicate that a vector does not fit within our simple model. It is easier to visualize that cecropia green has zero fraction of NPV and cecropia yellow has 0.11 NPV, rather than the reverse situation where cecropia yellow (as the GV endmember) has zero fraction of NPV and cecropia green has − 0.11 NPV.

SMA and the convex hull

The Fazendas image provides a good example of how difficult it is to define endmembers by those vectors on the convex hull of a data set. For example, from inspection of the scatter plot in Figure 7.13b, in order to encompass all of the data projected onto the plane of TM Bands 4 and 5, at least seven "endmembers" are needed: shade, riparian, cecropia green, kudzu, pasture, slash, and soil. (Soil is not shown in this plot, because it falls off the diagram to the right.) Requiring endmembers to lie on the convex hull creates several problems, aside from the fact that so many "endmembers" would be infeasible computationally. By selecting multiple vectors, we lose track of the fact that endmember spectra must be useful for describing mixtures. Spectra of discrete patches of different materials in the scene that do not mix, except at for a few pixels at edges, are not endmembers.

From the perspective of a field investigator, it would not make any sense to try to describe terra firme forest as a mixture of the seven hull endmembers. However, it does make sense to describe terra firme forest as consisting of mixtures of three basic spectral components, shade, green foliage and woody material, and to select endmembers that are proxies for these three components of the closed-canopy forest. Thus, we need to understand clearly that cecropia green, for example, is a spectrum that we use to represent green foliage, and it is NOT literally a patch of vegetation like the one shown in Figure 7.14.

Getting the shade out

We remove shade by re-summing GV and NPV fractions to 1 (Section 5.12). Normalization of shade can be visualized as a projection of all vectors onto the line that passes through GV and NPV (Figure 7.13). The normalized fractions of GV and NPV are shown in Table 7.2, in rank order, and the proportions of GV and NPV are shown in the fraction images in Figure 7.15. Now we can directly compare the NPV fractions in the primary forest types with those of the regeneration types. The relative ordering of the normalized fractions of NPV within

Table 7.2. *Normalized fractions of GV and NPV for the Fazendas sub-image using the cecropia-green endmember. Primary-forest types are italicized. Pasture is open canopy (soil exposed).*

Vegetation type	$\frac{NPV}{GV+NPV}$	$\frac{GV}{GV+NPV}$
Pasture	0.76	0.24
Varsea	0.25	0.75
Kudzu	0.20	0.80
Caatinga	0.15	0.85
Terra firme	0.13	0.87
Cecropia yellow	0.11	0.89
Mature regeneration	0.09	0.91
Cecropia average	0.05	0.95
Vismia	0.03	0.97
Cecropia green	0.00	1.00
Riparian	−0.07	1.07

the primary forest group and within the regeneration group is consistent with the ranking of Band 5 alone and with the ranking of the normalized difference of Bands 4 and 5.

Primary forest: attributes and ambiguities

Within the Fazendas sub-image there are two types of primary forest: terra firme, which occupies most of the image, and riparian forest, which occurs in narrow zones along the branching drainage patterns. Terra firme forest is defined by the combined spectral attributes of high shade (relative to all regeneration) and 0.13 NPV. Riparian has less shade (smoother canopy) than terra firme, but more shade than the regeneration, and it has the lowest NPV of all vegetation types.

Two additional types of primary forest occur outside of the Fazendas sub-image. Varsea, a seasonally flooded forest, and caatinga, a dark-foliage (high-shade) forest developed on nutrient-poor, sandy soils (Roberts *et al.*, 1998), resemble terra firme spectrally. However, the mean vector of varsea has more NPV (0.25) than terra firme (Fig. 7.10 and Table 7.2), and it can be mapped as a diffuse zone of high-NPV forest that is especially prominent along the north side of the Rio Negro (Figure 7.16). Caatinga has the lowest mean DN of all closed-canopy vegetation types in TM Bands 4 and 5 (Table 7.1), and it can be mapped by its high shade fraction relative to a background of terra firme forest (Figure 7.17).

(a)

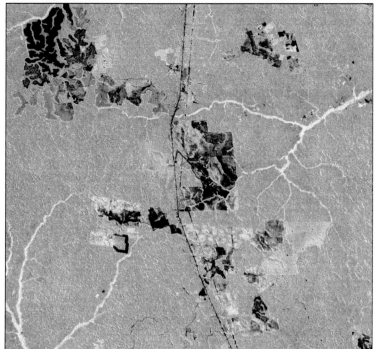

(b)

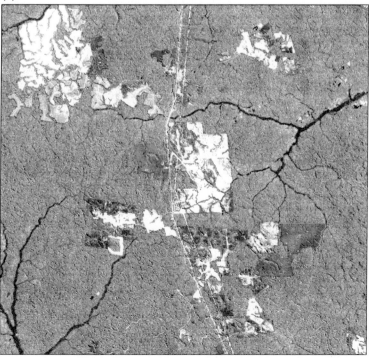

Figure 7.15. Complementary shade-normalized images of GV and NPV derived from the Fazendas sub-image. Endmembers are: GV = cecropia green; NPV = slash; shade = water. (a) Fraction image GV/(GV+NPV): lighter tones indicate higher fractions of GV; (b) NPV/(GV+NPV): lighter tones indicate higher fractions of NPV. The shade-normalized GV image is similar to the TM Band 4, 5 normalized-difference image in Fig. 7.9.

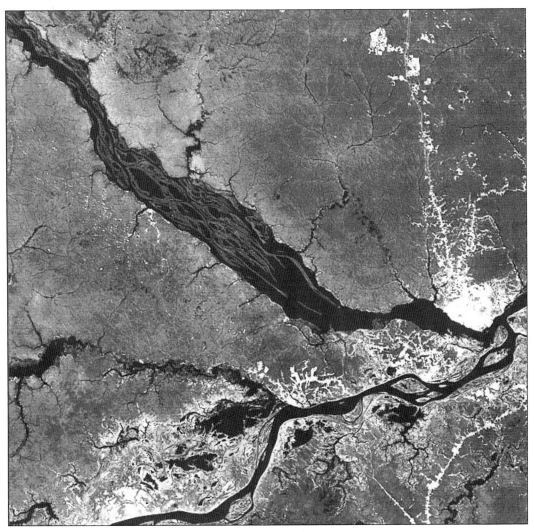

Figure 7.16. Normalized NPV fraction image for the August, 1991, Landsat TM image of Manaus, Brazil. See also Plate 3. Lighter tones indicate higher fractions of NPV. Varsea (seasonally flooded) forest along the banks of the Rio Negro has a higher NPV fraction than the terra firme forest, apparently due to partial leaf drop and exposure of woody material during times of seasonal flooding. The identity of the varsea forest is ambiguous where clearing of forest has exposed slash, dry grass, soil, or urban materials.

To be consistent with our rules for thematic mapping, we need to look carefully for possible ambiguities in these attributes that define the varieties of primary forest. Although riparian forest and terra firme forest have sharp boundaries along the main drainages, the two forest types become indistinguishable in the upper parts of tributaries. Apparently, these two types of vegetation grade into one another.

Figure 7.17. Dark, dendritic patches of caatinga forest north of the Rio Negro, Brazil. (1 – shade) image of the northwest corner of the Landsat TM image of Manaus, Brazil. See also Fig. 7.19, color Fig. 7.19 (see Web) and Plate 3. Darker tones indicate more shade. Arrows point to patches of caatinga forest. Shade is an image endmember derived from the Fazendas sub-image (Table 7.1).

Varsea forest is reliably mapped in the normalized NPV image of the full Manaus image in places where there has been little or no clearing of the primary forest (Figure 7.16). However, in developed areas, especially along the banks of the Rio Solimoes, the identity of varsea forest is ambiguous, because the spectrum of varsea mimics spectra of terra firme forest that contains small amounts of NPV as slash or dry grass in cleared areas (Figure 7.10). Furthermore, field observations suggest that the higher NPV fraction of varsea is related to leaf drop (deciduousness) that accompanies seasonal flooding of the main rivers. Therefore, an NPV-based thematic map of varsea should include the qualification that it may be valid only for a particular season. The identity of caatinga forest is potentially ambiguous on a per-pixel basis, due to mimicking by topographic-scale shadows in the terra firme forest. In this case, though, the ambiguity is relaxed by using photo interpretation. The patches of high-shade forest that correspond to known areas of caatinga (Figure 7.17) have a smooth surface texture, without the fine-scale drainage that is ubiquitous in the terra firme forest (see ahead to Figure 7.19, color Figure 7.19; see Web). This example emphasizes the importance of using shade in conjunction with other spectral attributes to define surface texture, as is discussed in the next section (Section 7.2.3).

7.2.3 Sub-pixel shade and surface texture

Sub-pixel shade consists of unresolved shading, shadows, and low-albedo objects within a pixel footprint. Shade at this scale is an important attribute for some kinds of thematic mapping, as we already have seen in the examples of the Gifford Pinchot (Figure 5.5 and color Figure 5.5; see Web) and Manaus images. The component of shade that is present below the scale of a pixel footprint must be inferred from information that is outside the image being analyzed. One kind of "outside" information would be elevation data that can be used to construct a digital-elevation model, which when subtracted from a shade-fraction image, leaves a residual that is the sub-pixel component. Another kind of outside information is knowledge that a given area has little or no topography. In some cases, it is possible to distinguish sub-pixel shade simply by photo interpretation, because the human eye–brain system is so adept at discounting shading patterns that are due to topography. In the example of the Gifford Pinchot forest, the component of sub-pixel shade was inferred from the residual after subtracting a DEM. In the Amazon example, we relied on knowledge that the area has subdued topography, and on photo interpretation, to distinguish the component of shade that provides information about vegetation texture and albedo.

Sub-pixel shade in the Manaus images

Shade-fraction images distinguish primary forest from regeneration, and the main types of primary forest from one another (Section 7.2.2). The low shade fraction of regeneration is interpreted to be the result of species compositions having relatively high NIR scattering, and canopy

Figure 7.18 and color Figure 7.18 (see Web). A (1 − shade) color image of the Fazendas sub-image. Red = (1 − shade); green = normalized GV; blue = normalized NPV. Regeneration vegetation is yellow (green + red) and exposed slash and dry grass are magenta (blue + red). In the terra firme forest, topography is mimicked in the unevenness of the closed canopy that is visually accentuated by the yellow (low-shade) and green (high-shade) areas. A complete key to interpreting the colors is given in Table 5.1.

structures that are relatively smooth. The canopies of the mature forests, in addition to having more diverse species composition, are relatively rough. Potentially confusing shading and shadows due to gentle topography in the upland areas can be distinguished by patterns that are consistent with illumination from the northeast.

Color composite images that combine shade and other spectral attributes such as GV and NPV are particularly useful for thematic mapping (Section 5.1.5). For example, in the Manaus image, if red = NPV, green = GV, and blue = shade, the varsea forest is mapped as red tones against the green terra firme forest, and shadows in the rough canopy are blue-green. Fine-scale topography (expressed in canopies) and sub-pixel shade are best mapped by rendering (1 − shade) in red, GV in green, and NPV in blue. A (1 − shade) color image of the Fazendas sub-image (Figure 7.18 and color Figure 7.18; see Web) clearly maps the regeneration vegetation in yellow against the green of the rougher-canopy terra firme forest. Riparian forest, however, is light green, because the shade fraction (canopy roughness and/or dark foliage) is intermediate between that of regeneration and terra firme. The (1 − shade) color image also shows clearly the magenta and white areas where substrate is exposed.

> **Reminder**
>
> A (1 − shade) image is an image of the mathematical complement of the shade fraction. In this image, high-shade areas are dark.

**Figure 7.19 and color Figure
7.19** (see Web). A (1 – shade)
color image of the caatinga
sub-image of the 1991 Landsat
TM image of Manaus, Brazil.
Red = (1 – shade);
green = normalized GV;
blue = normalized NPV.
Dark-green areas (high shade
fraction) are caatinga forest.
Yellow area (lower shade
fraction) is terra firme forest.
Orange (lower shade and
higher NPV fractions) indicates
varsea forest. The image has
been strongly stretched to
increase the color contrast. A
complete key to interpreting
the colors is given in Table 5.1.

A (1 – shade) color image helps us to resolve the potential ambiguity in mapping caatinga forest, as was discussed in the previous section. In Figure 7.19, the combined attributes for caatinga of high GV (green) and high shade fraction (low red) produce a dark-green color, whereas the surrounding terra firme forest that has a similar fraction of GV, but less shade than caatinga, is rendered as green + red = yellow. The result, dark-green caatinga set against a yellow background, makes it easier to see that the caatinga also lacks the texture of fine drainages that characterizes the terra firme forest. If the caatinga canopy is smoother than that of terra firme, it suggests that caatinga has a high shade fraction because the foliage itself has low albedo. This interpretation is consistent with field observations (Roberts *et al.*, 1998). Shade in this example, therefore, is a key attribute for thematic mapping, because it is part of the physical model that predicts the properties of the land surface.

Parsing vegetation shade with LIDAR
Most of the commercially available DEMs of terrestrial landscapes are based on stereoscopic measurements made from aerial photographs.

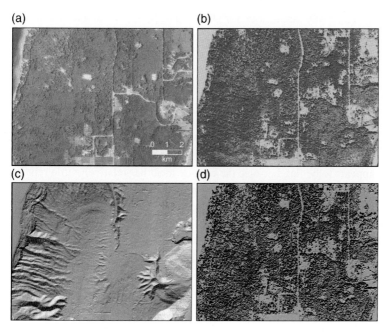

(a)   (b)
(c)   (d)

Figure 7.20. October 2002 aircraft-LIDAR images of a partially forested landscape near Seattle, Washington. See also Plate 2. (a) Aerial photograph. (b) Shaded-relief image calculated from the return time of the first part of the LIDAR pulse. The signal measures the elevation of the tops of the tree canopies. (c) Shaded-relief image calculated from the return time of the last part of the LIDAR pulse. The signal measures the elevation of the ground (hill shade). (d) Shaded-relief image of the difference between the early and late parts of the LIDAR returns (tree shade). The LIDAR images are by USGS and M. O'Neal.

These data encompass a wide range of resolutions, both horizontal and vertical. It is unusual, though, to find DEMs that provide topographic and roughness detail at finer scales than about 30 m. In the absence of direct measurements, photometric models and mixture models provide the main remote techniques to infer micro-topography and surface texture. However, remote measurements of topography at even finer scales are possible using optical-laser ranging, and aircraft light-detection and ranging (LIDAR) images are becoming available of a few areas (Figures 1.12 and 7.20).

The LIDAR images of landscapes with trees are particularly interesting, because it is possible to make separate images of the topography of the land surface and the micro-topography of the trees themselves. The method takes advantage of the fact that the return from each reflected laser pulse can be separated in time into the first part of the pulse that arrives from the tree tops and a later part that reflects from the ground, having passed through gaps in the trees. The first part of the signal contains information on the altitude of the trees plus the ground below. The last part of the signal is from the ground alone; therefore the difference between the two is a measure of the height of the canopy. An example is shown in Figure 7.20 from data acquired over a partially forested area on the Kitsap Peninsula near Seattle, Washington (Plate 2). The altitude measurements from the first and second arrivals, and the difference between them, are used to calculate shaded-relief models, in the same way that is done for DEMs, that is, by selecting a photometric function (Lambertian in this example) and a Sun azimuth and elevation.

The first-arrival image in Figure 7.20b is the calculated shaded-relief image of the trees plus ground. The second-arrival image in Figure 7.20c is the shaded-relief image of the ground (also known as "hill shade"), and it shows details of fine-scale topography, because it does not include the clutter of shading and shadow from the trees. The difference image in Figure 7.20d shows the calculated shading and shadows from the trees alone ("tree shade"); notice that the landscape appears to be flat, because the canopy itself has relatively little relief in comparison with the ground. At this point the shade is partitioned into two components: hill shade and tree shade. The information is limited to the effective spatial resolution of the laser beam; in this example, about half a meter. There is no information on shading, shadows, and albedo at the scale of leaves and branches; however, that information would be embedded in a shade-fraction image derived from a spectral image, and it potentially could be retrieved as the difference between a shade image and the LIDAR tree shade. Although LIDAR data are not available from most scenes, the above example illustrates the point that shade can be parsed into multiple components, each of which contains different information about a landscape.

Shade and volcanism on Mars

Mars Express	
Channel	*Wavelength* (μm)
1	0.45
2	0.54
3	0.75
4	0.95
Nadir	0.70

Surface texture can be an important attribute for mapping non-vegetated landscapes, as is shown in the next example that is drawn from images of Mars obtained in 2004 by the European Space Agency's Mars Express mission (Neukum *et al.*, 2004). The High Resolution Stereo Camera (HRSC) aboard the Mars Express orbiting spacecraft acquires images in four VIS-NIR spectral channels. The nominal size of the pixel footprint for these four channels is approximately 40 m, but it may be larger locally, depending on spacecraft altitude and channel-to-channel registration. A fifth, nadir channel has a pixel footprint of approximately 10 m.

Plate 7a is a color-composite image of part of orbit 097 from HRSC data acquired north of the Valles Marineris region of Mars. Red = channel 3, green = channel 2, blue = channel 1. Image spectra, calibrated to reflectance, for four regions in the image are displayed in Figure 7.21. The locations of the sampled image spectra are shown in the nadir-channel image in Figure 7.22. Exploration of the data space (Section 3.4.1) indicates that there are two basic spectral types in the orbit 097 sub-image, here designated simply as "blue" material and "orange" material (the descriptive names are derived from the colors in

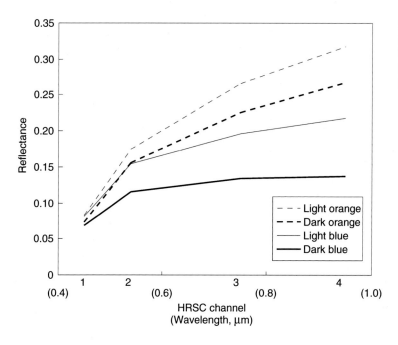

Figure 7.21. Reflectance spectra of representative areas in the orbit 097 Mars Express image. Areas are located in Figure 7.22.

Plate 7a). There also are light and dark varieties of each spectral type, as is shown in Figure 7.21.

An objective in analyzing this image was to understand the nature and origin of the blue material. Photo interpretation reveals that the darkest blue material is associated with possible volcanic landforms, including cones, aligned pit craters, and linear fissures. There is little evidence of lava flows; instead, the spatial evidence suggests that the darkest-blue materials consist of unconsolidated materials, possibly volcanic cinders and ash (tephra), with the light-blue materials being the finer ash. This interpretation is supported by the presence of dunes and streaks that imply mobility of the light-blue material with Martian winds. Additionally, pockets of the dark-blue material appear to have accumulated in low areas that would be protected from winds. To test the hypothesis that the blue material consists of mobile deposits, we sought evidence for differences in the texture of the surface, the idea being that the finest volcanic ash deposits might create relatively smooth, low-shade surfaces at a sub-pixel scale.

To assess sub-pixel texture, we need a shade-fraction image that excludes all of the other spectral components. However, there is a special problem in applying spectral-mixture analysis to this image, because one of the main scene components is the dark-blue material, and this dark material is a near mimic of shade. We addressed this problem by using a mixing model with two endmembers that were

Unusual endmember
The Mars image is an example where one of the non-shade endmembers (dark blue) is not the lightest sample of the spectral component.

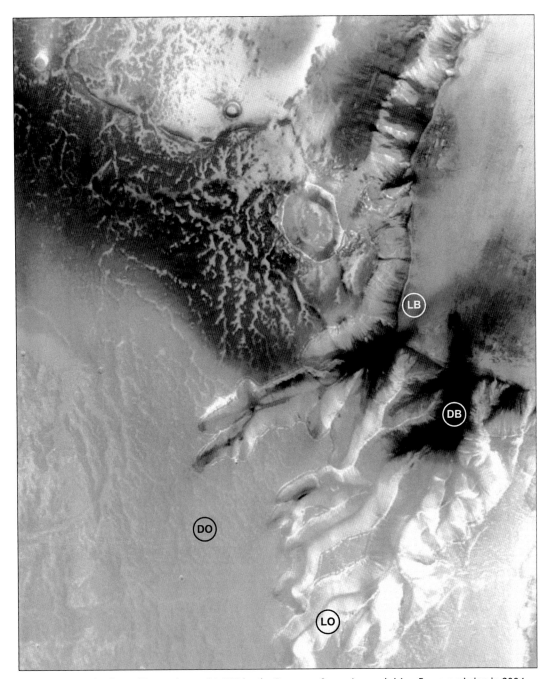

Figure 7.22. Nadir-channel image from orbit 097 by the European Space Agency's Mars Express mission in 2004. Illumination is from the upper left. North is at the top. See Plate 7a for location and scale. Circular structures are impact craters. A cliff extends north-northeast in the upper right corner, and the plain in the northeast corner is relatively low in elevation. To visualize the topography correctly it may be necessary to rotate the image (Section 1.3.3). Courtesy of the European Space Agency (ESA) and G. Neukum.

derived from the image: light orange and dark blue, and a reference endmember (0.01 reflectance) for shade (Section 4.4.1). The results are shown in the complemented-shade image in Figure 7.23 in which darker gray tones indicate higher fractions of shade. Notice that the complemented shade image captures the topographic shading of the large impact crater, and the northeast cliff and the canyons, but that the darkest patches in Plate 7a and in Figure 7.22 have only medium and light tones in Figure 7.23. In fact, the dark- and light-blue materials generally have the lowest shade fractions relative to the orange material. The only exception is in the cores of the darkest-blue patches where the nadir-channel image shows possible evidence of cones and elongated craters. The patterns of shading that are due to the recognizable topographic features are easily distinguished from the patterns that correspond to the orange and blue materials. The complemented-shade image indicates that the surfaces covered by the blue materials are smoother (less sub-pixel shading and shadow) than the surfaces of the orange materials. These results are consistent with the hypothesis that the darkest-blue patches consist of relatively coarse (higher fraction of sub-pixel shade) tephra, and that the medium- and lighter-blue materials are (lower-shade) deposits of volcanic ash.

We combined a $(1 - shade)$ image with the fraction images of the other two endmembers in Plate 7b: $red = (1 - shade)$, $green = light$ orange, and $blue = dark$ blue. As in previous examples of $(1 - shade)$ images, red indicates *less* shade. In the Mars image, green signifies the light-orange endmember that has relatively high shade, and yellow indicates areas where the light-orange endmember has low shade. Blue in the image signifies the dark-blue endmember that has relatively high shade. Magenta (red and blue) indicates the dark-blue endmember that has relatively low shade.

By combining information on surface texture with the compositional information of spectral attributes, we see, for example, that the plain in the upper right corner of the image is relatively smooth, and it consists of the "orange" material that is spectrally mixed with "blue" material. This is not evident from the color composite in Plate 7a that gives no hint of the presence of orange material in that area. The plain in the lower left of the image consists of the orange material that is relatively rough at the sub-pixel scale. At the diffuse boundary between this plain and the blue material, the $(1 - shade)$ image shows light-blue material, and no indication of the light-orange material. This is consistent with the interpretation that smooth, mobile blue material (volcanic ash?) has been deposited over the boundary and the edge of the orange surface, but it is inconsistent with the idea of smooth orange material (dust?) overlying the blue material. In Plate 7a the nature of the contact

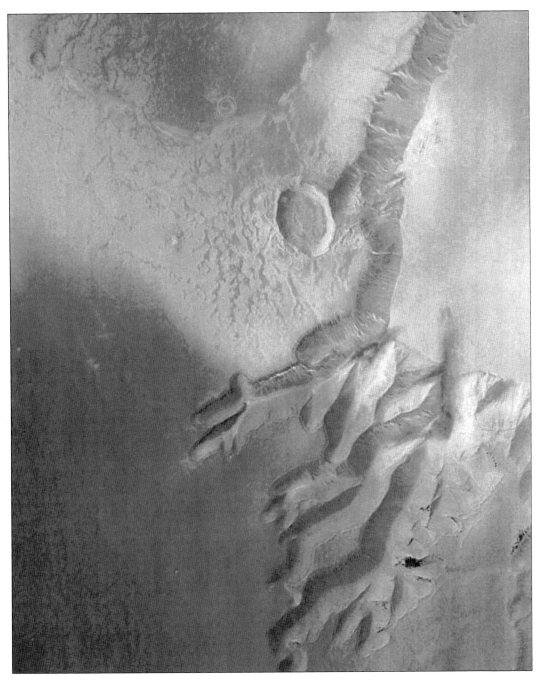

Figure 7.23. Complemented-shade image from the orbit 097 Mars Express sub-image. The mixing model consists of shade = 0.01 reflectance in all four channels, and light orange and dark blue from the reflectance spectra in Figure 7.21. Darker areas in the image have more of the shade endmember. The complemented-shade image correctly shows topographic shading. Notice, however, that the dark areas in Plate 7a and Figure 7.22 have relatively low shade, suggesting less shading and shadowing at sub-pixel scales. Illumination is from the upper left (northwest). See Figure 1.5 for help in visualizing the topography if it appears inverted. Derived from data supplied by ESA and G. Neukum.

between these materials is ambiguous. In this Mars example, sub-pixel texture is a key to understanding textural and compositional relations, that, in turn, are important for interpreting the geologic processes.

7.2.4 Unit maps

In Section 1.4.3 we distinguished between thematic maps that display attributes (properties) of an image and those thematic maps that have defined attributes *and* boundaries that constrain the attributes into units. Because attributes are measured on a per-pixel basis, it would be convenient if we also could use a per-pixel approach to define unit boundaries. Then, it would be possible to automate the process of making unit maps by setting attribute thresholds, and instructing a computer to collect all of the appropriate attributes into the desired units. The quest for an automated way to decide on unit boundaries brings us back to the many algorithms that already have been developed for classifying images. Unfortunately, though, as already discussed, classification algorithms by themselves do not produce thematic maps. First, most classification algorithms cannot take into account spatial context, and second, algorithms do not understand how to follow the rules for making thematic maps (Section 7.2.1). Let us return to each of the examples of attribute maps presented earlier in this chapter, and examine whether it is necessary to make units, and how to go about defining unit boundaries.

Death Valley TIMS image

The Death Valley TIR image in color Figure 7.1 (see Web) displays spectral attributes as colors. The colors have physical significance in terms of known rock types, but the boundaries between the colors, and, hence between the rock types, are left to the reader to interpret. For example, at the scale of the TIMS image, the boundary between the red (quartzite) and green (salt) is sharp, and we would have little trouble drawing a line on the image between red and green, or in finding this contact in the field. Other boundaries, such as between red (quartzite) and magenta (varnished quartzite) are less distinct, because the rock-varnish coating occurs in irregular and diffuse patches. Different readers might draw different lines to separate areas of varnish on the image. Do we need to make units here? Well, it depends on what information we are trying to convey. If we want to map a quartzite unit, we could ignore the varnish entirely, and just draw the boundaries between red (quartzite) and other rock types. Wherever the boundary is uncertain, we would use an appropriate symbol such as a dashed line instead of a solid one. Our red unit would have the label "quartzite," and we would expect that at the scale of the image a geologist in the field would reliably find quartzite (varnished and not varnished) within the unit boundaries. If we want to map varnish itself, though, we have a problem, because it is not possible to draw boundaries around one, or even a few, large patches that are mostly varnish. In our judgment, we would not try to define a varnish unit; instead, we would include varnish in the description of the quartzite unit, and use the color image to convey the idea that varnish occurs in small, discontinuous patches.

Seattle urban area

The attribute images of the Seattle urban area (Figures 7.2 and 7.3) do not meet our criteria for thematic maps. As discussed earlier, "urban" is a land-use name, and urban materials have a wide assortment of spectral properties, some of which mimic other materials such as soil, beach sand, and rock. Unless the measured attributes consistently match the name (urban) that we wish to give to a unit, there is no point in trying to define a boundary between urban and non-urban. Furthermore, the spatial patterns of urban materials are diffuse and highly complex. Materials that occur in the urban core also occur in urban-residential and industrial areas. Other materials such as road surfaces occur throughout the image. No useful information would be conveyed by drawing boundaries between urban and other materials. Nevertheless, the attribute images of Seattle can be useful to someone who understands that *they* must provide the interpretation, and who does not expect the interpretation to be the same for every highlighted pixel.

Amazon vegetation types

All of the types of closed-canopy vegetation in the Manaus, Brazil image that we selected as our themes could be distinguished from one another by their *mean* fractions of shade, GV, and NPV. These mean values, that are derived from all of the Landsat TM channels, are plotted on a ternary diagram in Figure 7.24. However, we have not yet evaluated the overlap of attributes among the vegetation types. An initial estimate of attribute thresholds can be made by grouping the mean values of fractions in Figure 7.24. One group consists of the primary forest types, terra firme, riparian, varsea, and caatinga, which differ from other types of closed-canopy vegetation by having high fractions (>0.30) of shade. (Recall that the shade fraction is measured relative to our image endmember, and it only applies to the illumination conditions in this image.) The other group comprises the closed-canopy regeneration types which have zero or negative shade relative to the GV endmember, cecropia green.

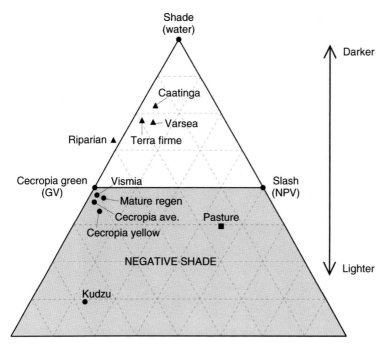

Figure 7.24. Ternary diagram showing mean fraction values of shade, GV, and NPV for the main vegetation types in the Fazendas image. Channel-DN values for the vegetation types are given in Table 7.1. Normalized fractions (without shade) are given in Table 7.2. The shade-normalized fractions (not shown) would plot along the line connecting GV and NPV. Negative fractions are a consequence of the choice of image endmembers (Figure 7.13), and simply mean that those vectors have less shade than the GV endmember (cecropia green). For a reminder about how to interpret fractions from a ternary diagram see Figure 4.3.

Figure 7.24 illustrates that there is a clear gap between primary-forest types and regeneration in terms of shade. If topographic shading is not an important factor, or if it can be removed by modeling (e.g., Figure 5.4), we can define a boundary between a primary-forest unit and a regeneration unit by selecting a shade-fraction threshold. In Figure 7.25 this threshold is set at a fraction of 0.20. In image form, all pixels having a shade fraction > 0.20 are rendered black, and the remainder are white. The threshold was determined by displaying the shade-fraction image and moving the slider bars on the histogram-stretch tool to extinguish pixels at various shade-fraction values, as was described in Section 5.1.4.

Variability and overlap of attributes

The caption to Figure 7.25 includes an explanation of the two units that we defined. The intent of the description is to make clear how one unit is different from the other, and to point out ambiguities. The units differ from classes (as in classification algorithms) in two important ways. First, we used the image context to interpret the shade threshold inter-actively. (We did not use an algorithm to select a threshold based on the statistics of the image data.) Second, we addressed the variability within each unit by alerting the reader to possible ambiguities, such as light,

---

Image unit

An aggregate of contiguous pixels within a specific interpretive context. The name of a unit applies to the aggregate, not necessarily to each pixel.

Image class

A population of pixels based on data attributes alone. Membership in a class is decided on a per-pixel basis.

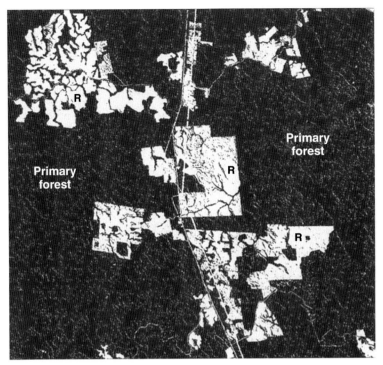

Figure 7.25. Unit map of primary forest and regeneration in the Fazendas sub-image. Explanation: *primary forest (PF)*. Black. Closed-canopy forest in which most pixels have a shade fraction > 0.20 owing to shading and shadowing by a rough canopy and/or intrinsically dark (low-albedo) foliage. Scattered pixels of Sun-facing crowns in the canopy have shade fractions < 0.20 and may mimic regeneration vegetation. Other land-cover types having high shade may mimic the PF unit, but usually can be distinguished by context. Boundary between PF and R generally is sharp on the scale of tens of meters. Boundary is more diffuse where regeneration is mature. *Regeneration (R)*. White. Closed-canopy regeneration in which most pixels have a shade fraction < 0.20 owing to low-to-moderate shading and shadowing in the canopy and/or intrinsically light (low-albedo) foliage. Locally, especially along ravines, shading and shadowing by topography in this unit may mimic PF. Occurs in irregular to rectilinear patches, but not as scattered single or few pixels within PF. Boundaries generally are sharp on the scale of tens of meters, but may be more diffuse where regeneration is mature.

Sun-facing crowns in the dark forests. Although the mean values of the fractions of the two units are distinct in Figure 7.25, there always are a few pixels within the primary forest that have low-shade fractions, regardless of where the threshold is set. This is a natural consequence of the fact that we are imposing a simplistic interpretive framework on a naturally complex scene. When classifying images using algorithms, analysts typically try to minimize or eliminate the effects of variability by smoothing the data or employing tools that take into account spatial statistics (e.g., Gillespie *et al.*, 1995; Gu *et al.*, 1999). Units, however can be internally heterogeneous, as long as they are consistent with the rules for making thematic maps (Section 7.2.1). A unit name (and the description) applies to an aggregate of pixels within an interpretive context, but that name may not be appropriate or useful on a per-pixel basis. The author of a unit map is able to convey the intended information about the landscape without being responsible for placing the correct name on every pixel.

Subdividing the primary forest and regeneration units

The initial organization of the Fazendas sub-image into primary forest and regeneration is convenient in terms of interpretation, and it also exploited a large difference in the shade fraction. In this image it helps to start with these two main units, and to consider whether to subdivide each one further, depending on our objectives. In Figure 7.24, with the exception of kudzu, regeneration types cluster near the endmember cecropia green. In Chapter 6 we used kudzu as an example of a target that could be detected based on spectral-length (SL) contrast in TM Band 4 (color Figure 6.17b; see Web). In Figure 7.24 the high albedo of kudzu is expressed as a high value of negative shade. If we were interested in making a unit map of kudzu, we would set a suitable unit boundary based on the fractions of shade and/or NPV, using the same interactive process described above.

The tight clustering of the other regeneration types presents a more difficult problem, because the natural variability within each type overlaps the mean values of all of the other types. There are no sharp boundaries between types when we manipulate the thresholds in the fraction images (or any other images). We can, however, select some boundaries that define units well enough to convey information about their general character. For example, cecropia green and cecropia yellow overlap in their fractions of GV, NPV, and shade; nevertheless they are visually distinct as an aggregate of pixels in fraction images and in other attribute images. In Figure 7.26 we have drawn boundaries on the normalized GV image to define cecropia green and cecropia yellow units in a patch of regeneration in the lower-right corner of the Fazendas sub-image (Plate 4). The boundaries are visually subjective, based on contrast-stretching the normalized GV image and the shade image.

Once we have started drawing the lines that form boundaries around units, we have to finish the job for each designated map area. In Figure 7.26 each of the cecropia units also is in contact with primary forest; therefore, we need to mark the boundary between each cecropia unit and primary forest. To complete the unit map, we also need to label the unit on the left side of Figure 7.26 that is not primary forest and that is not one of the cecropia units. If we do not know what that unit consists of, we can simply name it "un-mapped" or "un-differentiated land

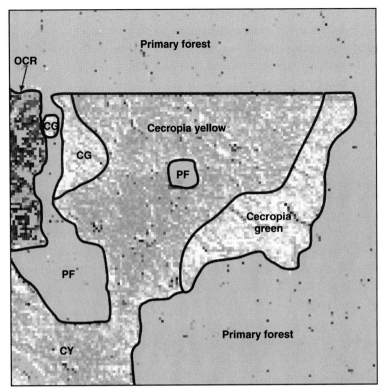

Figure 7.26. Unit map of vegetation types in a portion of the Fazendas sub-image. The map is approximately 4.5 km across, and is an enlargement of the patch of cecropia regeneration in the lower-right corner of the GV/(GV + NPV) image (see Plate 4 and Figure 7.15a). Explanation: *primary forest (PF)*. See caption for Figure 7.25. *Cecropia green (CG)*. Closed-canopy regeneration characterized by low NPV and low shade. Canopy is smooth as evidenced by a low-shade fraction and observations in the field (see Figure 7.14). Sharp contact with primary forest and open-canopy regeneration on the scale of tens of meters. Diffuse, irregular contact with cecropia yellow. Isolated patches of CG on the scale of one or a few pixels (30 m) occur within the cecropia yellow unit. *Cecropia yellow (CY)*. Smooth, closed-canopy regeneration characterized by higher NPV and lower shade than cecropia green. Higher NPV is interpreted as more exposed woody material, and is supported by observation from a low-altitude aircraft of wind damage in the canopy. Lower shade (higher albedo) than CG is interpreted as the result of less green foliage and/or more leaf senescence. Sharp contact with primary forest on the scale of tens of meters. Diffuse, irregular contact with cecropia green. Isolated patches of CY on the scale of one or a few pixels (30 m) occur within the CG unit. *Open-canopy regeneration (OCR)*. Undifferentiated vegetation and exposed soil and/or slash. Sharp contacts with PF, CG, and CY on the scale of tens of meters.

cover." But it does have to have a name. In this case we have attached the name "open-canopy regeneration," because we have field and image evidence that substrate is exposed.

In preparing the unit map of cecropia green and cecropia yellow we found that the attributes of the two units overlap, although we were able to define them by spatial context in a small, known area. The other types of closed-canopy regeneration, vismia, mature regeneration, and cecropia average, have attributes that are intermediate between cecropia green and cecropia yellow (e.g., Figures 7.10 and 7.25), indicating that we cannot distinguish any of these types of vegetation from one another by their spectral attributes, except in local areas where we have field data or other contextual information. This is a good example of a situation where training areas that appear to be distinct in a known part of an image are not really sufficiently different in their attributes to be used confidently for thematic mapping in unvisited areas. To make a complete map of closed-canopy vegetation types in the Fazendas sub-image, we would lump cecropia green, vismia, mature regeneration, cecropia average and cecropia yellow into a single unit.

Subdividing primary forest into units presents two additional kinds of problems that commonly are encountered when we evaluate whether to make unit maps. Riparian forest differs from terra firme forest in its NPV and shade fractions, and there is little overlap in these attributes. However, it is evident in Figures 7.7 and 7.15 that riparian forest is restricted spatially to narrow zones along streams, and at the scales of the Fazendas sub-image and the full Landsat TM scene it is not practical to draw boundary lines. The same is true of caatinga forest that in many places occurs in highly irregular, dendritic patterns (Figure 7.17). It becomes more practical to define unit boundaries for these forest types when we zoom in on individual patches, as is illustrated for caatinga forest in Figure 7.19.

Varsea forest presents a different but equally difficult problem. It overlaps with terra firme forest in its characteristic attribute, NPV, and the broad, diffuse gradation between varsea and terra firme forests extends over tens of kilometers (Figure 7.16). It is not clear where to draw a boundary line at any scale. Given the data available, an acceptable solution is just to make the attribute map, and to explain the spatial pattern in words.

In the above examples we have seen that some attribute maps do not translate well into unit maps. The decision as to whether or not to make a unit map depends on the spatial scale and what information we are trying to convey. When we make a thematic map in the field we usually select one spatial scale, and organize information accordingly. When we analyze images it is tempting to jump from one scale to another. As we discussed at the beginning of this chapter (Section 7.1.1), it is useful to remember that when we change scales we often have to change the way we extract, organize, and present information.

Next

Once we have made thematic maps from spectral images, how can we use them to infer processes and to detect change? Can we predict some of the spectral changes for those processes that we understand well? What gets in the way of monitoring change?

Chapter 8
Processes and change

Seattle, Washington State. Photograph by E. Evans

In this chapter

Surface processes that shape landscapes commonly can be identified within a single image by spectral expressions of stages along a few pathways through an image-data space. When images can be inter-calibrated, and we understand spectra in terms of land cover, spectral changes in a time series can be used to monitor surface processes.

8.1 Process pathways in spectral images

8.1.1 Photo interpretation of processes and change

Photo interpretation of spectral images requires experience (Chapter 1) – not just the experience needed to recognize familiar objects and patterns – but also the experience to understand what is "going on" in an image or from image to image over time. An experienced interpreter can infer processes that operated in a scene before an image was made, as well as processes that may have continued afterward. Inferring processes is more than just imagining a story that fits a scene. Images commonly contain evidence for different stages of processes. Once a

Process

A process is a continuous action, operation, or series of changes taking place in a definite manner.

few stages are recognized, a skilled observer can infer the nature of a complete process.

Consider the familiar example of a single image of an agricultural scene that includes expressions of various stages of the process of cultivating crops. Even if we are not farmers, we can recognize what is "going on" in the image, based on the presence of regular shapes of bare ground, others that are partially green, others that are entirely green, and still others that are brown. Our conclusion that crops are "being cultivated" in the image is based partly on our experience interpreting similar patterns, and, perhaps more importantly, on understanding the processes of tilling, planting, growing, and harvesting. In other words, we have a physical model in mind, and we compare evidence from image(s) with that model. In this example, the more experience we have about farming in the imaged area, the more specific our model can be, and the more information we can extract. However, the processes that shape landscapes do not have to be recent. Photo interpretation is used routinely to find evidence for very ancient processes. For example, geologists who are interested in the glacial moraines in Figure 1.2 seek information about deposition and erosion that occurred tens of thousands of years ago. Planetary scientists who study impact craters on the Moon (Figure 1.6) are observing evidence for excavation and deposition that took place billions of years ago.

> **Trading time for space**
> If a time series is not available, we may infer how a landscape changes spectrally by interpreting its different stages of evolution.

If we have access to a time series of images of a scene, we can observe changes, rather than inferring them. The more complete a time series is, the more photo interpretation approaches a record of continuous action, like watching a video of a landscape changing over time. Of course, landscape processes operate over a huge range of time scales – from seconds to millions of years. Furthermore, some processes ended long ago, and others are on-going. We do not need a time series of images to study craters on the Moon, because we missed the action there by several billion years. However, recent processes, ranging from floods and volcanic eruptions to urban development, can be documented by images that are closely spaced in time (hours, days, years). Each landscape expresses the net effects of all processes that acted on it, ancient and recent, and an image captures just a momentary manifestation of continuous action.

Process pathways

Processes can be inferred from spectral information, as well as from spatial information (Section 1.2). In our visual experience, this would be equivalent to observing color changes in an object of fixed size and shape. In a spectral image, for example, we might measure three agricultural fields side by side, one soil, another green vegetation, and the third, senescent (brown) vegetation; or we might observe one field change spectrally from one image to another in a time series. Each type of field is a spectral *sample* of the processes of the greening and senescence of crops. If one image (or a time series of images) contains many spectral samples, they define a process pathway in the data space of the spectral channels. Process pathways are models of physical processes. Some pathways are one-way, others represent processes that are reversible. Some pathways are simple, others complex. Spectral images, though, typically sample the expression of processes at just a few points, and when we connect those points, we only approximate the actual pathways. Furthermore, spectral pathways (as defined by image data) do not contain information about *rates* of the processes that they represent. Thus, when we see a field that consists of bare soil and another that is covered with green vegetation, we can infer the process of crop growth, but we do not have information about how long that growth took, unless we have many samples from images closely spaced in time or additional information. If the green vegetation is plowed under, the spectral sample would move back to soil – a reversal of direction that occurs nearly instantaneously.

An additional characteristic of process pathways is that they are not predictive. The field of bare soil might turn green, but we cannot be sure of that; for example, the pathway followed will depend on whether it is seeded and whether water is available. Similarly, we cannot be sure that the green field will turn brown, as it could be plowed under or harvested. These considerations have led us to adopt the term "spectral pathway," rather than the term "spectral trajectory," although "trajectory" has precedence in the literature. "Trajectory" has meaning that long pre-dates remote sensing, and brings to mind mathematical descriptions of planetary orbits or the flights of baseballs. Unlike pathways, trajectories have regular patterns of rates and rate changes, and their paths are predictable.

Pathway

Pathway: a path, course, route or way.

Path: a route, course, or track along which something moves; a course of action, conduct, or procedure; a continuous curve that connects two or more points.

Trajectory

Trajectory: the curve described by a projectile, rocket, or the like in its flight.

8.1.2 Process pathways in spectral data space

Vegetation indices and pathways

In the 1970s, images from the Earth Resources Technology Satellite (ERTS) Multi-Spectral Scanner (MSS) became available, and image analysts quickly realized that natural process operating on the ground had predictable responses in the spectral data space of the four MSS Bands. Initial interest was focused on vegetation indexes and applications for agriculture. The NDVI (Rouse *et al.*, 1973) and other vegetation indices were developed as ways to estimate cover of green vegetation in single images and changes in vegetation cover over time. Considerable subsequent work has led to improvements in vegetation indices, for example those that take into account variability in soil brightness and other factors (e.g., Huete *et al.*, 1994).

The idea of process pathways is implicit in "mapping" soil and green vegetation in the data space defined by red and NIR reflectance. We illustrate this in a diagrammatic sketch that shows three spectra, green vegetation (GV), light soil (LS), and dark soil (DS) in Figure 8.1. The heavy lines with arrows represent process pathways. One can easily visualize that increasing the cover of green vegetation will move the vectors from bare soil toward green vegetation. Process pathways, therefore, tend to be expressed in images as spectral-mixing trends. Pixels having partial cover of vegetation are spectral mixtures that plot between a soil vector and a vector for green vegetation. It is understood that the spectral data are derived from pixel footprints that are too large to resolve individual leaves and individual patches of soil that are exposed between leaves. Otherwise, with, say, centimeter-sized pixel footprints, the data would cluster around the "pure" leaf spectra and the "pure" soil spectra, with fewer data points between. Although vegetation indices such as the NDVI are widely used to measure vegetation cover, it is not generally appreciated that they, in fact, describe spectral mixing along process pathways.

If we consider Figure 8.1b as a mixing diagram, we can treat some of the basic spectral components of the scene (green vegetation, soil, and shade) as endmembers. The area that is enclosed by the endmembers contains all possible mixtures. Although all endmembers can mix with one another, this does not mean that all of the mixing lines between endmembers are equally realistic process pathways, or that the process rates are similar. For example, the pathways from soils to green vegetation represent the relatively slow process of growth of crops, but the reverse paths from green vegetation to bare soils (that could represent harvesting or plowing under), represent more rapid processes that entail a "jump" from one state to another. By the same logic, we would not expect a jump directly from soil to green vegetation. If we observed such a jump we would know that we had missed the intermediate stages.

A line between GV and shade is the pathway that describes shading of GV, which, of course, can trend in both directions. In Figure 8.1b DS, whether it is dark because it contains dark organic material or because it is wet, is spectrally indistinguishable from a mixture of LS and shade. As discussed in Section 4.2, shade is ambiguous in terms of what it represents on the ground, because it could mean shading, shadows, intrinsically dark soil or wet soil, or combinations of these. If we know that there is only flat terrain in a scene, we can discount the influence of shading on bare soil by topography and roughness

Vegetation indices
Vegetation indices describe spectral mixing along process pathways.

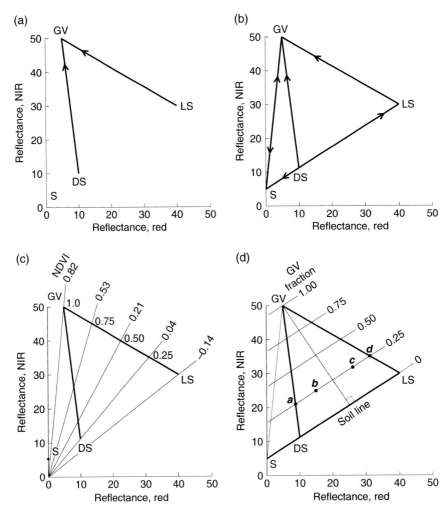

Figure 8.1. Scatter plots for red and NIR wavelengths showing the relationship between vegetation indexes and process pathways. (a) Increasing cover of green vegetation is modeled by process pathways from light soil (LS) and dark soil (DS) to green vegetation (GV). (b) Mixing space defined by endmembers GV, LS, and shade (S). The shade endmember has non-zero reflectance in the NIR. For a linear mixing model, the fraction of each endmember is proportional to the distance along a pathway. Variations in illumination are modeled by two-way pathways between GV and S, and between LS and S; DS mimics a mixture of LS and S. Other pathways are possible within the mixing area; for example, less illumination would move an un-shaded mixture of GV and LS toward S. (c) Lines of equal NDVI values radiate from the origin. Fractions of GV are marked where the lines of equal NDVI intersect the GV, LS pathway. To convert NDVI to fraction of vegetation cover, field calibrations are needed for each different soil type. (d) The perpendicular vegetation index (PVI) is based on the distance between the soil line and a parallel line through the vegetation vector. The PVI is equivalent to a shade-normalized mixing model, and is valid for both light and dark soils, as is demonstrated by the fact that vectors *a, b, c,* and *d* all represent a GV fraction of 0.25.

elements, although in vegetated areas we expect shading and shadows to be created by the vegetation itself. But with only red and NIR channels, we cannot eliminate the ambiguity between shading, shadowing, and dark soil, unless we have additional information.

How much vegetation cover?

From the perspective of spectral mixing, we can use the distance along a pathway between two vectors to estimate the amount of material on the ground that each vector represents. To a first approximation, the spectra of patches of green vegetation and the spectra of surrounding patches of substrate mix linearly (Chapter 2); therefore, the vector [0.5 soil, 0.5 GV] should be half way along a pathway between (shade-normalized) soil and GV. Although this is the case for a linear-mixture model, ratios such as the NDVI do not produce comparable results, as can be seen in Figure 8.1c where the intercepts of the ratio lines with the pathways vary with the orientation of the pathways. This means that the NDVI and similar indices must be calibrated using information from the ground. Furthermore, the amount of vegetation on dark soils will be underestimated relative to that on lighter soils; therefore, different ground calibrations are needed for each soil type.

| **Index calibration** |
| Indices such as NDVI must be calibrated against ground measurements ("vicarious" calibration) if they are to be interpreted thematically. |

An additional complication is that ratios are not accurate unless all offset terms have been removed. Instrumental and atmospheric offsets can be removed by calibrating, but this does not account for the offset that is caused by the illumination of shadows by NIR light that is transmitted by green leaves or illumination that is reflected from adjacent hillsides. Thus, for partially vegetated scenes, shade does not reside at the origin in the red–NIR scatter plot, even after the data are calibrated to reflectance. Offsets usually are ignored, though, when applying ratio-based vegetation indices. For many applications it appears to be sufficient to produce a simple, relative measure of vegetation cover. Besides, regardless of offset problems, accurate measurements using ratio-based indices still require calibration to field data.

Various methods have been devised to estimate vegetation cover that do not have the drawbacks of ratios. One of the earliest was the perpendicular vegetation index (PVI) (Richardson and Wiegand, 1977), which, in essence, is a mixing model, as can be seen in Figure 8.1d. Lines of equal fractions of GV in the mixing area [GV, soil, shade] are parallel to the line connecting soil and shade (the "soil line"). The soil line itself is the locus of all vectors having zero GV. The line parallel to the soil line and passing through the GV vector defines GV fractions of unity. The distance between the two parallel lines (it is perpendicular to the soil line; hence the name "PVI") is proportional to the fractions of soil and GV, regardless of the amount of shade. The

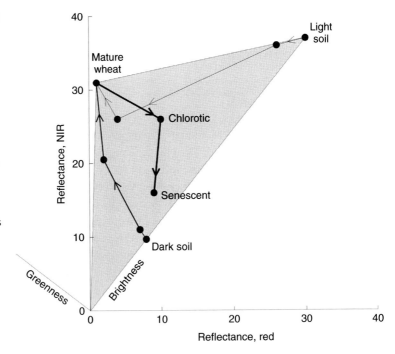

Figure 8.2. Pathways projected onto the plane of red and NIR wavelengths for growth and senescence of wheat. Data are from Landsat MSS images of agricultural plots (Kauth and Thomas, 1976). Growth pathways from soils to mature wheat lie close to the red–NIR plane. A rotation in this plane approximates the tasseled-cap transformation (TCT) plane of "greenness" and "brightness." The pathway from mature wheat to senescent wheat does not lie in the red–NIR plane.

same result is achieved using spectral-mixture analysis, deleting shade and normalizing the fractions of GV and soil (see Section 4.3.2 and Figure 4.7).

The tasseled-cap process model

Perhaps the earliest and clearest description of process pathways was made by Kauth and Thomas (1976) in the context of four-channel Landsat MSS images of agricultural crops. Their original tasseled-cap transformation (TCT) was based on a physical model for how vectors changed in the data space as crops grew, matured, and senesced. Later work by others further developed the TCT and extended it to Landsat TM images (e.g., Crist and Cicone, 1984; Crist and Kauth, 1986). As we discussed in Section 3.5.2, the TCT creates fixed, sub-space projections that show maximum spectral contrast between green vegetation and soil, and between green vegetation plus soil and senesced vegetation (NPV). Kauth and Thomas (1976) also placed their process model in the familiar two-dimensional framework of red and NIR wavelengths, and it is instructive to examine the shapes of these pathways in comparison with vegetation indices and mixture models. The Kauth–Thomas model that is shown in Figure 8.2 is based on the phenology of wheat on a light soil and on a dark soil. One pathway extends from light soil to mature wheat, another pathway extends from dark soil to mature wheat, and a third goes from mature wheat to senescent wheat. We have added lines connecting light soil, mature wheat, and the origin to facilitate comparison with Figure 8.1, and to make the point that all three pathways are within the mixing triangle that is defined by the endmembers light soil, mature, wheat, and shade (the origin).

Why are the pathways in the Kauth–Thomas red–NIR diagram more complicated than those in Figure 8.1? The main reason is that Kauth and Thomas measured two intermediate stages in the growth of the vegetation, and these stages show the effects of shadows cast by the vegetation itself. In the case of the pathway starting at light soil (LS), the progressive increase in shadows cast by the vegetation deflects the pathway toward shade. (Absorption of red light by chlorophyll in green vegetation moves the pathway toward the left side of the figure.) Eventually, shadowed areas are largely replaced by green vegetation, and the NIR reflectance increases. In the case of dark soil (DS), increasing the cover of green vegetation causes the pathway to move toward lower red reflectance, but the largest change is the increase in NIR reflectance. The shadows that are cast by the vegetation do not lower the overall reflectance, because the soil already is dark. By comparison, the simplistic model implied by Figure 8.1a,b ignores intermediate stages where the effects of shadowing are revealed.

The pathway from mature to senescent wheat in Figure 8.2 is the same for light soils and for dark soils, because there is still complete cover by vegetation. The first stage ("chlorotic") entails partial loss of chlorophyll that causes a decrease in red absorption. The second stage ("senescent") records the degradation of leaf cellular structure that is accompanied by a decrease in NIR scattering. This pathway is the "tassel" of the tasseled cap. As a reminder of the structure of the tasseled cap itself, we have placed the greenness and brightness axes of the TCT on Figure 8.2. Within the TCT framework, mature wheat is at the peak of the cap. It may help in visualizing the cap to turn back to Figure 3.4.

Although the senescence pathway projects onto the mixing triangle in Figure 8.2, it does not lie in the red–NIR plane, as can be shown by making scatter plots with other combinations of channels or applying the TCT. Vegetation indices such as NDVI and PVI that are confined to the red–NIR plane, only can measure relative amounts of *green* vegetation. They are not able to distinguish senescent or woody vegetation (NPV) from soil, because the spectra of these components mimic mixtures of soil, green vegetation, and shade. This is an advantage if the objective is to measure the amount of photosynthetically active vegetation, and it explains the widespread use of the NDVI for that purpose. Confinement to the red–NIR plane is a disadvantage, however, if the objective is to measure the total amount of vegetation that includes senesced leaves and/or woody material. For this purpose, additional channels are needed, as Kauth and Thomas proposed.

Images of vegetation pathways

The purpose of identifying pathways is to be able to construct images that can be interpreted in terms of processes on the ground, so let us review some of the ways that vegetation pathways can be displayed in image form if we have a physical model in mind. In the NDVI image of Seattle, for example, progressively lighter tones (higher NDVI) indicate pixels that are farther along pathways from unspecified "other" materials toward the spectrum of green vegetation (Figure 8.3). The context of the image determines specific interpretations. In the agricultural area (upper right corner of Figure 8.3), higher NDVI values suggest more crop coverage, whereas in the urban and suburban areas the higher NDVI values generally indicate closed-canopy trees or irrigated lawns and golf courses. Without context, intermediate or low NDVI values are ambiguous, because we cannot tell whether ratio values lie on pathways of greening or senescence, and we cannot differentiate among materials other than green vegetation (see Figures 8.1 and 8.2).

We can gain information about the substrate or background (soil, NPV, urban, water, etc.) in the Seattle image by displaying a color composite where red = Band 5, green = Band 4, and blue = Band 3 (Plate 2 and the reference images on the Web). With these color assignments, the end of the pathway that represents soil and/or dry vegetation is relatively red (Band 5), and progression along the pathway from soil/NPV to green vegetation is expressed by color changes from red to orange to yellow to yellow-green to green. In contrast, urban and built-up areas are relatively blue. A TCT color image of Seattle (Figure 8.4 and color Figure 8.4; see Web) can be interpreted in much the same way as the Band 5, 4, 3 color image, although the channels have been "tuned" by assigning weightings that are based on a physical model for growth and senescence of vegetation.

A geologic pathway

After volcanic flows erupt onto the surface of the Earth, they begin to react chemically and physically with the atmosphere. In some areas, such as the semi-arid parts of the island of Hawaii, USA, basalts that have been exposed for a few tens to a few hundred years accumulate thin (< 1 mm) veneers of opaline silica (Farr and Adams, 1984; Curtiss *et al.*, 1985). In the early stages of development, these coatings are not visible to the eye; however, the coated rocks can easily be distinguished from un-coated

Figure 8.3. NDVI sub-image of the 1992 Landsat TM image of Seattle. See Plate 2b for location. Lighter tones indicate higher NDVI values.

ones at thermal IR wavelengths (8–13 µm) (Figure 8.5). Kahle *et al.* (1988) obtained images with NASA's airborne TIMS of basalts on Mauna Loa, Hawaii, and compared the spectra of flows that had been dated historically or by radiocarbon. They demonstrated that basalt flows (erupted between about AD 450 and 1935) followed a pathway in the spectral data space that expressed the accumulation of silica with age (Figure 8.6). Therefore, TIMS images could be used to determine the relative ages of lava flows to assist in reconstructing the geologic

Figure 8.4 and color Figure 8.4 (see Web). A TCT sub-image of the 1992 Landsat TM image of Seattle. See Plate 2b for location. Red $= (1 - TCT_3)$; green $= TCT_2$; blue $= TCT_1$. Red pixels have relatively high DNs in TM Band 5 and correlate with the low water absorption of NPV, primarily dry grass in this August scene. Green pixels indicate GV, and blue corresponds to urban materials, roads and bare soil. Yellow pixels are a mixture of GV and NPV.

history of this area. Kahle *et al.* (1988) were able to define the age pathway for coated Hawaiian basalts using only three TIMS channels; therefore, they could represent the pathway in a color image (Plate 6 and Figure 8.7). Temperature effects were suppressed to enhance spectral emittance in each channel by applying a decorrelation stretch (Section 3.3.1). The TIMS color image was made by normalizing radiance for the three channels, and assigning red to channel 5 (10.3 μm), green to channel 3 (9.3 μm) and blue to channel 1 (8.3 μm). Thus, the youngest basalts at the beginning of the pathway in the ternary diagram in Figure 8.6 correspond to relatively blue areas in the TIMS image. With increasing age and development of silica coatings, the basalts become relatively red in the image, and the oldest coated rocks become green.

8.1.3 Illumination and atmospheric pathways

When we think of process pathways in the data space, we generally think first of those that represent physical processes on the ground. However, we also can apply the concept of pathways to variations in illumination and atmospheric conditions. In Figure 8.1b, for example, the pathways between GV and shade, and LS and shade, describe

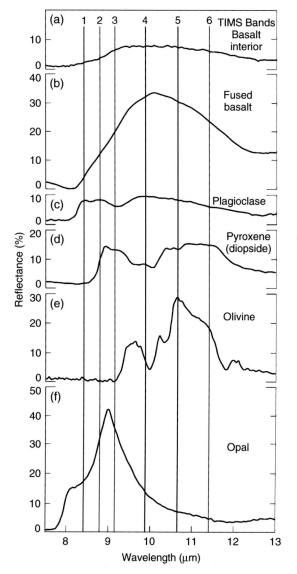

Figure 8.5. Laboratory thermal IR reflectance spectra for Hawaiian basalt and opal. (a) Crystalline basalt; (b) basalt glass; (c) plagioclase; (d) diopside (pyroxene); (e) olivine; (f) opal (amorphous silica). Emissivity of these samples can be estimated by applying Kirchhoff's law (Section 2.1.1); thus, for example, the emissivity of opal is relatively low in TIMS channel 3 where the reflectance is relatively high. Adapted from Kahle *et al.* (1988).

variations in illumination. Variations in illumination and atmospheric conditions cause vectors and pathways to shift within the data space. We illustrated this point in Figure 4.14 in the context of discussing how gains and offsets affect scatter plots of the synthetic image. Although the solar zenith angle generally does not change much within a single image, topography and view angle can cause significant variations in illumination from place to place, and, therefore, in the vectors that define pathways. Similarly, atmospheric variability within an image can shift vectors and pathways, although for many scenes this is a lesser effect.

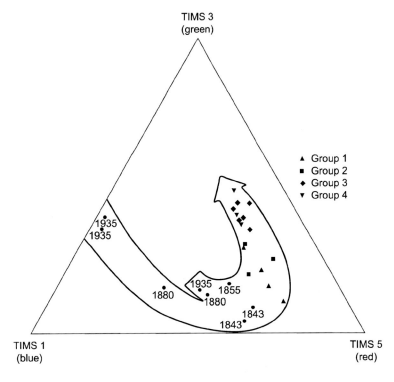

Figure 8.6. Ternary diagram for TIMS channels 1, 3, and 5, showing basalts of known ages. The data define a spectral pathway that expresses increasing thickness of amorphous silica coatings with age on pahoehoe (smooth-surface) basalts in a semi-arid part of Mauna Loa. After Kahle *et al.* (1988).

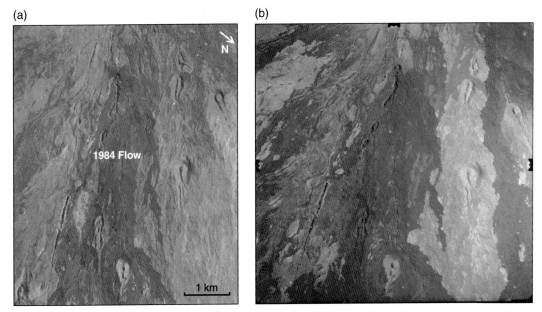

Figure 8.7 and color Figure 8.7 (see Web). See also Plate 6. (a) A TIMS image of part of Mauna Loa, Hawaii. In Plate 6 and in color Figure 8.7 (see Web): red = channel 5 (10.3 μm), green = channel 3 (9.3 μm) and blue = channel 1 (8.3 μm). Blue areas are youngest pahoehoe basalt flows with little or no silica coatings. Red areas are older flows with moderately developed silica coatings. Green areas are the oldest flows with the thickest silica coatings. Blue-green areas are aa (rough-surface) flows. (b) Color aerial photograph of the same region as the TIMS image. The TIMS image shows relative age differences that are not visible in the aerial photograph. NASA JPL images.

Sun elevation, topography, and texture

Predicting the effects of illumination variations on pathways in a single image is complicated by the fact that shading and shadowing occur over a wide range of spatial scales. If a landscape is flat, variations in solar illumination over time are described by a single gain that applies to all channels, unless the image has a very wide field of view. There is no effect on the shape of an image pathway, and we can adjust the absolute position of any pathway to a reference position in the data space by applying the gain that corresponds to a selected solar zenith angle. In contrast, topographic slope and aspect determine the *local* solar incidence angle, and they have the effect of differentially displacing spectral vectors along mixing lines that connect to the shade point. Pathways are difficult to interpret unless the topographic effects are suppressed by normalizing or by modeling (e.g., a DEM correction). Shading and shadowing at sub-pixel scales ("texture") has more complex and ambiguous effects on pathways. As we pointed out in Figure 8.2, the intermediate stage of wheat on light soil has more shade than would be predicted by a direct pathway from light soil to mature wheat, but the darkening could be due to topography or sub-pixel shading and shadowing. In this case, the ambiguity is resolved by knowledge that the scene has minimal topographic relief, and we conclude that the deflection of this vector toward shade is caused by sub-pixel shading and shadowing by the immature wheat itself – a important textural attribute that provides information about the stage of development of this crop. To predict the amount of sub-pixel shading/shadowing for a given solar zenith angle, we need to apply a model that takes into account the spatial structure (e.g., height, shape, spacing) of the surface elements (Section 1.3.3). Extensive work has been done on geometric models of shading and shadowing that are associated with different assemblages of plants, and there is a rich literature on the subject that dates back to the early 1970s (e.g., Suits, 1972; Verhoef, 1984; Rosema, 1992; Strahler, 1994; Hall *et al.*, 1997; Gu and Gillespie, 1998).

When we construct process pathways, we are not seeking information about solar zenith angle or macro-topography; indeed, we want to suppress their effects. However, as discussed in Section 7.2.3, sub-pixel texture can convey important information about the nature of a landscape surface, so we pay a price if we remove this information by normalizing. Generally, at relatively coarse spatial scales within a scene, we want to remove illumination effects to be able to compare image spectra with "reference" spectra, whereas at finer scales we may want to preserve illumination effects that reveal texture.

Visible–NIR measurements of surface texture are possible by making nearly simultaneous measurements of a scene from different view angles. For example, ASTER acquires two NIR images at view angles

(a)   (b)

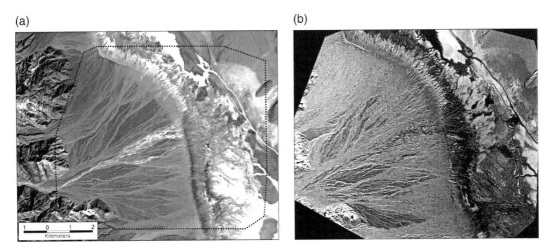

Figure 8.8. Sub-pixel texture from multi-angle, November, 2001 ASTER images of the Trail Canyon fan in Death Valley, California. See Plate 1 for location; also, Weeks *et al.* (1996). (a) Band 3 image viewed from nadir. Dark areas on the alluvial fan are coated with dark desert varnish. The alluvial fan and the valley floor in the upper right corner have low topography and little topographic shading relative to the mountains along the left side of the image. (b) Ratio image Band 3 / Band 3B. Mountains are masked in black, and are outside the dashed outline in image a. Band 3B views the scene about 55 s later than Band 3 from 27.6 degrees off nadir. Albedo variations are strongly suppressed. Dark areas indicate relatively more shading and shadow at all scales, and are interpreted as being relatively rough. Notice that the tongue of dark varnished rock along the bottom of image a is partly rough and partly smooth in image b. The high-albedo salt flats at the top of image a are seen to be very rough in image b. Images courtesy of A. Mushkin.

of $0°$ (toward nadir) and $27.6°$ (back-looking) separated in time by ~ 55 s. Depending on the season and the geographic location, the back-looking image may "see" more or less shade (shading and shadows) than the nadir image. The amount of the difference between the two ASTER images depends in part on the roughness of the terrain itself, and this can be modeled quantitatively if the illumination and acquisition geometry is known. A qualitative example of how textural information can be extracted from the two ASTER images of a desert surface in Death Valley, California is shown in Figure 8.8 (Mushkin and Gillespie, 2005).

8.2 Reference pathways

8.2.1 Reference spectra and calibration

Calibration

As long as we can convert image spectra from encoded radiance to reflectance or emittance, we can compare pathways from image to image, and we can compare image pathways with ones in the data spaces defined by laboratory and field measurements. To calibrate an image to corrected radiance or reflectance/emittance, we apply gains and offsets to all vectors (Section 3.2.1), and these alter the process pathways in predictable ways (Figure 4.14). It follows that if we know the location of a pathway in reflectance/emittance, we should be able to predict how the position and shape of that pathway would be altered in an image as a result of variations in topography or atmospheric conditions. Kauth and Thomas (1976), in fact, predicted how atmospheric haze would affect the tasseled cap. They further proposed that shifts in the pathways, because they are diagnostic of atmospheric processes, could be used to correct the image data for the effects of haze and viewing angle. We explored this subject of calibration feedback in Section 4.5.

Laboratory reference pathways

In Chapters 2 and 3 we emphasized the importance of laboratory spectra as the standard references that we use to interpret image spectra. Ideally, we would like to be able to define all process pathways using laboratory spectra; as was discussed in Section 2.2.2, however, we cannot measure laboratory spectra of many of the important components of landscapes (e.g., vegetation canopies, urban areas). Furthermore, pathways typically are mixing trends, and it is not feasible to collect the spectra of all possible mixtures of the materials that comprise landscapes. We can, however, select laboratory spectra that represent samples along a pathway, and infer the locations of the spectral mixtures between reference vectors.

To illustrate how laboratory spectra can help to define process pathways, let us examine several types of vegetation more closely. Figure 8.9a–e shows pairs of reflectance spectra that are sampled at Landsat TM wavelengths. The TM-equivalent spectra are sampled from laboratory spectra of Elvidge (1989) that cover the wavelength range from 0.4 to 2.2 μm.

It is instructive to review the physical processes that cause the spectrum of one type of vegetation to transform into the spectrum of another. The familiar visible yellowing of a mature green leaf during senescence is caused by a breakdown of chlorophyll; this removes the strong red absorption in TM Band 3, but it does not affect the absorption in TM Band 1 that is caused by other pigments that produce the yellow color. The increases in reflectance in TM Bands 4, 5, and 7 signal that the leaf has lost water, and that the cellular structure now scatters light more efficiently. This, of course, is a one-way path, as is shown in Figure 8.9f, because we generally do not expect yellow leaves to become green again. The transitions from a yellow leaf to a brown

Figure 8.9. Reflectance spectra of single leaves, sampled at Landsat TM wavelengths. (a–c) Spectra of California black oak leaves (*Quercus kelloggii*) derived from laboratory spectra of Elvidge (1989). (d) Estimated spectra of two closed canopies of green vegetation. (e) Spectra derived from laboratory spectra of Allen and Richardson (1968). (f) Vegetation-senescence pathways in TM Bands 3 and 4, based on the laboratory spectra of Elvidge (1989).

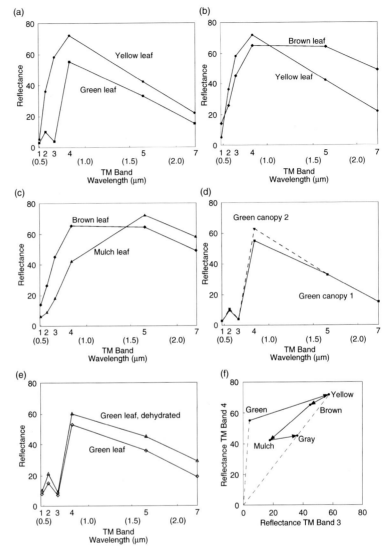

one, and from a (dry) brown leaf to a leaf that has begun to decompose in the ground litter, are accompanied by increased reflectance in TM Bands 5 and 7 as cellular water is lost. There also is decreased reflectance in TM Bands 1, 2, 3, and 4 as humic acid and other organic pigments become significant light-absorbing components (Figure 8.9b,c).

The senescence and decay pathway is shown in the plane of TM Bands 3 and 4 in Figure 8.9f, and in the plane of TM Bands 4 and 5 in Figure 8.10. We have added to both of these figures the vector for a decayed, gray-brown mulch leaf that has lost more of its pigments.

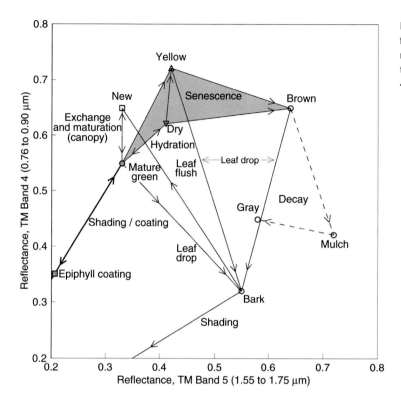

Figure 8.10. Process pathways for vegetation inferred from reflectance spectra projected in the plane of Landsat TM Bands 4 and 5.

Figure 8.9d,e includes spectra to illustrate the differences between mature and new leaves, and turgid and dehydrated leaves. The spectrum of a dehydrated oak leaf ("dry") is inferred, based on laboratory experiments on cotton leaves by Allen and Richardson (1968). Loss of water from leaf cells reduces absorption in Bands 5 and 7, and replacement of water by air increases scattering and reflectance in Band 4. The decreases in absorption in Bands 2 and 4 have the effect of "folding up" the wings of the chlorophyll band and shifting the "red edge" to shorter wavelengths. The dehydration pathway is not well expressed in the plane of Bands 3 and 4, because there is little change in Band 3; however, it is clear in the plane of Bands 4 and 5 (Figure 8.10). Holben et al. (1983) have shown, however, that at the scale of a vegetation canopy, wilted leaves can change the canopy architecture and cause a *decrease* in reflectance in Bands 4 and 5. Thus, laboratory-scale spectra do not give us the full picture of this pathway.

In general, new leaves of broad-leaf species tend to be thinner, smaller, and visibly lighter than mature ones, but it is difficult to make meaningful comparisons of reflectance spectra of single leaves. Measured in the laboratory, single mature leaves usually have higher reflectance than young leaves. However, a *canopy* consisting of young leaves typically is more reflective, because much of the transmitted light, especially in the NIR, is scattered back to the top of the canopy. Small leaves and/or an open architecture further enhance scattering by allowing more penetration of light to lower canopy levels (Roberts *et al.*, 1990). For Figure 8.9d we estimated the spectrum of a canopy of new leaves, based on calibrated TM and AVIRIS images. Thus, young leaves have higher reflectance in TM Band 4, owing to multiple scattering, but chlorophyll and water absorption remain strong, because the mean optical path is large. The spectrum of a closed canopy of mature leaves is represented by the spectrum of a single oak leaf, because it is relatively thick and non-transmitting in the NIR. These estimates of canopy spectra suggest pathways for leaf exchange and leaf maturation in Figure 8.10.

Figure 8.10 incorporates the spectra from Figure 8.9 and includes two more spectra. "Epiphyll coating" refers to dark epiphylls, primarily fungi, that accumulate on old green leaves, for example, in Amazon caatinga forest (Roberts *et al.*, 1998a). "Bark" is a representative spectrum of gray bark derived from the laboratory measurements of Elvidge (1989). The path from brown to mulch to gray illustrates the process of leaf decay and encompasses those spectra that represent leaf litter. The vectors for bark and litter define a distinct zone in the Band 4, 5 data space. We do not expect to find any spectra of green vegetation in this zone, although, from experience, we know that many types of non-photosynthetic vegetation and soils occur in the same zone. The process of leaf senescence is suggested by the path from mature to yellow or dry, and to brown. Again, this is a zone that is distinct from the zone of leaf decay and distinct from green vegetation.

For deciduous trees, leaf senescence typically (but not always) is accompanied by leaf drop that exposes bark and litter. Therefore, at remote-sensing scales, we infer that there must be leaf-drop pathways that describe mixing between senescing leaves and bark/litter. Leaf-drop is a one-way path, and when new green leaves appear, we expect a leaf-flush pathway to extend from bark/litter toward "new" (new green leaves). Spectra of species that do not exchange leaves all at once may follow a path between "new" and "mature" green leaves. Shading, shadowing, and dark coatings such as epiphylls, create pathways between all of the spectra and a point near zero reflectance.

Reference pathways improve with time

The pathways shown in Figures 8.9 and 8.10 are examples to illustrate the point that we can use laboratory spectra and calibrated image spectra as references to gain insight into the spectral behavior of vegetation as it undergoes various physical processes. The pathways that we have shown are greatly simplified, because they are based on limited vegetation types and only a few samples. A straight line drawn between two vectors may convey a sense of the overall spectral direction of a process, but in reality, many processes involve multiple stages and produce pathways that have irregular shapes. In Section 4.4.3 we pointed out that our choices of reference endmembers can be modified using feedback from modeling. Similarly, our original choices of a reference pathway can be adjusted as more information becomes available. Field observations may reveal the extent of variability of reference vectors, and this can affect the uncertainty envelope (the width or volume) of a reference pathway. Additionally, the detailed shapes of pathways may not be evident from initial image or field measurements. As observations of a given scene accumulate, it is possible to fill in missing stages. In Figure 8.2 the looping pathway from dark soil to senescent wheat has three intermediate stages (shaded soil, mature wheat, and chlorotic wheat). Measurements of additional stages would likely have revealed additional fine structure in this pathway.

Mixing and mimicking along reference pathways

Many pathways or segments of pathways describe spectral mixing between reference endmembers that represent basic scene components. However, in some instances these endmembers can mimic mixtures of other components. In Figure 8.2, for example, the pathway from chlorotic to senescent is defined by reference endmembers; however, as we pointed out earlier, in the data space of red and NIR channels, the vectors for chlorotic and senescent mimic mixtures of light soil, mature wheat, and shade. Ambiguities such as this often can be removed by introducing other channels. Indeed, the potential for mimicking emphasizes the need to use multiple channels to define pathways, rather than relying on simple two-channel projections of the data. In cases where mimicking cannot be avoided, it may be possible to remove ambiguity by image context. Thus, for example, Figure 8.2 suggests that pixels of senescing wheat will not be mimicked by soil as long as we know that the vegetation canopy is closed, or, if any soil is exposed, that the soil is dark.

Navigating in spectral data space

Despite the potential for mimicking, the most practical way to display pathways is as projections of the data onto the plane of two channels in which there is relatively large spectral contrast between the important scene components. At first glance it might appear that process pathways defined by laboratory spectra occupy much of a Cartesian data space.

Actually, there are large volumes of data space that are unoccupied by reflectance or emittance vectors of any known materials. These volumes are "out of bounds" for laboratory data, and if we find vectors there, we know that they are artifacts of measurement or calibration. (We can get a sense of the restricted nature of data volumes from two- and three-dimensional projections of the synthetic image in Figures 4.5 and 6.3.) Image spectra typically occupy even smaller data volumes than those for laboratory spectra. One reason is that landscapes do not contain the wide variety of materials that comprises spectral libraries. More importantly, though, most image spectra are mixtures of spatially unresolved "pure" endmembers, and mixing has the effect of shrinking the occupied portion of the data volume.

When we use laboratory spectra to define the outer limits of data volumes we have to be careful to specify the geometric properties of the measured samples. We know, for example, that the reflectance of green leaves varies with canopy structure, and that the reflectance of rocks and soils can vary with particle size. We can determine the extreme limits of real vectors from a combination of laboratory and calibrated image spectra. However, the most important task for image interpretation is to establish the locations of reference spectra that are typical of key scene components. If we define reference spectra and reference pathways using laboratory or imaging spectrometer data, we "map" their locations in a high-dimensional data space and, by extension, in any data space that is sampled by fewer channels. However, if we "map" a reference pathway using only a few channels, as was done in Figure 8.2, we have a limited sample that cannot be extended to images with more or different channels. There is another reason to base reference pathways on a combination of laboratory and image spectra. Pathways are not just the narrow lines that we draw between laboratory spectra on a scatter plot. Natural variability in the spectra of landscape materials tends to widen pathways.

We usually represent pathways as two-dimensional projections onto a computer screen or paper copy. With multiple channels, we can assemble many projections side by side (e.g., Figure 6.3), but this presents an incomplete picture of the data, and the result still is difficult for most of us to visualize. Nevertheless, pathways can be defined mathematically in higher dimensions, and this makes it possible to compare image data with reference pathways in much the same way that we compare single vectors. From an image-processing perspective, this is an advantage, because it allows us to classify image pixels according to how close they are to a given reference pathway. We return to this topic in Section 8.3.3.

8.2.2 Connecting pathways to field observations

Data visualization

To visualize pathways that are based on field observations, it is particularly convenient to plot fractions of surface components on ternary diagrams (e.g., Figure 5.12). A basic principle that guides how we make maps in the field is that the sum of the components of land cover on a landscape is unity (Albee's law). As long as components are aggregated into no more than three categories, they plot as fractions on a ternary diagram. But aggregates of attributes such as radiance, NDVI, greenness, etc. are not constrained to sum to 1; therefore, to plot them on a ternary diagram, their values must be normalized to sum to 1, and the results do not have the same meaning as fractions of land cover (Figure 8.6).

Although ternary diagrams commonly are used to show co-variance among chemical, mineralogical, and other materials, they rarely are used in remote sensing, because a data transformation needs to be made to convert spectral radiance (attribute data) into fractions of surface components. Spectral-mixture analysis solves this problem by using fractions of endmembers as proxies for materials on the ground (Chapter 4); SMA in its simplest form is limited to two to five endmembers, one of which is shade. After shade is removed, and the fractions of the remaining endmembers are re-summed to 1, any combination of three endmembers can be plotted on a ternary diagram. In the case of landscapes that are evenly illuminated, or where topographic effects on illumination have been removed, shade may be included as one of the endmembers on a ternary diagram to show the textural (sub-pixel) properties of the other two endmembers.

Aggregating land-cover components

From the perspective of a field investigator, three components often suffice to describe the most important variations in land cover. In the context of a semi-arid Australian range land, for example, Pech *et al.* (1986) found that the important land-cover components were green vegetation, dry vegetation, and soil. Each of these components was easy to recognize in the field and formed an intuitive framework to describe the landscape. Each component, of course, was an aggregate. "Green vegetation" included a variety of shrubs and grasses; "dry vegetation" included dry grass, leaf litter and woody material; and "soil" included the exposed soil and rock types. Pech *et al.* (1986) showed how the different proportions of the aggregated land-cover components that could be displayed conveniently on a ternary diagram (Figure 5.12) were related to image data through a linear-mixing model.

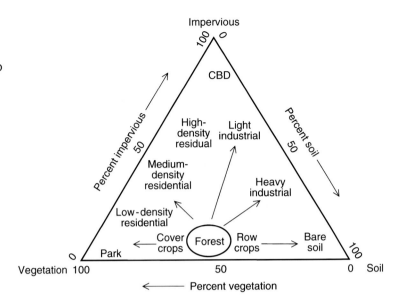

Ridd (1995) used green vegetation, impervious surface materials, and soil (termed the "VIS model") to describe the main land-cover components and process pathways associated with urban development (Figure 8.11). To those who study urban landscapes, the term "impervious" refers to a surface that is not penetrated by water – an attribute that is important for understanding landscape hydrology. Impervious surface materials include road materials (e.g, asphalt, concrete) and roofing materials (e.g., metal, tile, composite materials). In the field, where spatial attributes are so important (Chapter 1), "impervious surface materials" are easy to recognize. Ridd (1995) showed that the VIS model is useful for interpreting remote-sensing images of many urban areas, and Greenberg (2000) and Wu and Murray (2003) in their studies of urban areas recognized that a transformation from encoded radiance to fractions of endmembers established the needed proxies for the aggregates of components that comprise VIS land cover (Figure 8.12).

How can we use ternary diagrams to visualize the variations among more than three variables? A simple solution is to use more than one ternary diagram and different combinations of variables. Figure 8.13 illustrates a hypothetical (although realistic) case in which the land cover of a scene is modeled by five endmembers: green vegetation (GV), non-photosynthetic vegetation (NPV), soil, impervious materials (IMP), and shade. Three different models of land cover are shown in Figure 8.13 based on these endmembers. In the vegetation-cover model, shade has been deleted, and the remaining endmembers re-summed to 1; soil and IMP are combined to represent the substrate (SUB). (The

(a)

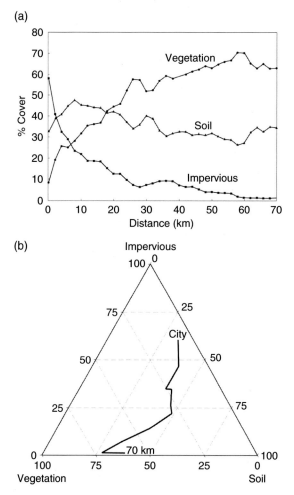

Figure 8.12. Spectral proxies for aggregated components of land cover and process pathways for the Seattle urban environments. (a) Mean fractions of endmembers (green) vegetation, impervious materials, and soil along radial distances from downtown Seattle. Data are from a spectral-mixture model applied to a 1991 Landsat TM image. (b) Data from a plotted on a ternary diagram. The pathway from 70 km (radial distance from city center) to city implies a process of urban development. From Greenberg (2000).

endmembers GV and NPV each could be aggregates of endmembers in a complex scene having multiple endmembers.) In the urban model, shade has been deleted and the remainders re-summed to 1; GV and NPV are combined (VEG), because the distinction between green and senescent vegetation is not necessary. The forest model includes shade, although *topographic* shading has been suppressed by applying a digital-elevation model. In this case, NPV, soil, and IMP are combined, because the objective is to emphasize the effect of shading and shadowing by trees.

Image pathways based on non-shade endmembers. Example: forest clearing for pasture

Figure 8.14 illustrates the use of normalized fractions of GV, NPV, and soil to describe process pathways in the Fazendas Landsat TM image.

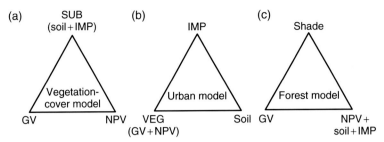

Figure 8.13. Ternary diagrams depicting three physical models for process pathways, based on five endmembers. Endmembers are green vegetation (GV), non-photosynthetic vegetation (NPV), soil, impervious materials (IMP), and shade. (a) Vegetation-cover model. Fractions of non-shade endmembers sum to 1; substrate (SUB) is the sum of soil and IMP. (b) Urban model, based on the VIS model of Ridd (1995). Fractions of non-shade endmembers sum to 1; vegetation (VEG) is the sum of GV and NPV. (c) Forest model. The topographic part of shade has been removed using a DEM. Remaining shade is below the DEM resolution.

Figure 8.14. Ternary diagram of mean fractions of shade-normalized endmembers GV, NPV, and soil for the Fazendas Landsat TM image. Categories are based on field observations. Pasture 1 is a good quality pasture for cattle; pasture 2 is a degraded pasture; OC and CC are open canopy and closed canopy, respectively. Arrows indicate process pathways that are inferred from patterns of land use in the local area.

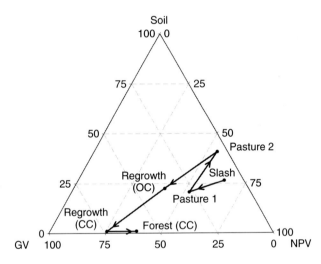

In this example, field observations established categories such as "pasture," "open-canopy regeneration," and "forest." The names of the categories reflect the known types of land cover *and* land use in the area. Data points on the ternary diagram are mean fraction values for the endmembers in the different categories (Adams *et al.*, 1995). The pathways between categories (arrows) are based on knowledge that mature forest has been cut and cleared to make pasture for cattle. Although several scenarios are possible, we have shown a common one at the time of image acquisition, which leads from good pasture to degraded pasture that subsequently is abandoned and reverts to forest.

Observations on the ground show, as expected, that cutting a patch of forest initially results in a cover of woody debris and dead leaves (NPV). The process corresponds to a one-way path (not shown) in fraction space from forest to "slash." In this area, the slash was cleared, piled, and burned and the area planted for pasture, thereby defining a pathway to "pasture 1." The diagram shows a straight arrow from slash to pasture 1, because the intermediate stages were not observed. However, we infer that, given sufficient time resolution, the slash vector actually would move toward soil as a result of clearing and burning, and then toward GV as grass became established, and then to pasture 1 that represents mature pasture. Keep in mind that a pasture that appears green when viewed horizontally on the ground actually exposes a significant fraction of senescent leaves (NPV) and some soil when observed from above, which explains why pasture 1 is not located closer to GV.

Two pathways describe the evolution of pasture that is not well maintained. Pasture 1 degrades to pasture 2, mainly by overgrazing, because soil increases at the expense of green vegetation. When degraded pasture is abandoned, it becomes populated by shrubs and trees. This change leads to open-canopy regrowth, and, eventually, to closed-canopy regrowth. Both pathways are reversible, although in this area clearing of regrowth was uncommon in the early 1990s. Finally, on the scale of decades, there is a pathway from closed-canopy regrowth to mature forest.

Image pathways that include shade. Example: timber harvesting and forest regeneration

To illustrate how field observations can be linked to spectral data, we return to the Landsat TM image of the Gifford Pinchot National Forest in Washington state. The area consists of coniferous forest and patches that have been cleared and are in various stages of regeneration. In this area, Landsat and AVIRIS images are well modeled by four endmembers, GV, NPV, soil, and shade that are proxies for the main land-cover components plus shade (Sabol *et al.*, 2002; Roberts *et al.*, 2004). The mean fractions of the endmembers are plotted in Figure 8.15 for five types of vegetation cover that correspond to stages of regeneration after removal of the forest cover. Shortly after clear-cutting, as the first green vegetation appears, there is still more NPV than GV, and the shade fraction is low. With further regeneration, the fractions of GV and shade increase at the expense of NPV until the canopy closes; then shade increases at the expense of GV, and there is a slight, but noticeable increase in NPV as more woody material is exposed in old trees. Soil is a minor component of the forested landscape, except along roads.

The normalized fractions of GV, NPV, and shade are shown on a ternary diagram in Figure 8.16. Data points correspond to patches of forest that were visited and photographed, and four examples of ground photographs are keyed to the ternary diagram. Ground observations at several

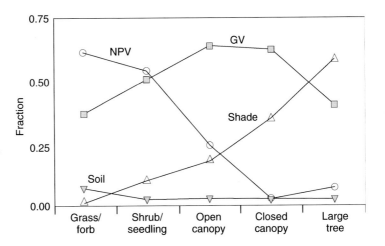

Figure 8.15. Mean fractions of reference endmembers for five land-cover types in a July 1991 Landsat TM sub-image of the Gifford Pinchot National Forest, Washington State. Soil is detectable only at the stage of regeneration of a clear-cut where green vegetation is sparse. From Sabol *et al.* (2002) with permission from Elsevier.

hundred sites confirmed the pattern of regeneration depicted by the end-member fractions. Based on these data, Sabol *et al.* (2002) defined a process pathway for regeneration of conifers that is shown in Figure 8.17.

8.3 Mapping changes in landscapes

The process pathways discussed in Section 8.2 were defined from single images. We can define process pathways from a momentary snapshot in time as long as an image contains samples of at least a few temporal stages; for example, agricultural scenes in which different areas are in different stages of crop cultivation. Using this approach, we infer temporal information from spatial information (substituting space for time). There are some processes, though, that only can be sampled using a time series of images. For example, one image of a deciduous forest cannot sample spring leaf flush, summer mature foliage, autumn senescence and winter leaf drop. Our next consideration is the analysis of multiple images of a landscape over time.

Given a time series of images, we naturally look first for obvious spatial and spectral differences. Most analyses do not extend beyond visual inspection, and this is consistent with the fact that photo interpretation alone is sufficient for a majority of applications of remote sensing (Chapter 1). Even monitoring of environmental changes, and the imaging requirements for surveillance or intelligence purposes usually can be satisfied by comparing spatial properties, with little or no input of spectral information. However, not all of the changes from one image to another are obvious enough to be detected by their spatial expression. In this section we focus on time-series spectral changes that

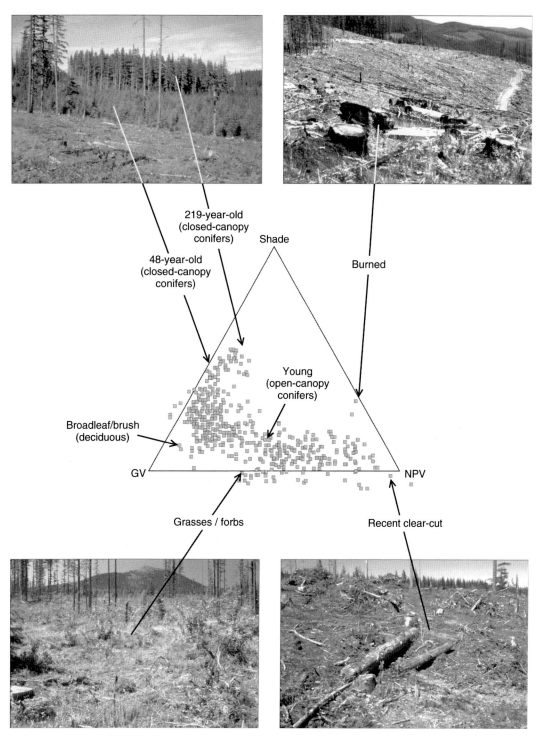

Figure 8.16. Ternary diagram of mean fractions of reference endmembers for land-cover types in a July 1991 Landsat TM sub-image of the Gifford Pinchot National Forest, Washington State. The endmember fractions are normalized to sum to 1 after deleting the soil endmember. Measurements sampled blocks of 3 x 5 pixels or larger at 495 sites. From Sabol *et al.* (2002) with permission from Elsevier.

Figure 8.17. Ternary diagram showing the inferred process pathway for conifer re-growth for the 1991 Gifford Pinchot sub-image. Numbers are calendar dates of clear-cutting. From Sabol *et al.* (2002) with permission from Elsevier.

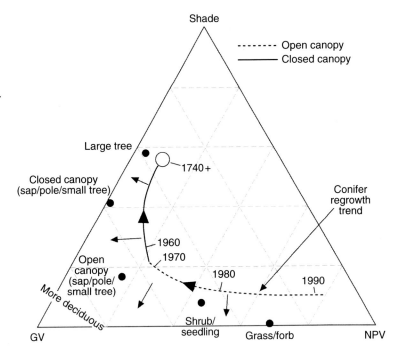

reveal information about surface processes that otherwise might be difficult to interpret by visual inspection alone.

8.3.1 Calibration and normalization

Image defects, miscalibration and variations in illumination geometry can masquerade as real changes in land cover, and in some cases it may not be possible to distinguish change from error (Chrisman, 1997). If we are looking for spectral differences in a series of images of a scene, we need to pay special attention to image quality and calibration. One essential task is to assure that each pixel is sampling the same footprint on the ground in all channels. The channels of each image must be correctly registered, and, then, the images being compared must be correctly registered together. Artifacts of misregistration propagate through spectral analyses (Section 3.1.5), and when comparing images acquired at different times, there is additional opportunity for interpretive error.

Misregistration

Misregistration in time series may lead to fictitious spectral pathways.

Normalization

If we are sure that images being compared sample the same parts of the ground, then each image needs to be calibrated and corrected for radiometric effects (Section 3.2) that might obscure real changes on the ground. The uncertainties of absolute radiometric corrections,

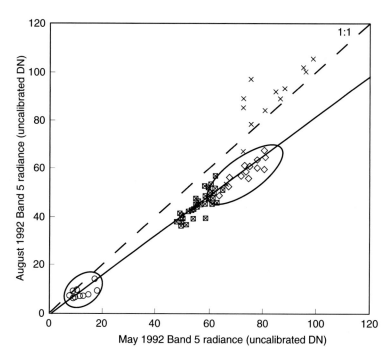

Figure 8.18. Inter-calibration of May 1992 and August 1992 Landsat TM Band 5 images of Marabá, Brazil. Dashed line (labelled 1:1) defines equal radiance values in May and August. Symbols: open circle = water; boxed cross = riparian and terra firme forests; open diamond = closed-canopy regenerating forest; cross = areas of pasture and land-use change. Water and regenerating forest were judged to be time invariant, based on field observations. The inter-calibration coefficients are consistent with the reported detector differences between Landsats 4 and 5. Vectors that plot above the inter-calibration line indicate areas where Band 5 radiance increased significantly from May to August with a seasonal increase in dry grass, primarily in pastures. Modified from Bohlman *et al.*, 1998.

especially when many images are being compared, commonly are bypassed by normalizing to an arbitrary reference image (e.g., Hall *et al.*, 1991; Schott, 1997, p. 222; Roberts *et al.*, 1998b). This process, also known as "relative radiometric calibration" or "inter-calibration," entails selecting representative DN values (channel by channel) of light and dark reference areas that are judged to be invariant from image to image over time. One image is selected as the reference image, and for each other image in a time series, the coefficients are calculated that "calibrate" them to the reference image. The success of the normalization method hinges on having appropriate invariant areas in the scene. Generally, the main uncertainty in comparing images in a time series is variability in the illumination of the scene; however, changes in detector responses and in atmospheric conditions also will be part of the derived coefficients.

Figure 8.18 is an example from Bohlman *et al.* (1998) that compares uncalibrated radiances of May 1992 and August 1992 Landsat TM Band 5 images of a partially forested area in the eastern Amazon Basin. The same areas of water and of closed-canopy regenerating forest in the two images were judged to be invariant over time, based on field experience. The August image was normalized to the May image by applying a gain and offset derived from the line fitted through the invariant data points. In this case, the main calibration difference between the two images turned out not to be due to illumination and atmospheric conditions; instead it is explained largely by differences in detector responses, because the May image was acquired by Landsat 4, whereas the August image was taken by Landsat 5.

Illumination

Normalization to a reference image may suffice to compensate for the effects of changing illumination, but for some vegetated landscapes, the problem can become more complicated, as we discussed in Section 8.1.3. If a landscape is covered with different surface materials, each having different photometric functions, an illumination correction using the gain derived from an invariant reference area may not be right for other parts of the image. In this situation, we need to apply separate relative calibrations to each part of the surface that has a different photometric function. The dilemma here is that we need to know the photometric response of each part of the scene to changing illumination to test for temporal changes on the surface, whereas we need to know that the surface is invariant to measure differences in the photometric function from place to place. One solution is to accumulate measurements over time of each invariant area, and to use this information to map the vector pathway in the data space that describes how each surface responds to changes in illumination. Alternatively, we can model the illumination changes, provided that we already understand each surface in terms of its composition and its physical structure at a range of scales. The objective of charting illumination pathways for specific surfaces is to be able to correct the DNs from any image of the same surface to a standard illumination geometry. Changes in the nature of the surface will be revealed if the corrected vectors do not coincide with the ones that are predicted from just the global variations in illumination.

8.3.2 Modeling temporal changes in vegetation

Connecting temporal pathways to process models

In Figure 8.10 we plotted the laboratory reflectance spectra of assorted kinds of vegetation in the reflectance space of TM Bands 4 and 5, and connected the vectors with lines to indicate probable process pathways. That simplistic diagram is based on only a few spectra, and, although we cannot use it as an accurate chart for all vegetation, we expect that the general trends of the pathways will be echoed in most time-series images of vegetation. As discussed in Section 8.2.1, laboratory spectra emphasize the importance of measurements of reflectance/absorption at 1.6 μm (TM Band 5) or 2.2 μm (TM Band 7), and these wavelengths are

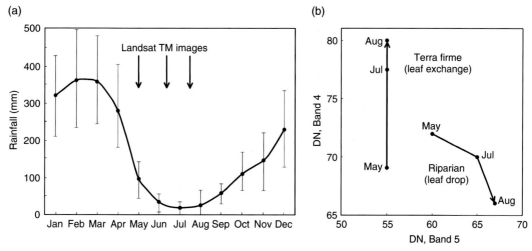

Figure 8.19. Seasonal changes in closed-canopy Amazon forest interpreted from Landsat TM images. (a) Monthly rainfall for Marabá, Brazil from 1973 to 1985. Landsat TM images (arrows) were acquired for May, 1992, July 1993, and August, 1992. (b) Mean vectors for terra firme forest and riparian forest for the three Landsat images, plotted in the plane of Bands 4 and 5. The temporal trend for terra firme forest is consistent with the trend for leaf exchange in Figure 8.10. The trend for riparian forest is consistent with the trend for partial leaf drop that exposes woody material. The processes inferred from the images were confirmed by field observations. Modified from Bohlman *et al.* (1998).

not sampled by all of the remote-sensing systems that acquire time-series data, for example, AVHRR and MODIS. The most accessible images for time series that sample at 1.6 and 2.2 μm are Landsat and ASTER.

The study by Bohlman *et al.* (1998) of Marabá, Brazil illustrates how the reference pathways of Figure 8.10 can assist in interpreting temporal changes in a tropical forest in Landsat images. The objective was to test for deciduousness of the vegetation cover, part of which is primary forest. Phenologic changes in closed-canopy tropical forest are especially difficult to measure for large areas on the ground, because of the high species diversity and problems of access and sampling. Even optical remote sensing is limited, because cloud cover obscures tropical forests, including this scene, for much of each year. Three images were found that spanned the dry season (Figure 8.19a), and although the full year was not sampled, the window from May to August showed changes in several vegetation communities. Figure 8.19b illustrates the changes in the terra firme forest and in one type of riparian forest. The images were intercalibrated to the May image (Figure 8.18), therefore, they are not calibrated to reflectance. Nevertheless, the trends of the time changes in the vectors can be compared with the reference spectra in Figure 8.10. The mean DN of terra firme forest shows an increase in Band 4 from May to July to August, with no change in Band 5. This trend is consistent with leaf flush where new, more reflective leaves replace older, darker leaves. This process of leaf exchange apparently does not expose large amounts of branches, and does not entail yellowing and browning of leaves before they are dropped. The temporal trend for the riparian forest, however, shows a decrease in Band 4 radiance, accompanied by an increase in Band 5. The trend is consistent with the pattern in Figure 8.10 that is expected when some of the mature green leaves are shed, thereby exposing more woody material. Field work by Bohlman *et al.* (1998) confirmed the interpretations that were drawn from the Landsat images. Furthermore, the authors were able to map the spatial patterns of these processes by assigning colors to each of the three images. For example, if May Band 4 = blue, July Band 4 = green and August Band 4 = red, the terra firme forest appears yellow, because Band 4 DNs are highest in the July and August images. The same color composite renders the riparian forest blue, because Band 4 DNs are highest for May.

Combined models

In Figure 8.2 we illustrated a pathway from light soil to mature wheat that showed an initial deflection toward shade, owing to shadows cast on the soil by the emerging vegetation. In that example, the global illumination did not change, but the photometric function of the surface did change with increasing vegetation cover. In order to predict pathways for vegetation in time, we need to be able to model both the global illumination *and* the photometric effects of each type of vegetation as it passes through its stages of growth and senescence, and as it undergoes phenologic changes. Hall *et al.* (1997) employed a strategy in which separate pathways (trajectories) were constructed for each vegetation type in a boreal forest. Pathways, based on measurements from images and physical models, were defined for different solar zenith angles and for different stages in regeneration of parts of the forest. For example, Figure 8.20 shows the modeled variation in red and NIR reflectance of aspen trees, on a boreal-forest landscape having minimal topography, through stages from no trees to spaced small trees to a closed-canopy forest. Process pathways are shown for five solar zenith angles, as calculated from hybrid models that take into account the type, number, and spacing of geometrical shapes (trees) on a given background. Using models of this type, Hall *et al.* (1997) defined pathways in the data space of red and NIR reflectance for the zenith angles encountered at the high latitudes of the Canadian boreal forest under study. Unknown vectors from images in a time series were evaluated by their Euclidean distances from the defined pathways. Vectors that fell on or near a pathway were judged to be members of the same class. The position of a vector

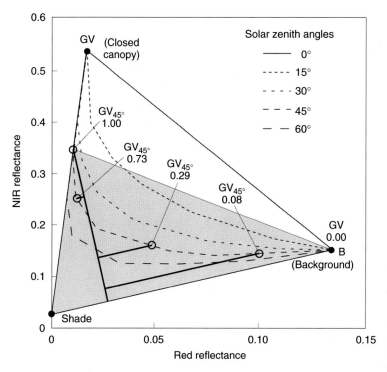

Figure 8.20. Modeled red and NIR reflectance of a landscape with variable cover of aspen trees (*Populus tremuloides*) and variable solar zenith angle. The amount of cover varies from background (B) to closed canopy (GV), following the solid and dashed lines. Each pathway from B to GV corresponds to a different zenith angle, as shown in the inset key. The pathways can be interpreted within a spectral-mixing framework when B, GV, and shade are considered as spectral endmembers. The shade fraction increases with increasing zenith angle. The GV fraction is measured as the proportional distance from the shade–background line to the point where a given zenith pathway intersects the shade–GV line. Fractions of GV are shown for a zenith angle of 45°, and the mixing space is gray for that illumination geometry. See also Figure 8.1. Based in part on data from Hall *et al.* (1997).

along a pathway reflected the fraction of the vegetation cover and its geometric properties, and this information was interpreted in terms of biophysical properties such as biomass.

Because the model of Hall *et al.* (1997) can be interpreted within the framework of spectral mixing between background (B), closed-canopy GV and shade, it can be extended to multiple channels. For example, the zero-degree zenith pathway from background to canopy describes mixing between two endmembers, and it is similar to the examples in Figure 8.1, except that in Figure 8.20 the red-reflectance scale is four times the NIR scale. At non-zero zenith angles, the pathways from B to GV are curved, because they describe the complex effects of shading and shadowing for the changing geometric shapes of the trees over time. The fraction of GV can be measured along a line from GV perpendicular to the line between shade (S) and B (see Figure 8.1d). With increasing fraction of GV, red reflectance decreases due to chlorophyll absorption. Simultaneously, the shade fraction increases as the trees grow larger and cast more shadows. Then, as tree canopies grow closer together, the fraction of shade decreases toward the amount for a closed canopy. The fractions of GV for a zenith angle of 45° are labeled in Figure 8.20, and the mixing space for that zenith angle is displayed as gray. Notice that each of the modeled pathways in

Figure 8.20 defines a separate mixing model having different image endmembers of GV. The zenith pathways in Figure 8.20 plot as curved lines on ternary diagrams such as those in Figures 8.16 and 8.17, and in that form it is easier to visualize the effects of illumination on the fraction of shade.

8.3.3 Thematic mapping with temporal images

Displaying and tracking changes

How do we display and organize information about the changes in an image time series when the number of images is greater than three? We can compare two images by making a difference image or a ratio image, and two or three images as color composites (e.g., Adams *et al.*, 1995; Bohlman *et al.*, 1998; Roberts *et al.*, 1998b); however, with four or more images, we either have to line them up and compare them visually, or we can view them as a video. The objective of displaying changes with images is to place information in a spatial context, which we know is essential for interpretation. However, simply viewing qualitative changes from image to image may not be sufficient, because we also may want to keep track of what vectors changed, the nature of the changes, and the amount of the changes. The problem of organizing and visualizing combined multispectral and multitemporal data is aggravated by the potentially huge amounts of data involved. Investigators who make thematic maps in the field cope with large amounts of information by being highly selective, that is, by deciding what is important and editing out what is unimportant (Chapter 7). A similar approach can be applied to temporal image sets.

Consider the hypothetical example of a set of ten multispectral temporal images, each having six or more channels. Further, let us assume that the set has been inter-calibrated, either by normalizing to a reference image or by calibration and correction to reflectance. The objective is to detect and track changes in the scene over the interval of measurement. If we consider one pixel address, x, y, it corresponds to one vector (in the data space defined by the number of channels) in each of the images of the time series. Scatter plots in two or three dimensions allow us to visualize the location of the x, y vector for each image individually. If we plot the locations of all of the vectors for the same pixel in the time series, and the scene has changed between each image, the scatter plot would have ten separated points. The ten points could be connected by arrows to indicate the pathway in time of the vector, following the basic idea of our previous examples (e.g., Figure 8.19). If each image were transformed into fractions of endmembers, the dimensionality of the vector for pixel x, y would be reduced to the number of endmembers (although information from all of the channels would be retained), and the temporal changes in three of the fractions could be displayed on a ternary diagram. It might be possible to interpret the temporal data for one pixel in terms of process pathways, but we certainly would not want to base conclusions on a single pixel. However, if we add the data even from just a few other pixels to the scatter plot or the ternary diagram, the displays become too cluttered to be interpretable. At the same time, the data space enlarges by ten more dimensions, thereby posing an additional computational burden.

One solution to the problem of organizing temporal image data is to select training areas from a reference image, and to compare the mean vectors of those spatially contiguous pixels for each time period sampled, as was done in Figure 8.19. Changes are interpreted according to the locations of the mean vectors and their variation from image to image with respect to process pathways. This approach has the advantage that we only need to compare the locations of a few vectors for each training area, but we do not have to keep track of all of the pixels that they represent.

A different approach entails classifying all pixels in each image of a time series. In this case, the objective is to compare the class membership of each pixel over time. We already have considered ways to classify image data that facilitate tracking temporal changes. For example, Hall *et al.* (1997) classified pixels according to their distance from process and illumination pathways (e.g., Figure 8.20). In Section 5.2 we described how to classify fractions based on image context and the proportions of basic scene components. Images of fraction classes from a four-image time series of part of the Landsat Fazendas subimage are shown in Plate 8. In another study, Roberts *et al.* (1998b) used endmember fractions and RMS error to classify a time series of five Landsat TM images of an area in the eastern Amazon Basin. In their analysis, membership in each class was determined by a decision-tree algorithm that split larger training sets into progressively more homogeneous groups.

Thematic maps of class changes

The class membership of each pixel over time can be displayed in a series of images, one for each time sampled. For a few images such as the four in Plate 8, we can interpret the changes in the scene by direct comparison; however, this can become difficult with a large number of images and/or classes. Furthermore, displaying images sequentially leaves the interpretation to the viewer. As we emphasized in Chapter 7, thematic maps, by definition, select certain themes, and they present an interpretation of the data. We can make thematic maps from

Figure 8.21 and color Figure 8.21 (see Web). Temporal thematic images of Fazenda Dimona, Brazil showing selected changes in land cover. See Plates 3 and 4 for location. Data are from Landsat TM images for August 1988, 1989, 1990, and 1991 (Plate 8). Each image in the time series was classified according to fractions of endmembers, based on a physical model of the spectrally dominant components in the scene. The classified images were edited to show those pixels having common histories of class membership. (a) Removal of forest. Red = cut primary forest. Blue = cut regeneration. Green = no change. White = other. The gray background is primary forest. (b) Pasture maintenance. Yellow = unchanged (maintained) pasture. Green = regeneration (pasture overgrown with shrubs). White = other. The gray background is primary forest. From Adams *et al.* (1995) with permission from Elsevier.

(a)

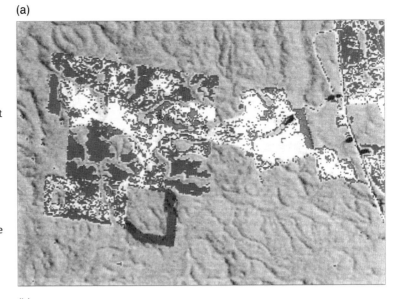

(b)

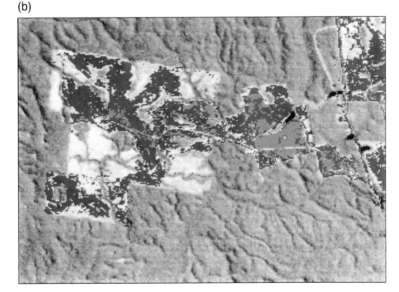

temporal data by selecting and editing pixels according to their historical attributes. Using this approach we can create images that display those changes that coincide with process pathways of interest. For example, Adams *et al.* (1995) used class changes over four years to display two types of thematic maps of Fazenda Dimona – one a map of forest removal, and the other a map of pasture conversion to regeneration (shrubs) (Figure 8.21 and color Figure 8.21; see Web). The history of forest removal was established by selecting pixels that: (1) started as primary forest and changed to slash, (2) started as regeneration forest and changed to slash, and (3) pixels that underwent no change. The second thematic map selected those pixels that (1) started as pasture and ended as regeneration (shrubs), (2) remained pasture, and (3) underwent no change. Classification was based on the complete temporal record, in this case four images, not just on the first and last samples. By assigning colors to each of the new "historical" classes, the viewer is presented with an interpretation that is easily understood. Furthermore, the image data also contain quantitative information about the area of the scene that is affected by a given change.

8.3.4 Accumulating reference information

We find the thematic approach to temporal data especially valuable, because, as is the case with field work, we can greatly simplify the complexity of the data by presenting one (or a few) ideas at a time. This means that we need to be clear about what information we are trying to extract from an image set. In the examples above, the objectives varied from study to study. Bohlman *et al.* (1998) wanted to test for phenological changes in known forest types. Roberts *et al.* (1998b) focused on mapping forest types correctly, in order to detect changes in land cover over time. When looking for seasonal changes in vegetation, the changes in cover such as cutting or regrowth are distractions that need to be ignored or "corrected." Conversely, when looking for changes in land use, seasonal changes in vegetation are "noise" that needs to be removed, perhaps by normalizing to a standard condition. Both studies share a need to normalize their data to a standard illumination geometry.

Perhaps efforts to monitor changing landscapes in the future would benefit from a more systematic accumulation of the measurements necessary to extract thematic information from spectral images. As just one example, it is widely recognized that there is a need to make temporal measurements of the Amazon forests at landscape scales, for reasons ranging from evaluating deforestation to assessing carbon sequestration by regrowth. Using a combination of measurements and models, it may be feasible to define standard corrections for illumination of the main forest communities. Eventually, enough measurements might be accumulated to model the seasonal changes in the same communities.

Inter-calibration of images poses a more formidable challenge, but the problem may not be intractable. The key may lie in the fact that the more measurements we accumulate about surface materials and their illumination and process pathways, the easier it is to interpret another image. Monitoring, by definition, means that we return to the same scene again and again. Scenes that deserve that much attention also are likely to be studied on the ground. However, analysts commonly approach spectral images "cold," without the benefit of a rich data base about the known surface materials and how a particular scene changes naturally over time. In the same sense that we proposed that it is feasible to define "universal" endmembers (Section 4.4.3), it might be useful to define "standard" images of those scenes that we revisit frequently, and "reference" areas within those scenes. Inter-calibrating to a standard image (acquired at a particular time) would have the advantage that it would facilitate comparison of results among different investigators. Also, calibrations and corrections applied to a reference image would carry over to those images that were normalized to the reference image. Within-scene atmospheric variations may pose the most difficult obstacle, but even here, if we have defined process pathways well, it may be possible to identify those vectors that do not fit known pathways, and to test whether they can be corrected by applying spatially variable atmospheric models.

The idea of connecting spectral images to the properties of the ground is as old as spectral remote sensing, and there are excellent examples of individual studies that have used a physical-modeling approach to advantage. Nevertheless, many investigators who are engaged in remote sensing with spectral images continue to be focused primarily on manipulation of the numerical image data. Apparently the quest is still on for "killer" algorithms to automate the extraction of information while minimizing the user's participation. In this book we have made a case for shifting the emphasis toward physical modeling that is guided by those who want specific information about a scene. This shift in emphasis is especially needed to understand the rapidly changing landscapes of Earth.

Glossary

A

Absolute measurements   Measurements that have been calibrated to an accepted reference and are in physical units such as meters, seconds, or $W\,m^{-2}\,sr^{-1}\,\mu m^{-1}$.

Absorption   The process by which photons transfer their energy to matter in their path.

Absorption band   Spectral wavelength region in which absorption occurs.

AIS   Airborne Imaging Spectrometer: an early experimental NASA sensor.

Albedo   The fraction of light that is reflected by a body or surface.

ASTER   Advanced Spaceborne Thermal Emission and Reflection Radiometer: a high-resolution imager on NASA's Terra satellite.

Attribute   A quality, characteristic, or property.

Attribute image   An image that portrays the properties of either the data or the physical scene (see *data attribute*).

AVHRR   Advanced Very High Resolution Radiometer: the primary sensor on NOAA polar-orbiting satellites.

AVIRIS   Airborne Visible-Infrared Imaging Spectrometer: a NASA hyperspectral (224-channel) airborne imager.

B

Background   The scene against which a sought-for target or scene component is viewed.

Band   In the context of spectroscopy, "band" describes a limited wavelength region in which radiant energy interacts with a substance; e.g., absorption band. In the context of remote-sensing instruments, "band" is equivalent to "channel," and means the wavelength interval in which a radiance measurement is made. In this book "channel" is used as the general term applied to instruments, except for specific names such as Landsat TM Band 4, where "Band" is capitalized.

Basic spectral components   Spectrally distinctive and commonly occurring materials that dominate the spectra of landscapes (see *universal endmembers*).

Bi-directional reflectance distribution function (BRDF)   The ratio of reflected to incident radiance at every possible direction.

C

Calibration   Strictly, a term that applies to defining the response of the imaging system to radiance, or determining the mathematical function that transforms measured data into data having standard physical units; casually, the application of that transfer function to the measured data. Even more casually: used to describe corrections to data, e.g. "calibration to reflectance."

Canopy   The upper surface of a community of vegetation, smoothed to the scale of measurement or observation.

Channel   The data plane devoted to radiance measurements in a given wavelength range (see *band*).

Channel ratios   The quotient of data from one image channel divided by another.

Channel residual   In spectral-mixture analysis, what is left over when the endmember spectral vectors are multiplied by the calculated endmember fractions, summed, and subtracted from the measured spectral vector.

Checkerboard mixing   Spectral mixing due to the spatial integration of light from unresolved opaque objects in a landscape (see *intimate mixing*).

Class   A group of spectral-data attributes according to specified thresholds. In common usage, a group of things having common attributes.

Classification   The process of grouping spectral vectors on the basis of common attributes.

Classifier   A mathematical algorithm that measures similarity between data vectors.

Classify   To arrange or organize by classes; to organize numerical data into groups; to make classes from digital-image data.

Clutter   A term for spectral-background variability due both to measurement error and to the natural complexity of landscapes.

Color-composite fraction image   A false-color image in which the three input images depict the endmember fractions calculated from spectral-mixture analysis.

Complement   In imaging, the number which, added to a measured DN, equals the maximum value in the dynamic range: e.g., in an eight-bit image ($0 \leq DN \leq 255$), the complement of 64 is 191. This has the effect of making a "negative" image (see *inverse*).

Complemented-shade image   An image of the complement of the shade fraction calculated in spectral-mixture analysis, in which high-shade areas are dark and low-shade areas are light.

Continuum   That smooth and continuous part of a spectrum lacking distinctive absorption or emission features.

Contrast enhancement   Adjustment of the gray-level distribution to make better use of the black/white or color dynamic range of a display medium.

Contrast stretch   A modification of image contrast.

Convex hull   The smallest convex set containing a cluster of data vectors, visualized as the shape of a balloon minimally enclosing the cluster. Vectors touching the balloon have extreme values for the cluster and are sometimes confused with spectral endmembers.

Correction   Adjustment of modeled data to minimize the residual with the measured data; adjustment of measured data to fit a model (see *calibration*).

CSIRO   Commonwealth Scientific and Industrial Research Organisation: the national government body for scientific research in Australia.

D

Data attribute   A quality, characteristic, or property of data.

Data transformation   Application of mathematical equations to express data in a new coordinate system.

Decorrelation stretch   Image enhancement that exaggerates the poorly correlated data in an image by equalizing the variances in the decorrelated image, and then re-transforming the data back to their original coordinate system.

Density slice   A DN range in an input image that is set to a single value in an output image.

Detectability   Qualitatively, the ease with which a target or sought-for object can be seen against the background of the landscape. Quantitatively, a measure such as the spectral angle between the target spectrum and the average background spectrum, divided by the standard deviation of the background spectral angles measured against the average.

Detection   Discovering something, or the process of discovering something, by distinguishing it from its background (see *identification*).

Detection limit or threshold   The quantitative value of the detectability below which the target cannot be recognized.

DEM   Digital elevation model: an image in which the measured variable is the height of the landscape above some datum.

DN   Data number: a commonly used term synonymous with gray level, a unit of encoded (uncalibrated) radiance or any other data displayed as an image.

E

Emissivity   The efficiency with which a surface radiates thermal energy. A blackbody or ideal cavity has an emissivity of unity.

Emittance   The ratio of thermal-infrared light emitted from a natural surface to the light emitted by an ideal blackbody at the same temperature. Differs from emissivity in the way that reflectance differs from reflectivity: reflectivity and emissivity are properties only of matter (ideal surfaces) whereas reflectance and emittance account also for multiple scattering and shadowing due to the shape and roughness of the surface.

Empirical   An observed relationship for which the physical cause is not understood.

Empirical calibration   Image calibration by fitting a line (light-transfer curve) through a cluster of data points relating image DN and field-measured radiances. Also called "empirical-line calibration" or "virtual calibration."

Encoded radiance   Radiance measurements represented as DN before calibration to radiance units (e.g., $W m^2 sr^{-1} \mu m^{-1}$).

Endmember   A spectral vector characteristic of a significant scene component that, when mixed with other endmember vectors in various proportions, models the measured spectra of a scene.

Error   The difference between a computed, estimated, or measured value and the true value (see *residual*).

Euclidean distance   The distance between two vectors in a Euclidean space.

F

Fit   The amount of the data described by a model, measured by the difference between the measured and modeled data (residual).

Foreground   An arbitrary scene or landscape component that is of key interest in the investigation of a problem. A target (see *background*).

Foreground–background analysis (FBA)   Parsing of an image into foreground components and background components, which together comprised the measured spectral data.

Fraction images   Images made of the fractions of endmembers calculated in spectral-mixture analysis.

G

Gain   A multiplicative factor in the encoding of radiance as DN, usually a setting in the amplifier of the imaging system.

GIS Geographic information system.

Green vegetation (GV) The chlorophyll-rich component of vegetation, such as leaves. A basic spectral component of many landscapes that commonly is used as an endmember in spectral-mixture analysis.

Ground truth The informed opinion of a field researcher about what actually is found in a scene, as opposed to what is inferred from remotely sensed data.

H

Hemispherical reflectance The ratio between the radiance reflected from a scene at all angles (over 2π steradians) and the incident irradiance, both in units of W m^{-2} μm^{-1} (see *reflectance*).

Hill shade A term coined for geographic information systems for use instead of shaded relief. Both terms refer to the predicted amount of radiance that would be measured from an ideal surface at specified viewing and illumination geometries and described by a photometric function.

Histogram A plot of the number of occurrences (or the normalized number of occurrences) in a scene or image of an attribute, against the value of the attribute. For example, the histogram of an image commonly is used to show the range of DNs that is encountered and the shape of the distribution.

Hyperspectral Spectral data sampled at high resolution in contiguous spectral channels and described by spectral vectors of high dimensionality. Spectra plotted from hyperspectral images resemble continuous spectra.

I

Identification In spectral-image analysis, the association of a spectrum or set of spectra unambiguously with a reference spectrum; the association of a spatial pattern with an object of known shape (see *detection*).

Illumination geometry The geometric positions of the Sun and sensor relative to a scene or specific scene element; also, the angles between the rays from the scene element to the Sun and sensor relative to the local surface normal, or the angle between the rays (phase angle).

Image Spatial distribution of scene attributes, especially radiance. The "scene" is an element of the landscape under observation or study (see *scene*).

Image endmember   An endmember spectrum derived from an image instead of a reference spectral library (see *reference endmember*).

Imagery   A term used in literature to describe fanciful mental or poetic descriptions or visions; sometimes used in image processing to indicate a general class or set of images.

Instantaneous field of view (IFOV)   The field of view of the imaging-system optics and individual detector when making a single measurement (pixel). In terms of pixel size projected onto the scene, it is the pixel footprint.

Intimate mixing   Nonlinear spectral mixing in which light is transmitted through materials such as leaves or mineral grains and is then reflected from other materials.

Inverse model   A mathematical model, generally under-determined, that is run "backwards" to estimate the value of scene or image attributes from the measured data.

Inversion   Turning something upside down. In image processing, "inversion" may mean taking the reciprocal or complement of a data value. In image modeling, inversion is the process of back-calculating or estimating important scene characteristics from the measured radiances.

J

JPL   Jet Propulsion Laboratory, a NASA laboratory in Pasadena, California, operated by the California Institute of Technology.

L

Landsat   A series of NASA Earth Satellites that aquire images in VIS, NIR, and TIR wavelengths.

Landscape   A portion of the land that can be seen from a single vantage point, usually on the ground. A term used in this book to distinguish the typical scales on which detailed field measurements are made, and contrasted with continental and global scales.

Leaf Area index (LAI)   The total area of all the leaves in a plant or canopy relative to the ground surface area covered.

LIDAR   LIght Detection And Ranging: the laser equivalent of RADAR.

Lightness   In color theory, the intensity of light reflected from a scene.

Linear mixing   In spectral-mixture analysis, "checkerboard" mixing of light from different surface elements by spatial integration, in proportion to their areal cover.

Longwave infrared (LWIR)   Thermal-infrared light in the 8–14 μm terrestrial atmospheric window.

M

Matched filtering (MF)   In signal theory, a measure of similarity between a measured signal and a known one, for example a sine wave. In remote sensing, generally a measure of similarity between a known and unknown spectrum, but it has come to be used in the specialized sense of foreground–background analysis in which the goal is to measure the distance of an unknown spectrum along a line connecting the cluster of image spectra (the "background") to the spectrum of a real or suspected scene component (the "foreground"). The measure is the distance of the projection of the unknown onto the line.

Mimics   Identical or similar spectral vectors (or attribute sets) for different scene components.

MISR   Multi-angle Imaging SpectroRadiometer.

Mixing model   A mathematical model describing measured image spectra in terms of linear or nonlinear mixtures of spectral endmembers.

Mixture   An aggregate of two or more substances that are not chemically united and that exist in no fixed proportion to each other.

Model   In remote sensing, a simplified description of a landscape, image, attribute, or pathway. A model may be qualitative or quantitative (mathematical) and empirical (statistical) or physical (physics-based), but all models should have predictive capability.

MODIS   Moderate Resolution Imaging Spectrometer, a 50-channel imager on the Terra spacecraft.

MODTRAN   A numerical radiative transfer model that is used to predict atmospheric effects such as transmissivity.

MSS   MultiSpectral Scanner: a four-channel imager on early Landsats.

MSSS   Malin Space Science Systems.

Multispectral   Spectral data sampled at low resolution in non-contiguous spectral channels and described by spectral vectors of low dimensionality.

N

Nadir   A point on the global surface directly below and opposite to the zenith.

Nadir view   A camera or detector orientation pointed at the nadir, i.e. "straight down" normal to the mean global surface.

NASA   National Aeronautics and Space Administration.

Near infrared (NIR)   Spectral wavelengths from 0.7 to 3.0 μm. NIR light consists primarily of reflected sunlight but is beyond the limit of human vision.

NEΔT   Noise-equivalent delta temperature (precision).

Noise   Data components unrelated to the imaged scene, and usually random. Noise can be introduced by electronic interference during measurement, amplification, and transmission of data.

Nonlinear mixing   Blending of spectra from different materials in which absorption or other nonlinear processes dominate simple spatial (checkerboard) integration.

Non-photosynthetic vegetation (NPV)   Non-chlorophyll-bearing plant matter, such as dry leaves and woody material. A basic spectral component of many landscapes that commonly is used as an endmember in spectral-mixture analysis.

Normalization   Division of an attribute by its maximum value such that its range is between zero and one; division of one attribute by another with the same goal.

Normalized-difference vegetation index (NDVI)   The difference between radiances measured at NIR and red wavelengths, divided by their sum. The NDVI is taken to be proportional to the amount of vegetation in an area (e.g., canopy percent or leaf area index).

O

Offsets   A shift in image DN unrelated to scene illumination or reflectance, for example due to atmospheric path radiance or to amplifier artifacts in the imaging system (bias).

Optical path length   The distance through a substance over which light has traveled before it arrives at the sensor, usually expressed mathematically as the physical distance normalized by the absorption coefficient (Bouguer–Beer–Lambert law).

Overflow fractions   Values of endmember fractions outside of the range of 0 to 1. Overflow fractions occur when a spectral vector lies outside the volume in a DN data space defined by the endmembers in spectral-mixture analysis.

P

Path, pathway   A route along which something moves; a series of spectral vectors in a data space for a scene element observed over time.

Path radiance An additive factor due to scattering of sunlight into the detector, or thermal-infrared light emitted directly by the atmosphere.

Photometric function A mathematical equation predicting the radiance from a surface under specified viewing and illumination geometries.

Physical model A mathematical or numerical model relating observations or measurements from a scene.

Pixel footprint The outline of the area in a scene over which radiance reported in an image pixel is integrated.

Pixel-to-pixel (PP) spectral contrast A measure of the difference between two image spectra or spectral vectors. Spectral-length contrast is the difference between the lengths of two spectral vectors; spectral-angle contrast is the difference between the directions in which they point (see *spectral contrast*).

Process A continuous action, operation, or series of changes taking place in a definite manner.

Process pathway The spectral pathway along which spectral vectors will evolve for a surface element affected by some significant process, such as seasonal growth and senescence.

R

Radiance The directional flux of radiant energy from a surface that is measured by an imaging system, usually in units of $W\ m^{-2}\ sr^{-1}\ \mu m^{-1}$.

Reference endmember An endmember spectrum derived from laboratory or field measurements that are corrected for shading.

Reference pathway A pathway through a data space that depicts the idealized or typical spectral evolution of a scene component.

Reference spectrum The spectrum of a known substance used to interpret spectra of unknown scene components from an image.

Reflectance The ratio between measured radiance from a scene and the radiance reflected from an ideal, smooth diffuse reflecting surface oriented at the same mean incidence angle to the Sun. For natural, rough surfaces shadows and shading darken the scene such that reflectance is never higher than reflectivity, unless the natural surface is a specular reflector.

Reflectivity The reflectance for the case in which the surface is smooth and diffuse. In practice, this means that the surface has been prepared and measured in a laboratory rather than in the field.

Residual The difference between predicted and actual measurements (see *error* and *fit*).

Residual image   An image of the difference between a predicted and measured parameter.

Resonance   Absorption or emission at specified wavelengths due to specific physical interactions between light and matter, including electronic and vibrational processes.

Resonance band   The neighborhood of a spectrum around a central wavelength at which resonance significantly affects the measured values.

RMS residual   Root-mean-squared (RMS) difference for a set of predicted and measured values.

Rule image   A data-attribute image showing the values of a calculated parameter and used to assign pixels to classes on the basis of spectral similarity.

S

Scatter plot   A Cartesian space in which the axes are the DNs of the different channels of a spectral image. A spectral vector from the image plots as a point in this space. For display purposes, it is common to visualize just a two-channel slice through this space, e.g., red vs. NIR.

Scene   An area within a landscape. In field investigation, the scene might be on the scale of what the human observer can see (horizon to horizon), but in image analysis it is the part of the landscape encompassed by an image (e.g., it is common to speak casually of "TM scenes").

Shade   An endmember spectrum describing the darkening in an image due to the combined effects of albedo, shading, and resolved and unresolved shadows.

Shading   A reduction in the amount of light reflected from a surface that accompanies a departure from normal illumination.

Shadow   A dark image cast on a surface by a body intercepting light.

Spectral angle (SA)   The angle between a spectral vector in a Cartesian space and a reference vector.

SA contrast   The angular difference between two spectral vectors.

Spectral component   A distinctive element in an image. A spectral component differs from an endmember in that it need not mix with other components, and may be a characteristic mixture of endmembers.

Spectral contrast   A measure of the difference between two spectral vectors or spectra. Wavelength-to-wavelength contrast within a single spectrum is a measure of variations in absorption or emission; pixel-to-pixel contrast is a measure of the overall difference between two spectra, and includes the lengths of the vectors (spectral length) or the angle between the vectors (spectral angle).

Spectral endmember   See *endmember*.

Spectral image   An image or set of images that is produced by radiant energy in specific, commonly narrow, wavelength regions for the purpose of making compositional inferences about the imaged scene.

Spectral length (SL)   Euclidean length of a spectral vector.

Spectral mixture   An aggregate of the spectra of two or more substances.

Spectral-mixture analysis (SMA)   The description of spectra or spectral vectors as mixtures of spectral endmembers; SMA includes both the selection of the endmembers to construct the mixing model and the inverse modeling (unmixing) to estimate, pixel by pixel, the fractions of the endmembers that are present in a spectral image.

SPOT   Système Probatoire d'Observation de la Terre. A commercial spaceborne imager.

Supervised classification   Classification in which the analyst defines the class attributes.

T

Target   A specific area or type of scene component sought for in image analysis.

Tasseled-cap transformation (TCT)   A data-space rotation in which axes of radiance are linearly combined to form axes of "greenness" and "brightness," forming a suitable framework for describing the growth and senescence of vegetation.

Texture   In this book, a term used to describe roughness inferred by shading and shadowing at sub-pixel scales.

Thematic map   A representation of an area of the land surface showing one or more themes.

Thematic Mapper (TM)   The imaging system carried on recent Landsats.

Theme   A unifying, dominant, or important idea or motif.

TIMS   Thermal-Infrared Multispectral Scanner, an airborne six-channel scanner.

Top-of-atmosphere radiance   The radiance reflected or emitted from a landscape and measured by a satellite, after it passes through and is affected by the atmosphere (also "exoatmospheric radiance").

Training areas   Areas, defined in an image and in the landscape, in which radiance measurements or image attributes and field measurements are compared for the purpose of adjusting the performance of an algorithm; a specific type of region of interest, or ROI.

Trajectory   The curve described by a projectile, rocket, or the like in its flight (see *pathway*).

U

Unit   Any group of things interpreted as an entity; the basic organizational entity used in thematic mapping on the ground and in image interpretation. Units have boundaries, labels, and defined physical attributes. In general use, any thing interpreted as an entity.

Universal endmembers   Generalized endmembers that, when mixed, describe most of the spectral variability found in a wide variety of scenes or landscapes: for example, green vegetation, soil, and shade.

Unmixing   Inverse spectral-mixture analysis, in which the fractions of endmember mixtures necessary to describe the maximum amount of image variance are calculated.

Unsupervised classification   Assignment of image pixels to classes or groups without the use of training areas or other human guidance. Classes are determined by the statistical properties of the spectral data and without consideration of image context.

USGS   US Geological Survey.

V

Validation   Quantitative demonstration by measurement in the field that an interpretive model is accurate.

Vegetation canopy   The volume above a landscape occupied by trees and shrubs.

Vegetation index   Any simple measure calculated from radiance data that is proportional to the amount of vegetation in a scene, such as leaf area or percent cover; also, the ratio of NIR and red radiance data.

Vibrational resonance   The wavelength at which interaction between light and matter causes strong vibration in the latter.

Virtual cold   An endmember in thermal-infrared images, analogous to shade in VNIR images. If radiance were linearly related to temperature, virtual cold would be the origin for calibrated radiance data.

Virtual endmember   A spectral endmember that is inferred from the image data, but is not actually present in the image.

W

Wavelength-to-wavelength (WW) spectral contrast   Normalized variation in reflectivity or emissivity from one part of the spectrum to another; from the bottom of an absorption feature to its sides in the continuum.

Z

Zenith   The point on the celestial sphere vertically above a given point on the surface; the opposite to nadir.

Zenith angle   The angle on the ground between the normal to the mean global surface and the direction to the Sun.

References

Adams, J. B. (1974). Visible and near-infrared diffuse reflectance spectra of pyroxenes as applied to remote sensing of solid objects in the solar system. *Journal of Geophysical Research* **79**, 4829–36.

Adams, J. B. and Adams, J. D. (1984). Geologic mapping using LANDSAT MSS and TM images: removing vegetation by modeling spectral mixtures. Third Thematic Conference, Remote Sensing for Exploration Geology, *ERIM* **2**, Michigan, Enivironmental Research Institute of Michigan, pp. 615–22.

Adams, J. B., Sabol, D. E., Kapos, V. *et al.* (1995). Classification of multispectral images based on fractions of endmembers: application to land-cover change in the Brazilian Amazon. *Remote Sensing of Environment* **52**, 137–54.

Adams, J. B., Smith, M. O., and Johnson, P. E. (1986). Spectral mixture modeling: a new analysis of rock and soil types at the Viking Lander I site. *Journal of Geophysical Research* **91**, 8098–112.

Allen, W. A. and Richardson, A. J. (1968). Interaction of light with a plant canopy. *Journal of the Optical Society of America* **58**(8), 1023–8.

Anderson, J. R., Hardy, E. T., Roach, J. T., and Witmer, R. E. (1976). A land use and land cover classification system for use with remote sensor data. US Geological Survey Professional Paper 964, Washington, DC, Department of the Interior.

Berk, A., Bernstein, L. S., Anderson, G. P. *et al.* (1998). MODTRAN cloud and multiple scattering upgrades with application to AVIRIS. *Remote Sensing of Environment* **65**, 367–75.

Bierregaard, R. O., Jr., Gascon, C. G., Lovejoy, T. E., and Mesquita, R. C. G. (2001). *Lessons from Amazonia: the Ecology and Conservation of a Fragmented Forest*. New Haven, Yale University Press.

Boardman, J. W. (1993). Automated spectral unmixing of AVIRIS data using convex geometry concepts. In *Summaries of the 4th JPL Airborne Geoscience Workshop*. Pasadena, CA: Jet Propulsion Laboratory, California Institute of Technology, JPL Pub. 93–26, vol. 1, pp. 11–14.

Bohlman, S. A. (2004). The relationship between canopy structure, light dynamics and deciduousness in a seasonal tropical forest in Panama: a multiple scale study using remote sensing and allometry. Ph.D. dissertation, University of Washington, DC.

Bohlman, S. A., Adams, J. B., Smith, M. O., and Peterson, D. L. (1998). Seasonal foliage changes in the eastern Amazon basin detected from Landsat Thematic Mapper satellite images. *Biotropica* **30**, 376–91.

Burns, R. G. (1970). *Mineralogical Applications of Crystal Field Theory*. Cambridge, Cambridge University Press.

Chrisman, N. R. (1997). *Exploring Geographic Information Systems*. New York, John Wiley & Sons.

Clark, R. N. (1999). Spectroscopy of rocks and minerals and principles of spectroscopy. In *Remote Sensing for the Earth Sciences: Manual of Remote Sensing*, 3rd edn., ed. A. N. Rencz. New York, John Wiley & Sons, vol. 3, pp. 3–58.

Clark, R. N. and Roush, T. L. (1984). Reflectance spectroscopy: quantitative analysis techniques for remote sensing applications. *Journal of Geophysical Research* **89**, 6329–40.

Clark, R. N., Swayze, G. A., Livo, K. E. *et al.* (2003). Imaging spectroscopy: Earth and planetary remote sensing with the USGS Tetracorder and expert systems. *Journal of Geophysical Research* **108**(E12), 5131. http://speclab.cr.usgs.gov/PAPERS/tetracorder.

Conel, J. E. (1990). Determination of surface reflectance and estimates of atmospheric optical depth and single scattering albedo from Landsat Thematic Mapper data. *International Journal of Remote Sensing* **11**(5), 783–828.

Crist, E. P. and Cicone, R. C. (1984). Comparisons of the dimensionality and features of simulated Landsat-4 MSS and TM data. *Remote Sensing of Environment* **14**, 235–46.

Crist, E. P. and Kauth, R. J. (1986). The tasseled cap de-mystified. *Photogrammetric Engineering and Remote Sensing* **52**(1), 81–6.

Crist, E. P., Laurin, R., and Cicone, R. C. (1986). Vegetation and soils information contained in transformed Thematic Mapper data. In *IGARSS '86*, Zurich, ESA Publications Division, vol. ESA SP-254, pp. 1465–72.

Curtiss, B., Adams, J. B., and Ghiorso, M. (1985). Origin, development and chemistry of silica–alumina rock coatings from the semi-arid regions of the island of Hawaii. *Geochimica et Cosmochimica Acta* **40**, 49–56.

Ehrlich, P. R. and Ehrlich, A. H. (2004). *One With Nineveh*. Washington, Island Press.

Elvidge, C. D. (1989). Vegetation reflectance features in AVIRIS data. Proceedings of the sixth Thematic Conference on Remote Sensing for Exploration Geology, Houston, Texas, May 16–19.

Farr, T. and Adams, J. B. (1984). Rock coatings in Hawaii. *Geological Society of America Bulletin* **95**, 1077–83.

Gillespie, A. R. (1992). Spectral mixture analysis of multispectral thermal infrared images. *Remote Sensing of Environment* **42**(2), 137–145.

Gillespie, A. R., Adams, J. B., Smith, M. O. *et al.* (1995). Forest mapping potential of ASTER. *Journal of Remote Sensing Society, Japan* **15**(1), 62–71.

Gillespie, A. R., Kahle, A. B., and Palluconi, F. D. (1984). Mapping alluvial fans in Death Valley, California, using multichannel thermal infrared images. *Geophysical Research Letters* **11**, 1153–6.

Gillespie, A. R., Kahle, A. B., and Walker, R. E. (1986). Color enhancement of highly correlated images, I. Decorrelation and HSI contrast stretches. *Remote Sensing of Environment* **20**, 209–35.

Goetz, A. F. H. and Billingsley, F. C. (1973). Digital image enhancement techniques used in some ERTS application problems. In *Third Earth Resources Technology*

Satellite-1 Symposium. Washington, DC, National Aeronautics and Space Administration, NASA SP-351, pp. 1911–93.

Goetz, A. F. H., Vane, G., Solomon, J. E. and Rock, B. N. (1985). Imaging spectrometry for Earth remote sensing. *Science* **228**(4704), 1147–53.

Greenberg, J. D. (2000). Analysis of urban–rural gradients using satellite data. Ph.D. dissertation, University of Washington, Seattle, WA.

Gu, D. and Gillespie, A. (1998). Topographic normalization of LANDSAT TM images of forests based on subpixel sun-canopy-sensor geometry. *Remote Sensing of Environment* **64**, 166–75.

Gu, D., Gillespie, A. R., Adams, J. B., and Weeks, R. (1999). A statistical approach for topographic correction of satellite images using spatial context information. *IEEE Transactions on Geoscience and Remote Sensing*, **37**(1), 236–46.

Hall, F. G., Knapp, D. E., and Huemmrich, K. F. (1997). Physically based classification and satellite mapping of biophysical characteristics in the southern boreal forest. *Journal of Geophysical Research*, **102**(D24), 29567–80.

Hall, F. G., Strebel, D. E., Nickeson, J. E., and Goetz, S. J. (1991). Radiometric rectification: toward a common radiometric response among multidate, multisensor images. *Remote Sensing of Environment* **35**, 11–27.

Hapke, B. (1967). A readily available material for the simulation of lunar optical properties. *Icarus* **6**, 277–8.

Hapke, B. (1993). *Theory of Reflectance and Emittance Spectroscopy*. Cambridge, Cambridge University Press.

Harsanyi, J. C. and Chang, C. I. (1994). Detection of low-probability subpixel targets in hyperspectral image sequences with unknown backgrounds. *IEEE Transactions on Geoscience and Remote Sensing* **32**, 779–85.

Head, J. W., Pieters, C., McCord, T. B., Adams, J. B., and Zisk, S. (1978). Definition and detailed characterization of lunar surface units using remote observations. *Icarus* **33**, 145–72.

Holben, B. N., Schutt, J. B., and McMurtrey, J., III (1983). Leaf water stress detection utilizing Thematic Mapper bands 3, 4 and 5 in soybean plants. *International Journal of Remote Sensing* **4**, 289–97.

Hook, S. J., Abbott, E. A., Grove, C., Kahle, A. B., and Palluconi, F. (1999). Use of multispectral thermal infrared data in geological studies. In *Remote Sensing for the Earth Sciences: Manual of Remote Sensing*, 3rd edn., ed. A. N. Rencz. New York, John Wiley & Sons, vol. 3, pp. 59–103.

Huete, A., Justice, C., and Liu, H. (1994). Development of vegetation and soil indices for MODIS-EOS. *Remote Sensing of Environment* **49**, 224–34.

Huete, A. R., Liu, H. Q., Batchily, K., and van Leeuwen, W. (1997). Comparison of vegetation indices over a global set of TM images for EOS-MODIS. *Remote Sensing of Environment* **59**, 440–51.

Jensen, J. R. (1996). *Introductory Digital Image Processing: a Remote Sensing Perspective*, 2nd. edn. Upper Saddle River, NJ, Prentice Hall.

Kahle, A. B., Gillespie, A. R., Abbott, E. A. *et al.* (1988). Relative dating of Hawaiian lava flows using multispectral thermal infrared images: a new tool for geologic mapping of young volcanic terranes. *Journal of Geophysical Research* **93**, 15239–51.

Kapos, V., Wandelli, E., Camargo, J. L., and Ganade, G. 1997. Edge-related changes in environment and plant responses due to forest fragmentation in central Amazonia. In *Tropical Forest Remnants: Ecology, Management, and Conservation of Fragmented Communities*, eds. W. F. Lawrence and R. O. Bierregaard, Jr. Chicago, University of Chicago Press, pp. 33–44.

Kauth, R. J. and Thomas, G. S. (1976). The tasseled cap – a graphic description of the spectral-temporal development of agricultural crops as seen by LANDSAT. Symposium Proceedings, Machine Processing of Remotely Sensed Data. Purdue University, IN, pp. 41–51.

Kruse, F. A. (1999). Visible and infrared: sensors and case studies. In *Remote Sensing for the Earth Sciences: Manual of Remote Sensing*, 3rd edn., ed. A. N. Rencz. New York, John Wiley & Sons, vol. 3, pp. 567–611.

Li, X., Strahler, A. H., and Woodcock, C. E. (1995). A hybrid geometric optical radiative transfer approach for modeling albedo and directional reflectance of discontinuous canopies. *IEEE Transactions on Geoscience and Remote Sensing* **33**, 466–80.

Linthicum, K. J., Anyamba, A., Tucker, C. J. (1999). Climate and satellite indicators to forecast Rift Valley fever epidemics in Kenya. *Science* **285**(5426), 397–400.

Logsdon, M. G., Kapos, V., and Adams, J. B. (2001). Characterizing the changing spatial structure of the landscape. In *Lessons from Amazonia: the Ecology and Conservation of a Fragmented Forest*, ed. R. Bierregaard, Jr., C. G. Gascon, T. E. Lovejoy and R. C. G. Mesquita. New Haven, Yale University Press, pp. 358–68.

Lucas, R. M., Honzak, M., Foody, G. M., Curran, P. J., and Corves, C. (1993). Characterizing tropical secondary forests using multi-temporal Landsat sensor imagery. *International Journal of Remote Sensing* **14**, 3016–67.

Lucas, R. M., Honzak, M., Do Amaral, I., Curran, P. J., and Foody, G. M. (2002). Forest regeneration on abandoned clearances in central Amazonia. *International Journal of Remote Sensing* **23**(5), 965–88.

Mandelbrot, B. B. (1967). How long is the coast of Britain? Statistical self-similarity and fractional dimension. *Science* **156**, 636–8.

Mazer, A. S., Martin, M., Lee, M., and Solomon, J. E. (1988). Image-processing software for imaging spectrometry data analysis. *Remote Sensing of Environment*, **24**(1), 201–10.

McCord, T. B., Charette, M., Johnson, T. V. *et al.* (1972). Lunar spectral types. *Journal of Geophysical Research* **77**, 1349–59.

McCord, T. B., Pieters, C., and Feierberg, M. A. (1976). Multispectral mapping of the lunar surface using ground-based telescopes. *Icarus* **29**, 1–34.

Mushkin, A. and Gillespie, A. R. (in press). Estimating sub-pixel surface roughness using remotely sensed stereoscopic data. *Remote Sensing of Environment*.

Mustard, J. F., Li, L. and He, G.-Q. (1998). Nonlinear spectral mixture modeling of lunar multispectral data: implications for lateral transport. *Journal of Geophysical Research*, **103**(E8), 19419–25.

Mustard, J. F. and Sunshine, J. M. (1999). Spectral analysis for Earth science: investigations using remote sensing data. In *Remote Sensing for the Earth Sciences: Manual of Remote Sensing*, 3rd edn., ed. A. N. Rencz. New York, John Wiley & Sons, vol. 3, pp. 251–306.

Neukum, G., Jaumann, R., and the HRSC Co-Investigator and Experiment Team (2004). HRSC – the high resolution stereo camera of Mars Express. European Space Agency SP-1240, pp. 17–35.

Pech, R. P., Graetz, R. D., and Davis, A. W. (1986). Reflectance modeling and the derivation of vegetation indices for an Australian semi-arid shrubland. *International Journal of Remote Sensing* **7**(3), 389–403.

Pieters, C. M. (1993). Compositional diversity and stratigraphy of the lunar crust derived from reflectance spectroscopy. In *Remote Geochemical Analysis: Elemental and Mineralogical Composition*, eds. C. M. Pieters and P. A. J. Englert. Cambridge, Cambridge University Press, pp. 594.

Richardson, A. J. and Wiegand, C. L. (1977). Distinguishing vegetation from soil background information. *Photogrammetric Engineering and Remote Sensing* **43**, 1541–52.

Ridd, M. K. (1995). Exploring a V–I–S (vegetation–impervious surface–soil) model for urban ecosystem analysis through remote sensing: comparative anatomy for cities. *International Journal of Remote Sensing* **16**(12), 2165–85.

Roberts, D. A., Adams, J. B., and Smith, M. O. (1990). Predicted distribution of visible and near-infrared radiant flux above and below a transmittant leaf. *Remote Sensing of Environment* **34**(1), 1–17.

Roberts, D. A., Batista, G., Pereira, J., Waller, E., and Nelson, B. (1998b). Change identification using multitemporal spectral mixture analysis: applications in eastern Amazonia. In *Remote Sensing Change Detection: Environmental Monitoring Applications and Methods*, eds. C. Elvidge and R. Lunetta. Ann Arbor, MI, Ann Arbor Press, pp. 137–61.

Roberts, D. A., Gardner, M., Church, R. *et al.* (1998c). Mapping chaparral in the Santa Monica Mountains using multiple endmember spectral mixture models. *Remote Sensing of Environment* **65**(3), 267–79.

Roberts, D. A., Nelson, B. W., Adams, J. B., and Palmer, F. (1998a). Spectral changes with leaf aging in Amazon caatinga. *Trees* **12**, 315–25.

Roberts, D. A., Ustin, S. L., Ogunjemiyo, S. *et al.* (2004). Spectral and structural measures of northwest forest vegetation at leaf to landscape scales. *Ecosystems* **7**(5), 545–62.

Rosema, A., Verhoef, W., Noorbergen, H., and Borgesius, J. J. (1992). A new forest light interaction model in support of forest monitoring. *Remote Sensing of Environment* **42**(1), 23–41.

Rouse, J. W., Haas, R. H., Schell, J. A., and Deering, D. W. (1973). Monitoring vegetation systems in the great plains with ERTS. In *Third Earth Resources Technology Satellite-1 Symposium*. Washington, DC, National Aeronautics and Space Administration, NASA SP-351, pp. 309–17.

Rowan, L. C., Simpson, C. J., and Mars J. C. (2004). Hyperspectral analysis of the ultramafic complex and adjacent lithologies at Mordor, NT, Australia. *Remote Sensing of Environment* **91**(3–4), 419–31.

Sabins, F. F. (1996). *Remote Sensing: Principles and Interpretation*, 3rd edn. New York, Freeman.

Sabol, D. E., Adams, J. B., and Smith, M. O. (1992). Quantitative sub-pixel detection of targets in multispectral images. *Journal of Geophysical Research*, **97**(E2), 2659–72.

Sabol, D. E., Jr., Gillespie, A. R., Adams, J. B., Smith, M. O., Tucker, C. J. (2002). Structural stage in Pacific northwest forests estimated using simple mixing models of multispectral images. *Remote Sensing of Environment*, **80**(1), 1–16.

Salisbury, J. W., Walter, L. S., Vergo, N., and D'Aria, D. M. (1991). *Infrared (2.1–25 μm) Spectra of Minerals*. Baltimore, Johns Hopkins University Press.

Schott, J. R. (1997). *Remote Sensing: the Image Chain Approach*. New York, Oxford University Press.

Schowengerdt, R. A. (1997). *Remote Sensing: Models and Methods for Image Processing*, 2nd edn. San Diego, CA, Academic Press.

Shipman, H. and Adams, J. B. (1987). Detectability of minerals on desert alluvial fans using reflectance spectra. *Journal of Geophysical Research* **92**, 10391–402.

Smith, M. O., Adams, J. B., and Sabol, D. E. (1994a). Spectral mixture analysis – new strategies for the analysis of multispectral data. In *Imaging Spectrometry – a Tool for Environmental Observations*, eds. J. Hill and J. Megier. Dordrecht, Kluwer Academic, pp. 125–43.

Smith, M. O., Adams, J. B., and Sabol, D. E. (1994b). Mapping sparse vegetation canopies. In *Imaging Spectrometry – a Tool for Environmental Observations*, eds. J. Hill and J. Megier. Dordrecht, Kluwer Academic, pp. 221–35.

Smith, M. O., Ustin, S. L., Adams, J. B., and Gillespie, A. R. (1990a). Vegetation in deserts: I. A regional measure of abundance from multispectral images. *Remote Sensing of Environment* **31**, 1–26.

Smith, M. O., Ustin, S. L., Adams, J. B., and Gillespie, A. R. (1990b). Vegetation in deserts: II. Environmental influences on regional abundance. *Remote Sensing of Environment* **31**, 27–52.

Strahler, A. H. (1994). Vegetation canopy reflectance modeling: recent developments and remote sensing perspectives. *Remote Sensing Reviews* **15**, 159–94.

Suits, G. H. (1972). The calculation of the directional reflectance of a vegetative canopy. *Remote Sensing of Environment* **2**, 117–25.

Sunshine, J. M., Pieters, C. M., and Pratt, S. F. (1990). Deconvolution of mineral absorption bands: an improved approach. *Journal of Geophysical Research* **95**, 6955–66.

Tanner, E. V. J., Kapos, V., and Adams, J., 1998. Tropical forests – spatial pattern and change with time, as assessed by remote sensing. In *Dynamics of Tropical Communities*, eds. D. M. Newbery, H. H. T Prins, and N. D. Brown (37th Symposium of the British Ecological Society). London, Blackwell Science, pp. 599–615.

Tompkins, S., Mustard, J. F., Pieters, C. M., and Forsyth, D. W. (1997). Optimization of endmembers for spectral mixture analysis. *Remote Sensing of Environment* **59**(3), 472–89.

Tucker, C. J., Townshend, R. G., and Goff, T. E. (1985). African land cover classification using satellite data. *Science* **227**(4685), 369–75.

Ustin, S. L., Smith, M. O., Jacquemoud, S., Verstraete, M., and Govaerts, Y. (1999). Geobotany: vegetation mapping in Earth sciences. In *Remote Sensing for the Earth Sciences, Manual of Remote Sensing*, 3rd edn., ed. A. N. Rencz. New York, John Wiley & Sons, vol. 3, pp. 189–248.

Verhoef, W. (1984). Light scattering by leaf layers with application to canopy reflectance modeling: the SAIL model. *Remote Sensing of Environment* **16**(2), 125–41.

Weeks, R. J., Smith, M. O., Pak, K. *et al.* (1996). Surface roughness, radar backscatter, and visible and near-infrared reflectance in Death Valley, California. *Journal of Geophysical Research* **101**, 23077–90.

Wu, C. and Murray, A. T. (2003). Estimating impervious surface distribution by spectral mixture analysis. *Remote Sensing of Environment* **84**(4), 493–505.

Websites

http://speclib.jpl.nasa.gov/
http://speclab.cr.usgs.gov/

Index

*All numbers in italics indicate figures or plates.